I0072235

Traffic Engineering and Transport Planning

Traffic Engineering and Transport Planning

Edited by **Samuel Morgan**

Willford Press
New York

Published by Willford Press,
118-35 Queens Blvd., Suite 400,
Forest Hills, NY 11375, USA
www.willfordpress.com

Traffic Engineering and Transport Planning
Edited by Samuel Morgan

© 2016 Willford Press

International Standard Book Number: 978-1-68285-095-4 (Hardback)

This book contains information obtained from authentic and highly regarded sources. Copyright for all individual chapters remain with the respective authors as indicated. All chapters are published with permission under the Creative Commons Attribution License or equivalent. A wide variety of references are listed. Permission and sources are indicated; for detailed attributions, please refer to the permissions page and list of contributors. Reasonable efforts have been made to publish reliable data and information, but the authors, editors and publisher cannot assume any responsibility for the validity of all materials or the consequences of their use.

The publisher's policy is to use permanent paper from mills that operate a sustainable forestry policy. Furthermore, the publisher ensures that the text paper and cover boards used have met acceptable environmental accreditation standards.

Trademark Notice: Registered trademark of products or corporate names are used only for explanation and identification without intent to infringe.

Printed in the United States of America.

Contents

Preface

In my initial years as a student, I used to run to the library at every possible instance to grab a book and learn something new. Books were my primary source of knowledge and I would not have come such a long way without all that I learnt from them. Thus, when I was approached to edit this book; I became understandably nostalgic. It was an absolute honor to be considered worthy of guiding the current generation as well as those to come. I put all my knowledge and hard work into making this book most beneficial for its readers.

The increase in transportation systems has fueled the growth of traffic engineering. Traffic safety, counter-measures for road traffic accidents, etc. are some of the important areas wherein the focus of transport planning and traffic engineering lie. This book attempts to understand the multiple branches that fall under the discipline of traffic engineering and how such concepts have practical applications in the modern times. Included in this book are elucidations on important topics like traffic planning, control and management, traffic and transport safety, traffic policies, urban transit systems, traffic information engineering and control, etc. Students, researchers, experts and all associated with traffic and transportation engineering and allied branches of engineering will benefit alike from this book.

I wish to thank my publisher for supporting me at every step. I would also like to thank all the authors who have contributed their researches in this book. I hope this book will be a valuable contribution to the progress of the field.

Editor

1

Road traffic congestion measurement considering impacts on travelers

Liang Ye · Ying Hui · Dongyuan Yang

Abstract The article intends to find a method to quantify traffic congestion's impacts on travelers to help transportation planners and policy decision makers well understand congestion situations. Three new congestion indicators, including transportation environment satisfaction (TES), travel time satisfaction (TTS), and traffic congestion frequency and feeling (TCFF), are defined to estimate urban traffic congestion based on travelers' feelings. Data of travelers' attitude about congestion and trip information were collected from a survey in Shanghai, China. Based on the survey data, we estimated the value of the three indicators. Then, the principal components analysis was used to derive a small number of linear combinations of a set of variables to estimate the whole congestion status. A linear regression model was used to find out the significant variables which impact respondents' feelings. Two ordered logit models were used to select significant variables of TES and TTS. Attitudinal factor variables were also used in these models. The results show that attitudinal factor variables and cluster category variables are as important as sociodemographic variables in the models. Using the three congestion indicators, the government can collect travelers' feeling about traffic congestion and estimate the transportation policy that might be applied to cope with traffic congestion.

L. Ye (✉)
Transport Planning and Research Institute, Ministry of Transport of China, Room 1109, building 2, Jia 6 Shuguangxili, Chaoyang, Beijing 100028, China
e-mail: yel1231@gmail.com

Y. Hui · D. Yang
School of Transportation Engineering, Tongji University, Shanghai 201804, China

Keywords Traffic congestion indicator · Attitudinal factor variable · Linear regression model · Ordered logit model

1 Introduction

Traffic congestion is one of the worst problems in China, especially in those metropolises, such as Shanghai, Beijing, and Shenzhen. After long-time struggling with traffic congestion, most of researchers realize it is not easy to eliminate congestion but it is possible to relieve it. A number of traffic congestion studies [1–3] focused on improving transportation system but not transportation users' feelings. Presently, more and more researchers [4, 5] realize that it is not enough to just study transportation system capacity, and transportation users' feelings and reactions are also important to decide how to relieve traffic congestion. It is an important point to know transportation users' feelings and reactions about urban road traffic congestion, which can help decision makers to make more efficient and useful policies and strategies. A method should be found to quantify traffic congestion's impacts on travelers to help transportation planners and policy decision makers well understand congestion situations standing on travelers' side. Some prior studies [6, 7] revealed that traffic conditions especially traffic congestion may impact people's travel-related decisions and behaviors.

Under this background, we study the traffic congestion impacts on travelers and their reactions to congestion. A random sampling survey was taken in Shanghai, China during August 1st to August 31st in 2009 to collect data for this research, including transportation users' attitudes about road traffic congestion, baseline transportation characteristics of transportation users, their reactions to traffic congestion and sociodemographics. Totally, 274 valid samples were collected, covering most of districts of Shanghai.

In order to quantify traffic congestion impacts, we found a way to evaluate the service level of transportation system. It is a hotspot to study traffic congestion relieving policies in China. Most of these studies focus on seeking sources of congestion and qualitative analysis of policies to relieve congestion. However, study on quantitative indicators for congestion impacts is as important as study on congestion-estimating policies. Study of traffic congestion impacts on travelers and their reactions can provide some supports for setting the target of urban transportation system service level, also for choosing congestion policies.

Three travelers' feeling indicators, namely, transportation environment satisfaction (TES), traffic congestion feeling and frequency (TCFF), and travel time satisfaction (TTS) were selected to quantify congestion impacts on travelers. The "likert-type scale" is used to get data of TES and TTS. A series of questions were asked to get the information of travelers' feelings and the frequency they suffer congestion in a typical month about 9 traffic congestion situations which were designed based on previous studies and our hypothesis. A merged indicator was created based on both travelers' feelings and frequency they met from the nine congestion situations using factor analysis. Ordered logit models and linear regressive model were set up to analyze impact factors of the three indicators, respectively.

The remainder of the article is organized as follows. Section 2 briefly reviews previous related research. Section 3 describes the data collection and survey contents in this study. Then Sect. 4 presents the reason for select the three traffic congestion indicators and their values in Shanghai, China. Models were built to analyze the impact factors of each indicator in Sect. 5. Finally, Sect. 6 summarizes the study and suggests future research directions.

2 Literature review

Definitions of traffic congestion could differ with different organizations and purposes. The Federal Highway Administration [8] defines traffic congestion as "the level at which transportation system performance is no longer acceptable due to traffic interference." They also state that "the level of system performance may vary by type of transportation facility, geographic location (metropolitan area or sub-area, rural area), and/or time of day." The regional council of governments in Tulsa, Oklahoma [1] defines congestion as "travel time or delay in excess of that normally incurred under light or free-flow travel conditions." In Minnesota [8], when the traffic speed is below 45 mph in peak hours, freeway congestion could be defined. Michigan also defines freeway congestion using level of service.

By user expectation, "unacceptable congestion" was defined using travel time in excess of an agreed-upon norm, which might vary by type of transportation facility, travel mode, geographic location, and time of day. Lomax et al. [9] realized that "A key aspect of a congestion management strategy is identifying the level of 'acceptable' congestion and developing plans and programs to achieve that target." Pisarski [10] used the U.S. Census data to conduct the commuting patterns, and defined the unacceptable congestion as "if less than half of the population can commute to work in less than 20 min or if more than 10 % of the population can commute to work in more than 60 min." The Metropolitan Washington Council of Governments [11] developed a "user satisfaction" transportation system performance measure based on acceptable travel time and delay. The measure incorporated a set of curves that show the percentage of users satisfied for a given trip length and time.

Some more studies about traffic congestion indicators are listed in Table 1. In those studies, we can find that most of traffic congestion indicators are focused on transportation capacity, travel time, delay, travel speed, et al., which could be classified as transportation system performance indicators. A few indicators are based on user expectation and satisfaction, which concern users' acceptable travel time or delay.

Attitude data analysis in travel behavior researches were started from 1970s, and became more popular ever since [12]. Attitudinal surveys provide a means for measuring the importance of qualitative factors in travel behavior. Factor analysis was often used to collapse the questions into a smaller set of factors as explanatory variables in travel behavioral models [13]. A significant amount of studies used factor analysis, cluster analysis, and discrete model to study traveler's behavior under specific situations or policies. Redmond [14] used factor analysis to identify the fundamental dimensions of attitude, personality, and lifestyle characteristics; then used cluster analysis to group respondents with similar profiles. Mokhtarian [15] used the discrete model to describe the choice of increasing transit use during the Fix I-5 project. She also used the discrete model to estimate the preference to telecommute from home [16]. Factor analysis is performed on two groups of attitudinal questions, identifying a total of 17 factors in that article.

3 Data collection and survey

3.1 Data collection

A random sampling household survey was taken in Shanghai during August 1st to August 31st in 2009 to collect data for this research. The data were collected from

Table 1 Traffic congestion indicators in different research

Author/ organization	Years	Purpose	Indicators	Note
Texas transportation institute [11]	2007	Used in both the public and private sectors as a means of communicating the congestion trends in the larger U.S. urban areas	Roadway congestion index (RCI)	The RCI is an empirically derived formula that combines the indicator of urban area daily vehicle kilometers of travel (DVKT) per lane of roadway for both freeways and principal arterial streets
Chicago's freeway management system [11]	1996	Quantify freeway congestion	Lane occupancy rates	Using lane occupancy rates requires the installation of a freeway detector network
The metropolitan Washington council of governments [11]	1996	Measure transportation system performance based on acceptable travel time and delay	User satisfaction	The measure incorporates a set of curves that show the percentage of users satisfied for a given trip length and time
Herbertlevinson, timothyj. lomax [11]	1996	Consistent with the myriad analytical requirements	Delay rate index (DRI)	DRI combines the beneficial effects of using travel time and speed data with the ability to relate congestion and mobility information
Highway capacity manual [9, 11]	1985	Reflect traffic volume counts and peaking, roadway characteristics, and traffic signal timing	Level-of-service (LOS)	The LOS is defined in terms of density for freeways, average stopped delay for intersections, and average speed for arterials
Department of Transportation in UK [17]	2001	To well understand congestion and cope with it	Extra time taken compared with free-flow time risk of serious delay average speed on different road types amount of Time stationaryor less than 10 mph	Four measures people would find most helpful to measure congestion by publish information
The federal highway administration (FHWA) [3, 8]	2005	To measure travel time in a mobility monitoring program	Travel time index average duration of congested travel per day (hours) buffer index	They are trying to answer a mobility question: "how easy is it to move around?" and a reliability question": how much does the ease of movement vary?"

a mixed internet-based survey and mail survey in Shanghai. We sent 15,000 letters by mails to invite people taking part into the survey, and the survey website link was provided in the letter for those who were willing to attend the survey by internet. We also provided four ways for people to ask for the paper questionnaires: our survey service phone number, email address, text message to cell phone, and mail back the postcard which is paid by us. Totally, 274 valid samples were collected, covering most of districts of Shanghai, including 233 internet-based respondents and 21 paper questionnaire-based respondents.

Table 2 presents the sample statistics for some selected characteristics. A majority of the respondents (59 %) are less than 40 years old; 79 % respondents' education level is higher than high school graduate; company employees form the largest part in whole respondents, the proportion is 45 %; more respondents (34.4 %) have an annual income of 60,000–119,999 Yuan.

3.2 Survey contents

There are six parts in the survey:

Part *A* collects respondents' characteristics and attitudes, including satisfaction about current life, the city and neighborhood, the transportation system, personal characteristics, and general attitudinal statements.

Part *B* offers attitudinal statements to seek transportation-related attitudes under the traffic congestion.

Part *C* collects the information about most frequent trips of respondents, including trip purpose, travel mode, trip OD, departure time, frequency, and feeling about different traffic congestion statements.

Part *D* collects the general trip information of respondents, including trip purpose, travel mode, and total travel time per week.

Part *E* explores the active choices and reactions to traffic congestion.

Table 2 Selected characteristics of the sample

Characteristic	Number of cases	Percentage (%)	Sample sizes
Number of females	133	48.9	272
Age group			
16–20 years old	29	10.7	272
21–30 years old	106	39.0	
31–40 years old	54	20.0	
41–50 years old	33	12.1	
>50 years old	50	18.4	
Education background			
Doctoral degree	7	2.6	274
Master's degree	23	8.4	
Four-year college, university, or technical school graduate	115	42.0	
Some college or technical school	71	25.9	
High school graduate	29	10.6	
Some grade or high school	18	6.6	
Other	11	4.0	
Occupation			
Officer	28	10.2	274
Company employee	123	44.9	
Student	44	16.1	
Business man	6	2.2	
Teacher	14	5.1	
Retiree	30	10.9	
Production/construction/crafts	16	5.8	
Other	13	4.8	
Annual household income			
Less than 24,999 yuan	32	11.7	273
25,000–59,999 yuan	76	27.8	
60,000–119,999 yuan	94	34.4	
120,000–249,999 yuan	59	21.6	
250,000–399,999 yuan	8	2.9	
400,000–599,999 yuan	1	0.4	
600,000 yuan or more	3	1.1	

Part *F* collects information on sociodemographic characteristics, including age, gender, income, occupation, and education.

4 Road traffic congestion indicators based on the impacts by travelers

4.1 Road traffic congestion indicators selection

Based on previous studies [9, 18] and our hypothesis, three traffic congestion indicators based on the impacts by travelers were created in this article. They are

(1) TES. This indicator presents people's satisfaction of total transportation environment, not only for evaluating traffic congestion. However, we can set it as an indicator to show situations at the macro-level about transportation system.

(2) TCFF. It is a new indicator created by the author to present travelers' feelings of different traffic congestion situations by considering both frequency of congestion happening and travelers' feelings about the congestion. In our survey, we designed nine congestion situations[1] to present congestions in our daily life. TCFF integrated these 9 situations.

(3) TTS. It is a popular indicator in some previous studies based on traveler's feelings. In our survey, we also asked a question for traveler's satisfaction of their travel time. This indicator was also used to present travelers' particular feelings of travel time.

4.2 The value of congestion indicators in Shanghai, China

A question was asked in our survey about the TES: "How satisfied do you feel with your current life,..., and the transportation system?" One statement is "Travel environment in the city." The options are "Not satisfied at all," "Not satisfied," "Slightly satisfied," "Moderately satisfied," and "Extremely satisfied." About 30 % respondents presented their dissatisfaction of transportation environment, and 24 % respondents felt satisfied. The following question was asked about the TTS: "Are you satisfied with your usual travel time for your most frequent trips?" The options are the same as the former one. The information of travelers' most frequent trips were required. The most frequent trips could be a trip from home to work (or work to home), or a non-work trip, but it should always have the same trip purpose and the same (single) origin and destination. The reason to ask for the most frequent trips information is that, we want to get more exact information like departure time, trip origin, and destination for a special trip which will not change by different purpose or trip distance. And the most frequent trip will be the most familiar trip in travelers' daily trips which impact them most. For this question, about 20 % respondents report that

[1] 9 congestion situations: (a) You are delayed about 30 min because of traffic congestion; (b) The traffic you are in basically stops for more than 5 min because of traffic congestion; (c) The traffic you are in always stops but restarts soon; (d) Your speed is slower than a bicycle; (e) Although you can move smoothly, the road is full of vehicles and people; (f) The trip takes longer than you expected; (g) It takes at least two green lights before you can get through the intersection; (h) You can't estimate travel time because of traffic congestion; (i) You are stacked behind people who are slower than you like.

they are satisfied or unsatisfied with their travel time for the most frequent trips, respectively.

TCFF is a new indicator which was not designed directly in the questionnaire. Instead, we set a series of situations (see the footnote on the last page) to describe traffic congestion, and ask for the frequency respondents meet the similar situation in a typical month, and how it makes them feel. Even if a certain event never happens, the respondent would be also asked to image the feeling. The statistical results indicate that most respondents (76.2 %) feel moderately bad or extremely bad when they are delayed about 30 min because of traffic congestion, which is consistent with the previous study results of Al-Mosaind [19]. However, in Shanghai, 14.3 % of respondents indicate that they meet this kind of situation more than once a week in a typical month. The situations that the speed is slower than a bicycle and cannot estimate travel time because of traffic congestion are the following two events which make respondents feel moderately bad or extremely bad, about 67.5 % and 66.7 % respectively. 24 % and 26 % of respondents said that they meet these two situations more than once a week in a typical month. The frequencies of the situations such as that taking at least two green lights to get through the intersection, being stacked by slower people, and travel time being longer than expected occur more often than other situations. More than 40 % of respondents suffered these three situations more than once a week in a typical month.

We hypothesize that the frequency of a congestion situation will impact travelers' integrate feeling about congestion. In other words, if two travelers have the same feeling to one congestion situation itself, such as slightly bad, but one traveler suffers it once a week and another one just meet it once a month, we assume that the traveler who suffers more often would feel worse than the low frequency one in their true life. Therefore, we set a integrate index to describe this relationship which we call as TCFF. The formula of TCFF is as follows:

TCFF = Traffic congestion frequency × Traveler's feeling

In order to calculate the index, in this study, we transferred the survey options of frequency to the exact number of value:

"Never" → 0 per month;
"Less than once a month" → 0.5 per month;
"1–3 times a month" → 2 per month;
"1–2 times a week" → 6 per month;
"3–4 times a week" → 14 per month;
"5 or more times a week" → 20 per month.

At the same time, we set the value of travelers' feeling as

"Not a problem" → 0;
"Slightly bad" → 1;
"Moderately bad" → 2;
"Extremely bad" → 3.

After calculated, the average value of TCFF is shown in Fig. 1. The value of the situation that traffic flow always stops is the highest one (10.66) in the 9 congestion situations, with high share rate of respondents who suffered it more than once a week and feeling moderately or extremely bad.

TCFF is a kind of indicator that combines the frequency of respondents suffered congestion and their feeling. It presents the real and integrated feeling of congestion situations in the true life. The value of this index can be used to evaluate travelers' feeling and their experiences of traffic congestion.

5 Models of road traffic congestion indicators

5.1 Methodology and variables

5.1.1 Methodology

The purpose of this study is to estimate how traffic congestion impacts travelers' feeling. The relationship of congestion indicators and impact factors needs to be studied through models to help understand which make travelers feel bad or not. As the type of data for TES and TTS are ordered data, the ordered logit model is selected to analyze the relationship between impact factors and indicators. The linear regression (LR) model is used for TCFF calculation.

5.1.2 Dependent variables

Two dependent variables—TES and TTS—are created from the survey question which asks "How satisfied do you feel with your current life,..., and the transportation system?" One statement is "Travel environment in the city." And the question asks "Are you satisfied with your usual travel time for your most frequent trips?" The options are the same: "Not satisfied at all," "Not satisfied," "Slightly satisfied," "Moderately satisfied," and "Extremely satisfied."

The dependent variable of TCFF model is calculated from the integrated value of the index in 9 congestion situations. The factor analysis is used to obtain the integrated value by setting just one factor number. The 9 congestion situations can be set as 9 statements in factor analysis after the value of statements are standardized by dividing 10 (from 0–60 to 0–6). The principal components analysis (PCA) is used in this study to derive a small number of linear combinations of a set of variables that retain as much of the information in the original variables as possible, using the SPSS statistical software package. For the result, the main factor explained 62 % of the total variance in the statements which could be seemed as a high value and able to present most of information for those variables [20]. The factor score was used in the subsequent model as the dependent variable.

5.1.3 Explanatory variables

Based on literature review and previous empirical studies [2, 6, 9, 15, 19, 21–23], the explanatory variables obtained from the survey fall into five main categories, each described as below.

General attitude and transportation-related attitude: in survey Part *A* and Part *B*, we asked a series of general attitude and transportation-related attitude statements on a 5-point scale from "strongly disagree" (1) to "strongly agree" (5). Common factor analysis was used to extract the 4 general attitude factors and 6 transportation-related attitude factors. Table 3 presents the factor loadings by general attitudinal statements, and Table 4 presents the factor loading by transportation-related attitudinal statements.

General attitude and transportation-related attitude cluster variables: a cluster analysis was used to classify the categories of respondents based on their general attitudinal factors and transportation-related attitudinal factors. We produced solutions for predefined cluster numbers of 2 and 3. For the criteria of interpretability and maintenance of statistically robust segment sizes, we selected the two-cluster solution. Table 5 presents the cluster results for each of them.

Baseline travel characteristics: Part *C* and Part *D* of the survey collected the information of respondents about their general trips and the most frequent trips including trip purpose, travel mode, travel time, and so on.

Other traffic congestion indicators: other traffic congestion indicators were added to estimate the relationship between them and the dependent variable.

Sociodemographic characteristics: Part *F* of the survey captured an extensive list of sociodemographic variables such as gender, age, educational background, household income, household size, and so on.

5.2 Model results

5.2.1 TES model results

Due to missing data, the final TES model (Table 6) has 239 respondents. The ρ^2 goodness-of-fit measure [24] with the market-share model as base is 0.145, which shows that the true explanatory variables add 0.145 to the goodness-of-fit.

Nine variables besides the constant are retained in the model: three sociodemographic variables, three additional factors, and three other congestion indicators.

Three sociodemographic variables are gender, owning the current residence, and annual household income. Women are more likely to be satisfied with transportation environment than men, which could be explained using

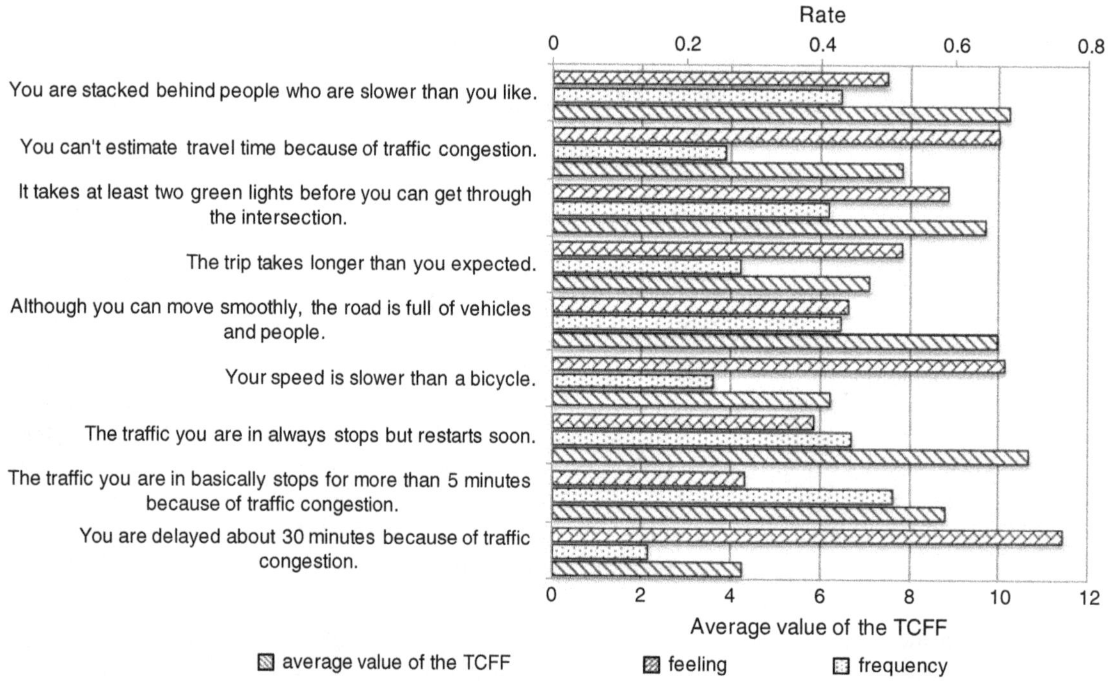

Fig. 1 Average value of TCFF and share rate of respondents for congestion frequency and feeling. Note The "frequency" bar presents the share rate of respondents who suffered the situation more than once a week week; the "feeling" bar presents the share rate of respondents whose feeling to the situation is moderately bad or extremely bad

results of the previous study of Mokhtarian [15] that women are easier to adjust themselves to the external changes. Respondents who own the current residence are more likely to be satisfied with transportation environment, for they may have more acceptances with the city when they decided to buy the house or apartment. Respondents with higher income show their less satisfaction with transportation environment. Maybe it is because people with more money will have higher requirements to the city.

We got attitudinal factors from a series of statements using factor analysis, and more details could be seen in the author's another article [25]. Three significant attitudinal factors are hates wasting time, contend with travel conditions, and dislikes travel. It is easy to understand that people who hate wasting time will be more likely to feel dissatisfied with transportation when they are stacked on the road. People who can contend with travel conditions are more likely to feel satisfied with transportation. If people dislike travel, then it means there are some aspects with trips which make them uncomfortable, and so they will feel less likely to be satisfied with transportation environment.

Other congestion indicators also involved in the model to estimate the relationship between TES and other congestion indicators. Three other congestion indicators are significant in the model. TTS is a major index to present whether travelers are satisfied with their travel time. Respondents who are satisfied with their travel time are more likely to be satisfied with the total transportation environment. The 30-min-delay frequency and feeling and slower than bicycle frequency and feeling are indicators presenting the frequency and respondents' feelings with two congestion situations. If travelers meet these two congestion situations more frequently or they feel worse than other people, then they will less likely to be satisfied with the urban transportation environment. The results also indicate that travel time and travel speed are the two important aspects for travelers when they do the daily trips, which will impact their feeling to the total transportation environment.

5.2.2 *TTS model results*

TTS model (Table 6) has 235 valid respondents. The ρ^2 goodness-of-fit measure with the market-share model as base is 0.271, which shows that the true explanatory variables add 0.271 to the goodness-of-fit. Eleven variables

Table 3 Rotated factor loadings (pattern matrix) by general attitudinal statements ($N = 271$)

Survey statement	Hates wasting time	In a hurry and out of control	Confident	Likes quiet living	Communalities
Even if I have something else pleasant or useful to do while traveling for routine activities, it often bothers me if the trip takes a long time	0.579	–	–	–	0.352
In my daily life, I have to spend too much time waiting	0.435	–	–	–	0.412
I make productive use of the time I spend on daily traveling	−0.393	–	–	–	0.249
If the line is moving, waiting is OK for me	−0.357	–	–	–	0.144
In general, waiting is unpleasant even if I have an interesting way to pass the time	0.351	–	–	–	0.142
Work and family do not leave me enough time for myself	0.268	–	–	–	0.129
I'm often in a hurry to be somewhere else	–	0.703	–	–	0.461
I have to admit that sometimes I make other people wait for me	–	0.499	–	–	0.267
I will do something humiliating, if you give me enough money	–	0.439	–	–	0.271
I often feel like I don't have much control over my life	–	0.325	–	–	0.423
In choosing where to live, there are many factors much more important than transportation conditions	–	0.224	–	–	0.065
It is understandable for someone to be a bit late	–	0.180	–	–	0.043
I am confident that I can deal with unexpected events effectively	–	–	0.583	–	0.323
I can always rely on my own ability to handle difficult situations	–	–	0.500	–	0.261
Even when I have a lot of things to do, I seldom feel pressure	–	–	0.254	–	0.093
I like living in a small and quiet city instead of a bustling city	–	–	–	0.617	0.393
I like the idea of having different types of businesses (such as stores, offices, post office, bank, and library) mixed crowdedly in with the homes in my neighborhood	–	–	–	−0.392	0.289
I like to live in a crowded neighborhood with lots of people	–	–	–	−0.357	0.212

Table 4 Rotated factor loadings (pattern matrix) by transportation-related attitudinal statements ($N = 271$)

Survey statement	Contend with travel conditions	Likes driving	Travel planner	Transportation aware	Travel constraint	Dislikes travel	Communalities
Thinking about both good and bad aspects, overall the public transportation system is pretty good	0.736		–	–	–	–	0.532
It's convenient to travel from one place to another in my city	0.609	–	–	–	–	–	0.411
Getting stuck in traffic doesn't bother me too much	0.494	–	–	–	–	–	0.257
Some amount of traffic congestion is inevitable, no matter what we do	0.358	–	–	–	–	–	0.153
I prefer to drive rather than travel by any other means	–	0.731	–	–	–	–	0.520
I like driving itself, without having any other reason	–	0.584	–	–	–	–	0.404
To me, a car is a status symbol	–	0.436	–	–	–	–	0.247
I like the idea of walking or biking as a means of transportation	–	−0.430	–	–	–	–	0.249
I get where I'm going more quickly than other people because I know how to choose my departure time and route to avoid congestion	–	–	0.747	–	–	–	0.463
It is important for me to organize my errands so that I make as few trips as possible	–	–	0.542	–	–	–	0.433
I really need to get more information about traffic conditions before I make a trip	–	–	0.416	–	–	–	0.362
Even though I'm only one person, my actions can make a difference to the transportation system	–	–	–	0.519	–	–	0.246
Transportation condition plays an important role when I choose my job	–	–	–	0.504	–	–	0.187
I like the idea of using public transportation whenever possible	–	–	–	0.443	–	–	0.231
When I choose the means of transportation for a certain trip, I consider traffic congestion	–	–	–	0.369	–	–	0.417
It's unfair to expect me to sacrifice to help reduce traffic congestion, if other people aren't doing it too	–	–	–	0.215	–	–	0.229
It's really hard to estimate my travel time before leaving because of congestion	–	–	–	–	0.522	–	0.365
I know very little about the transportation system of this city	–	–	–	–	0.433	–	0.341
The only good thing about traveling is arriving at your destination	–	–	–	–	0.393	–	0.185
I generally know when and where Congestion will happen in the city	–	–	–	–	−0.383	–	0.306
The traveling that I need to do interferes with doing other things I like	–	–	–	–	0.282	–	0.101
Sometimes I would enjoy staying at home for the whole day and not having to go anywhere	–	–	–	–	–	0.607	0.366
I want to go somewhere at least once a day, even if I have nothing particular to do	–	–	–	–	–	−0.569	0.357
I prefer to shop near where I live, in order to make fewer trips	–	–	–	–	–	0.322	0.212

Table 5 Cluster centroids and between-cluster mean sum of squares ($N = 274$)

General attitudinal factor	Cluster centers		Between-cluster MSS
	Stressed	Executive	
Hates waiting time	0.363	−0.502	50.029 (HH)
In a hurry and out of control	−0.392	0.543	58.342 (HH)
In control	0.073	−0.010	1.986 (BB)
Likes quiet living	0.259	−0.357	25.295 (B)
No. (%) of observations in each cluster	159 (58.0)	115 (42.0)	–
Transportation-related attitudinal factor	**Savvy traveler**	**Travel planner**	**Between-cluster MSS**
Contend with travel conditions	0.193	−0.377	19.973 (B)
Likes driving	0.088	−0.171	4.130 (BB)
Travel planner	−0.356	0.694	67.653 (HH)
Transportation aware	0.362	−0.707	70.183 (HH)
Travel constraint	−0.153	0.298	12.475 (B)
Dislikes travel	0.039	−0.077	0.831 (BB)
No. (%) of observations in each cluster	181 (66.1)	93 (33.9)	–

The average BMSS of 33.913 for general attitudinal factors and 29.208 for transportation-related attitudinal factors. BB and B means much below and below, respectively; M means the value is about equal to the mean BMSS; H and HH means above and much above mean BMSS, respectively

besides constant variable are significant in the model, including three sociodemographic variables, four attitudinal factors, two trip characteristics, and two other congestion indicators.

Three significant sociodemographic variables are gender, government employee, and company employee. Inconsistent with the TES model, women are more likely to be unsatisfied with travel time for their most-frequent trips which is the same as previous study [15]. The reason could be due to gender differences in response style: women could be more inclined than men to use the extreme ends of a scale [26]. TES is a kind of overall indicator to describe the total transportation status of a city, however, TTS indicator more focuses on the most frequent trips. Therefore, they may have lower level acceptance in travel time than men but more of them like the total transportation system. Government employee and company employee are more likely to be satisfied with travel time which may be because generally, their-most frequent trips are commuted trips for which they are already used to the travel time. So they may be more satisfied with travel time than other respondents whose most frequent trips' purposes are not commuting.

Four attitudinal factors are residence satisfaction, satisfaction of urban transportation system, in a hurry and out of control, and likes quiet living. Respondents who are satisfied with their residence and transportation system will obviously more likely to be satisfied with the travel time of the most frequent trips. Respondents who are always in a hurry and out of control will be more likely to be unsatisfied with travel time. That is because these kinds of people do not have the ability to organize or plan their errands, and so they will more likely feel to be hurrying with everything including their trips. People who like quiet living are more likely to be unsatisfied with travel time either. The reason is that such people do not like the busy life and traveling itself, so they will be less likely to take long time on traveling.

The longer travel time of the most frequent trips is, the less likely the respondents are to be satisfied with the travel time. Accordingly, the longer the total travel time in a week is, the less likely the respondents are to be satisfied with the travel time. Two congestion indicators are also significant in the model. If the road is full of vehicles, then respondents will be less likely to feel satisfied with travel time. And if respondents need to wait for two green lights to go through the intersection, it means the travel time is longer than usual, so they will be less likely to feel satisfied with the travel time.

5.2.3 *TCFF model results*

The LR model was used here. In the model, 220 respondents are valid (see Table 7); the ρ^2 is 0.345, and the adjusted ρ^2 is 0.300, which could be deemed as acceptable [27].

There are fifteen variables significant in the model, including two sociodemographic variables, five attitudinal factors, one cluster category, and seven trip characteristics variables. Two sociodemographic variables are currently owning residence and annual household income. Different from the TES model results, respondents who currently

Table 6 Ordered logit models of TES and TTS (0 = strongly disagree, 1 = disagree, 2 = neutral, 3 = agree, 4 = strongly agree)

Variable name	TES		TTS	
	Coefficient 2.765	*P* value 0.001	Coefficient 7.503	*P* value 0.000
Socio-demographics				
Female (dummy variable-DV)	0.478	0.067	−0.643	0.030
Annual household income	−0.266	0.030		
Own the current residence (DV)	0.794	0.050		
Government employee (DV)			0.982	0.048
Company employee (DV)			0.948	0.005
Attitudinal factors				
Residence satisfaction			0.757	0.000
Satisfaction of urban transportation system			0.685	0.000
In a hurry and out of control			−0.584	0.000
Likes quiet living			−0.533	0.011
Hates wasting time	−0.287	0.059		
Contend with travel conditions	0.600	0.000		
Dislikes travel	−0.320	0.036		
Trip characteristics				
Travel time of the most frequent trips (minutes)			−0.163	0.001
Total travel time of a typical week for commuting (hours)			−0.228	0.019
Other congestion indicators				
TTS	0.459	0.016		
30 min delayed frequency and feeling	−0.257	0.026		
Slower than bicycle frequency and feeling	−0.218	0.057		
Full with vehicles on the road frequency and feeling			−0.527	0.000
Waiting for more than one green lights frequency and feeling			−0.405	0.001
Valid number of cases, *N*	239		235	
Final log-likelihood, $LL(\beta)$	−263.298		−188.877	
Log-likelihood for market share model, $LL(MS)$	−307.953		−259.095	
No. of explanatory variables, *K* (including constant)	10		12	
$\rho^2_{\text{MSbase}} = 1 - LL(\beta)/LL(MS)$	0.145		0.271	
χ^2 (between final and MS models)	89.310		140.436	

own their residences have higher value of TCFF. This may be due to the differences between these two indicators. TCFF presents the real statuses of the respondents in their most-frequent trips—frequency at which they meet the congestion situations and their feelings about these congestion situations. The same thing happens to the annual household income: respondents with higher income have less satisfaction of transportation environment but also meet less-frequent congestion situations or feel better with those congestion situations. The interpretation is that people with higher income levels have higher requirements with urban transportation system. At the same time, they also have higher ability to cope with the traffic congestion.

Five attitudinal factors are satisfaction of urban transportation system, hating wasting time, contending with travel conditions, disliking travel, and transportation awareness. Respondents who are satisfied with transportation system are less likely to meet the congestion situations or have better feeling with congestion. For those who hate wasting time, they are more likely to feel worse with congestion. Respondents who have higher awareness of transportation are more sensitive to congestion that makes them easier to point out congestion or feel worse about congestion. If travelers who can contend with travel conditions, then they will be less likely to suffer congestion situations or feel bad with congestion. And for those who dislike travel respondents, they will make as fewer trips as they can, and the frequency of meeting congestion will be less than others, and their TCFF value will be lower.

One cluster category variable became significant in the model which indicates that different people group will have

Table 7 LR models of TCFF

Variable name	TCFF index	
	Coefficient	*P* value
	−2.415	0.000
Sociodemographics		
Annual household income	−0.103	0.054
Own the current residence (DV)	0.482	0.005
Attitudinal factors		
Satisfaction of urban transportation system	−0.129	0.062
Hates wasting time	0.268	0.002
Contend with travel conditions	−0.401	0.000
Dislikes travel	−0.122	0.102
Transportation aware	0.232	0.001
Executive (DV)	0.606	0.000
Trip characteristics		
Total travel time of a typical week for commuting (hours)	0.091	0.021
Total travel time of a typical week for recreation or social activities (hours)	0.107	0.050
Trip purpose of the most frequent trips commute (DV)	1.479	0.017
Trip purpose of the most frequent trips work related (DV)	1.545	0.016
Trip purpose of the most frequent trips grocery shopping (DV)	1.704	0.012
Trip purpose of the most frequent trips recreation or social activities (DV)	1.961	0.003
Trip purpose of the most frequent trips picking up other people (DV)	1.914	0.007
Valid number of cases, *N*	220	
No. of explanatory variables, *K* (including constant)	15	
ρ^2	0.345	
Adjusted ρ^2	0.300	

different feelings of congestion. Executive travelers will meet more frequent congestion situations or feel worse about congestion than stressed people (cluster results shown in Table 5).

Different from TES model, several trip characteristic variables are significant in the model. Besides, two travel time-related variables—total travel time of a typical week for commuting and total travel time of a typical week for recreation or social activities, other five variables are all about the trip purpose of the most frequent trips. In general, if the travel time of respondents' daily trips is longer, they are more likely to suffer more congestion and feel worse. The significant variables of trip purposes are commuting, work-related trips, grocery shop, recreation, or social activities, and picking up other people. During trips with these five purposes, respondents will be more likely to meet more congestion or feel worse than those with other trip objectives.

6 Conclusions and suggestions for future research

The article uses three new congestion indicators to estimate urban traffic congestion based on travelers' feelings. They are TES, TTS, and TCFF. A survey was taken in Shanghai China to collect travelers' attitudes about congestion and trip information. Based on the survey data, we estimated the three indicators' value of travelers in Shanghai. About 30 % respondents showed they were unsatisfied with transportation environment and 23 % respondents said they were unsatisfied with the travel time of the most frequent trips. Nine congestion situations were designed in the survey to collect the frequency that travelers meet in their most frequent trips and the feelings when meet these situations. In the nine congestion situations, most respondents (76.2 %) feel moderately bad or extremely bad when they are delayed about 30 min. The situations that the speed is slower than a bicycle and cannot estimate travel time because of traffic congestion are the two events which make about 67.5 % and 66.7 % respondents feeling moderately bad or extremely bad, respectively. TCFF was created by multiplying the frequency with the feeling value.

Subsequently, in order to estimate the whole congestion status, the PCA was used to derive a small number of linear combinations of a set of variables. We set the factor as the dependent variable in TCFF model. The LR model was used to find out the significant variables which will impact respondents' feelings. The ordered logit model was also used to select significant variables of TES and TTS. Nine variables are significant in the TES model, eleven variables are significant in the TTS model, and fifteen variables are significant in the TCFF model. The results show that attitudinal factor variables and cluster category variables are as important as sociodemographic variables in models. Three congestion indicators can describe travelers' feelings of congestion from three different levels. Using these congestion indicators, the government can collect travelers' feelings about congestion besides traffic condition index.

Acknowledgments This study was supported by the Key Natural Science Foundation of China: Urban Transportation Planning Theory and Methods under the Information Environment, Grant No. 50738004/E0807. The authors gratefully acknowledge the help provided by Prof. Patricia L. Mokhtarian on the survey design and her suggestions for the whole study. Ke Wang, Weiqi Yao, and Chen Chen were essential to the data collection.

References

1. INCOG (2001) Congestion management system. FHWA, Tulsa, Oklahoma
2. Boarnet MG, Kim EJ, Parkany E (1998) Measuring traffic congestion. Transp Res Rec 1634:93–99
3. Texas Transportation Institute (2005) Traffic congestion and reliability: trends and advanced strategies for congestion mitigation. Cambridge Systematics, Cambridge
4. Taylor BD (2002) Rethinking traffic congestion. Access 21:8–16
5. Salomon I, Mokhtarian PL (1997) Coping with congestion: understanding the gap between policy assumptions and behavior. Transp Res D 2(2):107–123
6. Davis AF (2004) The impact of traffic congestion on household behavior: three essays on the role of heterogeneity. Dissertation, North Carolina State University, Raleigh
7. Cullinane S, Cullinane K (2003) Car dependence in a public transport dominated city: evidence from Hong Kong. Transp Res D 8(2):129–138
8. Bertini RL (2005) Congestion and its extent. Access Destin 398(1):1–28
9. Lomax T, Turner S, Shunk G (1997) Quantifying congestion. Texas Transportation Institute, National Academy Press, College Station
10. Pisarski AE (2006) Commuting in America iii: The third national report on commuting patterns and trends, transportation research board, 2006, http://onlinepubs.trb.org/onlinepubs/trnews/trnews247CIAIII.pdf. Accessed 20 Oct 2010
11. Levinson HS, Lomax TJ (1996) Developing a travel time congestion index. Transp Res Rec 1564:1–10
12. Bohte W, Maat K, Wee BV (2009) Measuring attitudes in research on residential self-selection and travel behaviour: a review of theories and empirical research. Transp Rev 29(3):325–357
13. Clifton KJ, Handy SL (2003) Qualitative methods in travel behavior research. In: Transport survey quality and innovation. Kruger National Park, South Africa
14. Redmond L (2000) Identifying and analyzing travel-related attitudinal, personality, and lifestyle clusters in the San Francisco Bay Area. Dissertation, University of California, Davis
15. Mokhtarian PL, Ye L, Yun M (2009) The effect of gender on commuter impacts and behavior changes in the context of a major freeway reconstruction. In: 4th international conference on women's issues in transportation, Irvine
16. Mokhtarian PL, Salomon I (1997) Modeling the desire to telecommute: the importance of attitudinal factors in behavioral models. Transp Res A 31(1):35–50
17. DFT (2001), Perceptions of and attitudes to congestion, United Kingdom Department of Transport http://www.dft.gov.uk/pgr/statistics/datatablespublications/att/perceptionsofandattitudestoc5124
18. Kikuchi S, Mangalpally S, Gupta A (2005) Precision of predicted travel time, the responses of travellers, and satisfaction in the travel experience. In: International symposium on transportation and traffic theory, University of Maryland, College Park, Maryland, pp 447–465
19. Al-Mosaind MA (1998) Freeway traffic congestion in riyadh, Saudi Arabia: attitudes and policy implications. J Transp Geogr 6(4):263–272
20. Garson GD (2013) Factor analysis. Statistical Associates Publishers, Asheboro
21. Choo S, Mokhtarian PL (2008) How do people respond to congestion mitigation policies? A multivariate probit model of the individual consideration of three travel-related strategy bundles. Transportation 35(2):145–163
22. Mokhtarian PL (2004) Reducing road congestion: a reality check—a comment. Transp Policy 11(2):183–184
23. Mokhtarian PL, Raney EA (1997) Behavioral response to congestion: identifying patterns and socio-economic differences in adoption. Transp Policy 4(3):147–160
24. Ben-Akiva M, Lerman SR (1985) Discrete choice analysis: theory and application to travel demand. The MIT Press, Cambridge
25. Ye L, Hui Y, Yang D (2011) Traffic congestion-related attitudes and segments of travelers in Shanghai, China. In: Transportation research board 90th annual meeting, Paper #11-1117. Transportation Research Board, Washington DC
26. de Jong MG, Steenkamp J-BEM, Fox J-P et al (2008) Using item response theory to measure extreme response style in marketing research: a global investigation. J Mark Res 45(1):104–115
27. Veall MR, Zimmermann KF (1996) Pseudo-r^2 measures for some common limited dependent variable models. J Econ Surv 10(3):241–259

2

Study on the fatigue and wear characteristics of four wheel materials

G. Y. Zhou · J. H. Liu · W. J. Wang · G. Wen · Q. Y. Liu

Abstract The fatigue and wear characteristics of four different steel wheel materials are investigated in detail by using rolling contact fatigue and wear bench tests on a JD-1 apparatus, analyzing chemical composition and hardness, and performing profile analysis and micro-morphology analysis. The wear and fatigue behavior of one of the materials under different operation speeds is also investigated. The results show that the wear resistance of the materials has a positive correlation with their carbon content, while fatigue resistance has a negative correlation. Based on hardness analysis as a function of depth into the specimen, the thickness of layers with a steep hardness gradient has a negative correlation with the initial surface hardness in the tests using different materials. The hardness increments, however, have a positive correlation with initial surface hardness. The rolling tests on one material using different rotation speeds show that the hardness increments and the thickness of layers with a steep hardness gradient increase with the rotation speed. The analyses and experimental results demonstrate that two of the four materials exhibit good wear resistance and rolling contact fatigue resistance, making them suitable for either high-speed or heavy axle railroad operations.

Keywords Wheel material · Fatigue · Wear · Hardness

G. Y. Zhou (✉) · J. H. Liu · W. J. Wang · G. Wen · Q. Y. Liu
State Key Laboratory of Traction Power, Tribology Research Institute, Southwest Jiaotong University, Chengdu 610031, China
e-mail: zhouguiyuangift@sina.cn

1 Introduction

Researching high-performance wheel materials is important in wheel/rail development in order to reduce the wear between wheel and rail and prolong service life [1]. In the USA, Robles Hernández et al. [2] developed a high performance wheel steel, called SRI wheel steel, and compared it with seven other high-performance wheels, six pearlitic and one bainitic, manufactured by different companies. In China, Mi et al. [3] researched the wear characteristics of two types of cast steel wheel materials, which were named B+ and B grades of steel. Experimental inquiry is very important in new wheel material research, such as the experiments done by Cvetkovski et al. [4] using low-cycle fatigue tests on a new wheel material for passenger trains.

Wear is the most critical factor in the replacement of rails and wheels in commercial railroad systems, and in restricting the service life of wheels. Enhancing the wear resistance of wheels, therefore, can bring economic benefit to railway operations, and a large amount of railway research is spent on reducing the wear between the wheel and rail by simply reducing the weight loss or by reducing special wear forms such as corrugation.

We know that the hardness of a material directly relates to wear, and that increasing the hardness of the steel can reduce the wear of wheels and rails. There is a limit to the benefit of increasing the hardness of the wheel and rail, however, and simply improving the hardness of material to reduce the wear is not an effective method. Many studies have shown that a plastic deformation layer on the surface of hard steel is formed during wearing, significantly increasing the hardness of the worn surface [5–8]. This forms a special material which consists of a hard external material and a tough internal material which is ideal for railway operation. However, depending upon plastic deformation to improve surface hardness is a limited technique.

The roughness of surfaces can increase considerably during rolling contact experiments, which can cause high contact pressure and lead to plastic deformation of the materials [9]. The plastic deformation can accumulate, called ratcheting, and can ultimately lead to failure due to cracking. However, the contact surface asperities between the wheel and rail contribute to the tangential friction force at the wheel/rail interface, resulting in an increase of the adhesion coefficient [10]. Therefore, if the interface roughness can be held at a suitable value, it would be beneficial for operation safety and maintenance.

Fatigue of a material results in conditions such as fracture toughness and fracture brittleness, but the level of fatigue of a material is always evaluated by cracks and service time. In the railway industry, fatigue cracks are produced by abnormal braking heat [11], and the service time of a material is defined as the period from initial use to the time of fatigue crack initiation [12]. Finite element analysis (FEA) can simulate fatigue crack initiation effectively and conveniently, but experimental methods can directly obtain fatigue characteristics and are more reliable.

In this paper, the wear and fatigue characteristics of four different wheel materials are investigated using the JD-1 wheel/rail simulation facility and special analysis methods. Material weight loss is measured by weighing, and measurements of the surface, hardness, and fatigue cracks are performed using a scanning electron microscope (SEM) and an optical microscope.

2 Experimental details

2.1 Materials

The four types of wheel steel in this study are intended for trains, and the number labeling and chemical compositions in weight percentage of these materials are shown in Table 1. The carbon content of materials #3 and #4 were slightly higher than common wheel materials [4, 7, 13]. The main difference in composition among the four materials was the amount of carbon, which varied from 0.51 % to 0.72 %. The steel of the rail roller used was U71Mn rail steel, which is discussed in [14].

The pearlite microstructure of the four materials is shown in Fig. 1, revealing the typical ferritic-pearlitic structure which is tough and ductile, but soft. As the carbon content was increased, the amount of pro-eutectoid ferrite decreased and the micro-hardness increased, as listed in Table 2. Although there was no specific measurement of the pearlitic grain size, the grain size of the #1 material was clearly the smallest, and the pearlitic grain sizes increase with the carbon content.

The surface hardness of the materials as listed in Table 2 was measured on a micro-hardness tester (MVK-H21, Japan) using a 200 g load. Each specimen was measured at five different points to reduce the errors due to material non-uniformity, and each point was measured five times.

The surface hardness of the materials relates to the carbon content and metallographic analysis, as discussed previously, with our #1 material exhibiting the smallest value of hardness and #4 exhibiting the greatest.

2.2 Methods

Prior to the rolling tests, the mass, roughness, and hardness of the specimens were measured. The rolling contact fatigue test was then carried out, and the weight, roughness, and hardness of the specimens were subsequently measured again. To analyze the wear and fatigue damage mechanisms in the wheel materials, SEM was used to observe subsurface cracks and the scar morphology, and an optical microscope was used to analyze plastic deformations near the surface.

All experiments were carried out on a JD-1 wheel/rail simulation facility apparatus [7], as shown in Fig. 2. The tester was composed of a small wheel which served as the locomotive or rolling stock (called the "wheel roller") and a large wheel which served as the rail (called the "rail roller"). The rail roller was driven by a DC motor, and an opposing torque unit generated an opposing torque against the rotation direction imposed on the wheel roller.

The geometric size of the rollers was determined using the Hertzian simulation rule, shown in Eqs. (1) and (2):

Table 1 Compositions of four wheel materials in mass fraction (%)

Number	C	Mn	Si	P	S	H	Cr	Ni	Mo	V	Cu
#1	0.51	0.75	0.28	0.016	0.002	1.4	0.22	0.01	0.01	0.01	0.02
#2	0.58	0.72	0.25	0.012	0.001	0.8	0.18	0.01	0.01	0.01	0.01
#3	0.62	0.79	0.82	0.013	0.012	1.2	0.17	0.01	0.01	0.01	0.03
#4	0.72	0.81	0.86	0.014	0.016	2	0.02	0.01	0.01	0.01	0.02

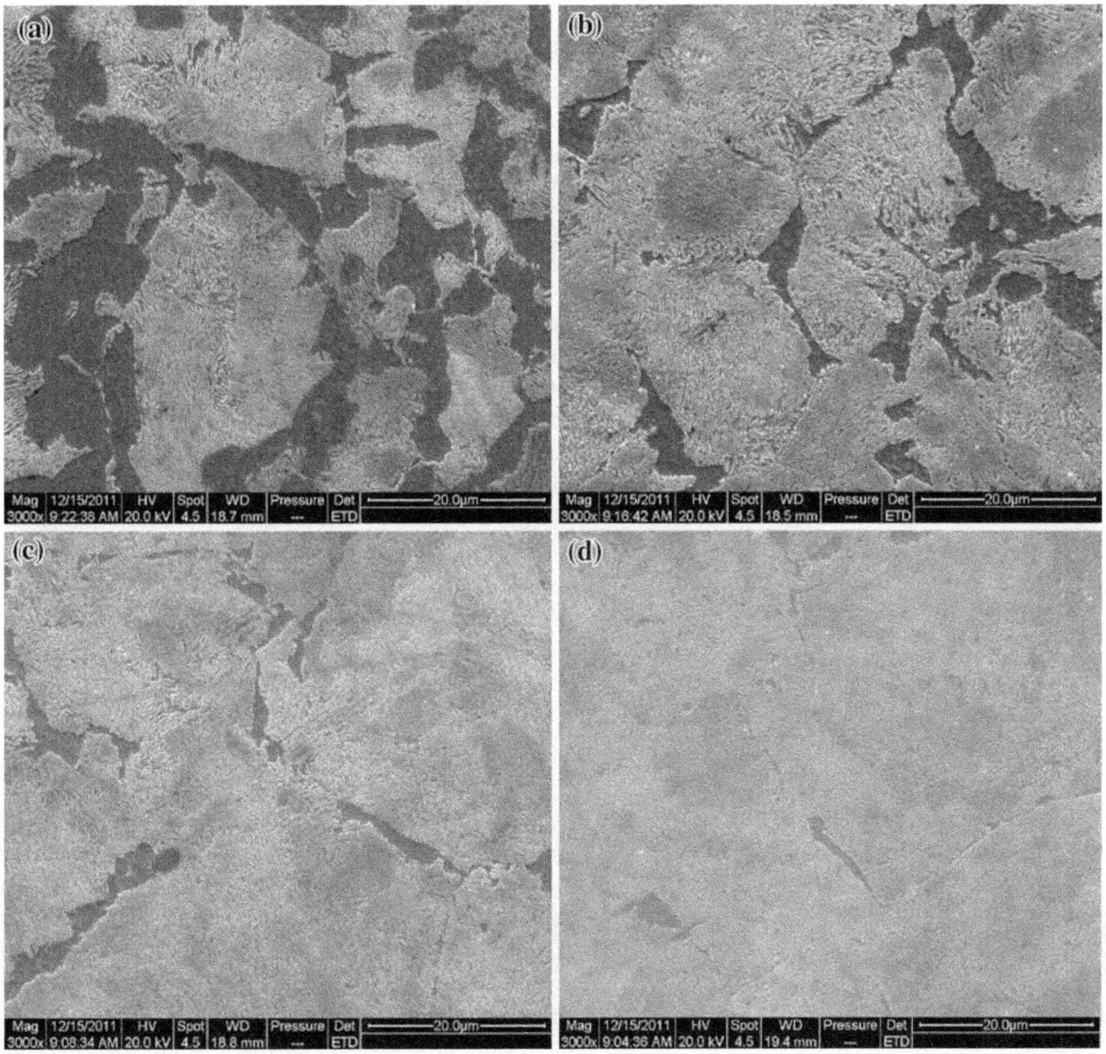

Fig. 1 SEM graphs of the pearlite microstructure (3,000×). **a** #1 material. **b** #2 material. **c** #3 material. **d** #4 material

Table 2 The surface hardness of materials

Number of material	#1	#2	#3	#4
Hardness (HV_{200g})	252.95	274.28	303.47	330.82

$$(P_{\max})_{\text{field}} = (P_{\max})_{\text{lab}}, \tag{1}$$

$$\left(\frac{a}{b}\right)_{\text{field}} = \left(\frac{a}{b}\right)_{\text{lab}}, \tag{2}$$

where $(P_{\max})_{\text{lab}}$ and $(P_{\max})_{\text{field}}$ are the maximum contact stresses in the laboratory and in the field, respectively; and $(a/b)_{\text{lab}}$ and $(a/b)_{\text{field}}$ are the ratios of the semi-major axis to the semi-minor axis of the contact ellipses between the wheel and rail in the laboratory and field, respectively. The schema of the rollers' geometric sizes as calculated by the above equations is shown in Fig. 3.

All experiments were conducted in dry and ambient conditions (temperature 18–23 °C, relative humidity 50 %–70 %), and all contact surfaces were cleaned with acetone prior to testing. All experimental parameters were determined by means of the Hertzian simulation [1], with the diameter of the rail roller set at 1,050 mm and the diameter of the wheel roller, which was cut from an actual wheel, set at 68 mm. The total number of cycles undergone by the wheel roller was 10^6, and the normal load used in the laboratory was 1,420 N, which simulated an actual field axle load of 19 t. The corresponding maximum contact stress as calculated by the Hertz formulae was 1,242 MPa. The rotation speeds of the wheel rollers were 32, 69, and 94 r/min, which simulated train speeds of 120, 250, and 350 km/h, respectively. Using an attack angle, which is the angle between the axes of the wheel and rail rollers, to simulate the curvature of an actual track, we used an attack angle of 0.3772° to simulate a curvature radius of 2,000 m. To simulate the traction condition, a force of 100 N was applied on the opposing torque unit to generate an

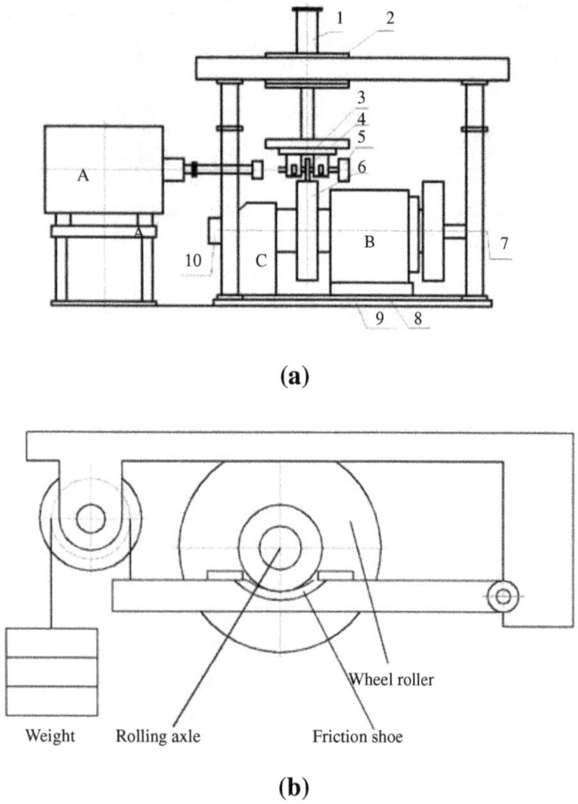

Fig. 2 Scheme of JD-1 wheel/rail simulation facility. **a** JD-1 wheel/rail simulation facility. *1* Normal loading cylinder, *2* loading carriage, *3* 3D loading sensor, *4* wheel roller, *5* opposing torque unit, *6* rail roller, *7* speed measuring motor, *8* turning plate, *9* base plate, *10* optical shaft encoder. **b** Opposing torque unit

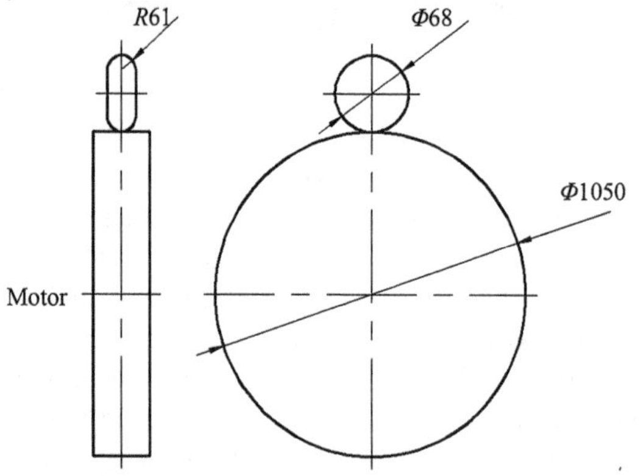

Fig. 3 Scheme of the wheel and rail rollers

opposing torque (Fig. 2b). The main parameters of all experiments are listed in Table 3.

3 Experimental results

A photograph of the four specimens after the rolling test for different materials (specimens A–D), together with a ruler, is shown in Fig. 4a. It is obvious that the width of the wear scars on the wheels decreases with the specimen series, from specimen A–D. Since all specimens have approximately the same original profile, it is logical to conclude that the wear resistance has a positive correlation with the carbon content of the material.

In addition, there exists an uneven wear phenomenon on all of the specimens, especially in specimens B, C, E, and F. Specimen E, obtained with rolling tests using different speeds, exhibits another unusual wear phenomenon of smooth areas, as pointed out in Fig. 4b. Also obtained in tests using different speeds, the corrugation pattern of specimen F is well-distributed along the circumferential direction, exhibiting smooth (trough) and unsmooth (crest) areas with a distance between of about 1 mm.

3.1 Scars of specimens under SEM

After rolling contact tests, small slices are cut from the specimens and cleaned using ultrasonic cleaning with alcohol and acetone before being observed with SEM (QUANTA200, FEI, England). The morphology of the scars evident on the specimen surfaces are shown in Fig. 5.

Analysis of the rolling surfaces shows the occurrence of flaking and adhesive wear for all specimens, including specimens E and F in Fig. 6. Surface ratcheting cracks are most evident in specimen D, and an example of one is shown in Fig. 5(d). Ratcheting cracks occur when the material loses its ductility due to plastic strain accumulation [15]. Specimens A, B, and C shows evidence of adhesive wear rather than flaking. This may indicate that the occurrence of the ratcheting phenomena in specimens A, B, and C is not as great as in specimen D, which may be due to the differences in carbon content and metallographic structure.

The worn surfaces of specimens E and F, tested using rotation speeds of 32 and 94 rpm, respectively, exhibit uneven wear phenomena (Fig. 6). Both the morphology of the scars and the wear mechanisms are different at different regions of uneven wear, as can be seen in Fig. 6b, c, e, and f. At the crest the surface is much rougher and ratcheting is the dominant wear mechanism, while in the trough the surface is smoother and adhesive wear is as significant as ratcheting effects. The special wear phenomenon called smooth area wear is observed macroscopically in Fig. 4b, and is observed at high magnification using SEM in Fig. 6a.

3.2 Weight loss and profile analysis

Figure 7 shows the material weight loss against the carbon content of all specimens during all tests. The weight loss is obtained using an electronic scale (JA4103, China) before

Table 3 Test parameters

Test type	Specimen series	Rotational speed of simulating rail (rad/min)	Material number
Test for different materials	A	69	#1
	B	69	#2
	C	69	#3
	D	69	#4
Test under different speed	E	32	#3
	F	94	#3

and after the rolling tests, and is a direct way to obtain the weight loss of specimens. The results indicate that the wear resistance is directly related to the carbon content, and specimens with a higher carbon content have greater wear resistance. Particularly, the least weight loss occurs in specimen D, which has the highest carbon content, and the greatest weight loss occurs in specimen a, which has the lowest carbon content. Specimens C, E, and F use the same material (#3) with different rotation speeds, and the weight loss is seen to have a negative correlation with the experimental rotation speed.

3.3 Results of test for different materials

The variation of surface hardness as a function of depth is measured to study the work-hardening effect, and the results are presented in Fig. 8. These plots show that the hardness of the specimens changes rapidly nearest the rolling surface (i.e., steep hardness gradient), and changes at a less rapid rate after a certain depth into the surface (i.e., gradual hardness gradient). The thickness of the hardened layer of all specimens is more than 200 μm. Comparing the post-experiment hardness values (Fig. 8) with the initial surface hardness of different materials (Table 2), we can see that the hardened layers are thicker than the plastic deformation layers observed in Figs. 9 and 10.

In the rolling tests of different materials (Fig. 8a), for the layers displaying a steep hardness gradient, the thickness of these layers has a negative correlation with the initial surface hardness, while the hardness increments have a positive one. This result is consistent with the results of the thickness of plastic deformation in Fig. 9.

For comparison, rolling tests with different rotation speeds using the #3 material show both the hardness increments and the thickness of layers displaying a steep hardness gradient of the three specimens to increase with rotation speed (Fig. 8b). These results are consistent with the results in Fig. 10. It is obvious that the dominant wear damage mechanism in the specimen with high surface hardness and with high experimental rotation speed is ratcheting rather than an adhesive wear mechanism. It is known that ratcheting can lead to fatigue cracks at the near-surface layer, so generating fatigue failure will therefore be easier in the cases involving high surface hardness and high experimental rotation speed.

Both the plastic deformation and the hardness increments have a positive correlation with the contact stress. Some research indicates that severe plastic deformations in the near-surface layer of the rail cross-section penetrate just a few tens of microns into the material [9, 11, 15]. In Fig. 8 we see that at thicknesses greater than about 100 μm into the material, the hardness gradient is quite gradual.

The results of the SEM observations are shown in Fig. 11, and the observation schematic is shown in Fig. 12. It is obvious that the fatigue resistance of specimen A is so perfect that no cracks can occur, and that the length of fatigue cracks in the other specimens increase as a function of increasing carbon content. In other words, the fatigue resistance decreases with carbon content in this rolling test. Moreover, all of the fatigue cracks propagate from the surface into the material at approximately the same angle.

Examining the cross sections of specimens from A to D with an optical microscope (OLYMPUS BX60M, Japan), we can observe significant plastic bands due to ratcheting,

Fig. 4 Macrograph of specimens. **a** For different materials. **b** For different speeds

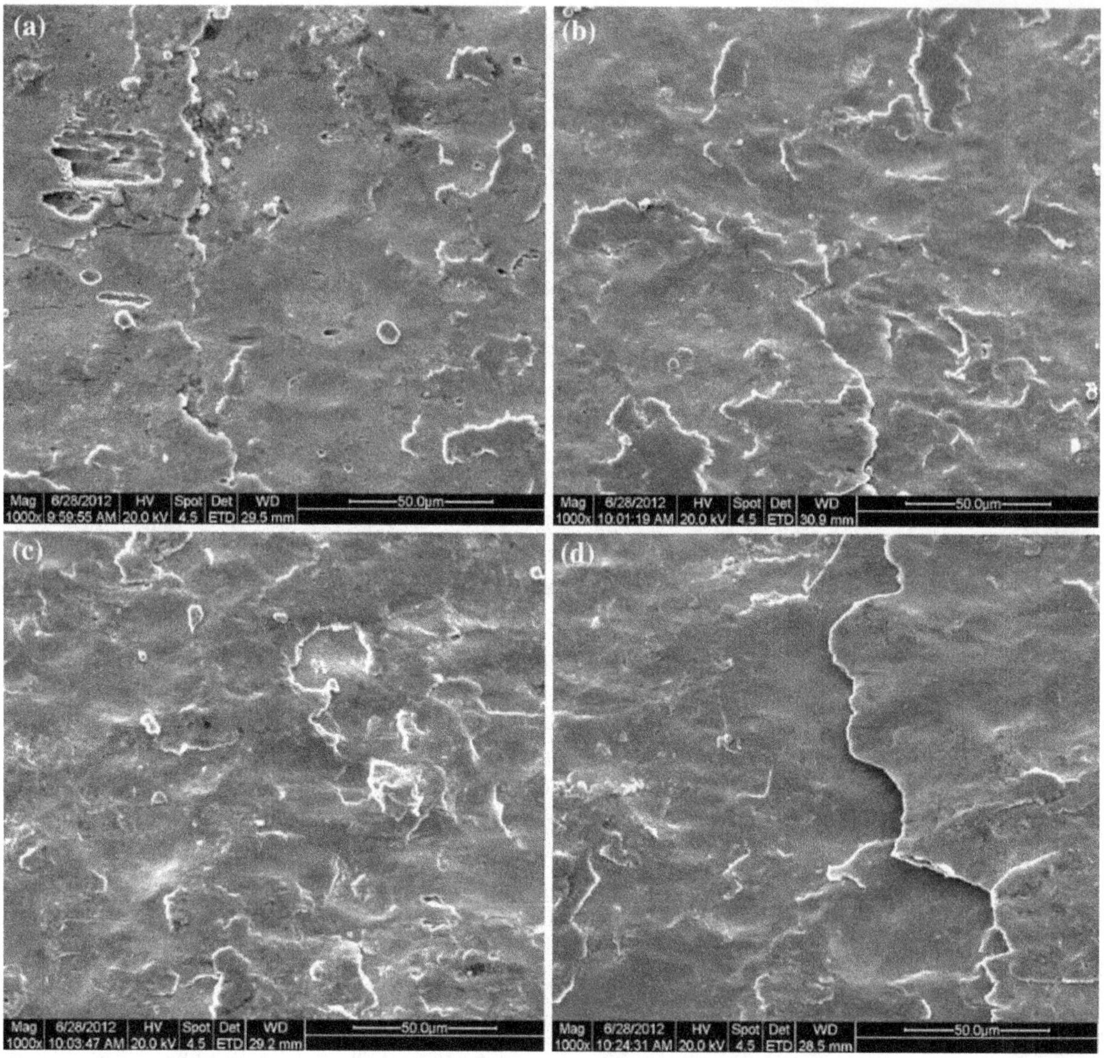

Fig. 5 Scars of specimens A to D under SEM. **a** Specimen A. **b** Specimen B. **c** Specimen C. **d** Specimen D

as shown in Fig. 9. The plastic deformation of specimens A and B are mild, while the deformations of specimens C and D are severe. In addition, both specimens C and D exhibit significant fatigue cracks.

3.4 Results of test for different rotational speeds

Optical microscope images of the cross sections of specimens E and F, un-etched after polishing, are shown in Fig. 13. These images show that longer fatigue cracks occur after the higher rotation speed is applied (Fig. 13b), but the amount of small cracks is greater at the lower rotation speed (Fig. 13a).

The thickness of the deformed layer is measured to be about 115 μm for specimen E and about 180 μm for specimen F (Fig. 10). The deformed layer of specimen F is much thicker than that of E, and is related to wear rate and ratcheting. The wear rate is greater at lower rotation speeds, but the effect of ratcheting is almost the same for all speeds in the test, so the deformed layer increases more slowly at a slower rotation speed.

4 Discussion

The wear process of materials is very complex. During the initial stages of the rolling–sliding process, the strain accumulation and hardening rate is strongest and oxidative wear is the main wear mechanism [16]. The initial hardening of the material plays an important role during the whole wear process of wheel and rail. If there is a hardening layer on the surface of the rollers and if the load is insufficient to generate severe flaking during the subsequent operation, the material, which has internal toughness and an induced external hardness, can experience reduced wear and fatigue damage and convey an extended service life for rollers composed of it.

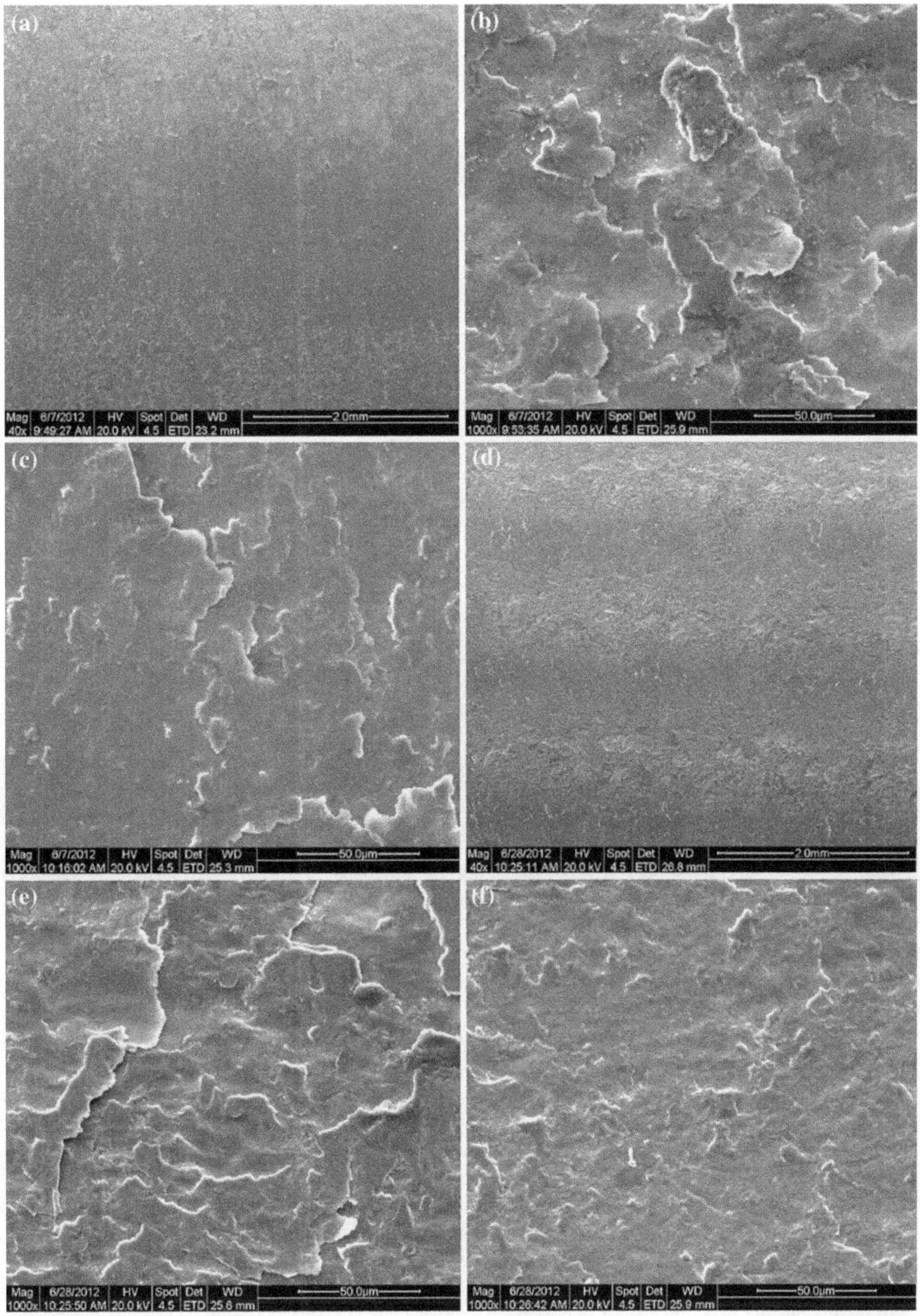

Fig. 6 Scars of specimens E and F under SEM. **a** Smooth area of specimen E. **b** Crest of specimen E. **c** Trough of specimen E. **d** Uneven wear pattern of specimen F. **e** Crest of specimen F. **f** Trough of specimen F

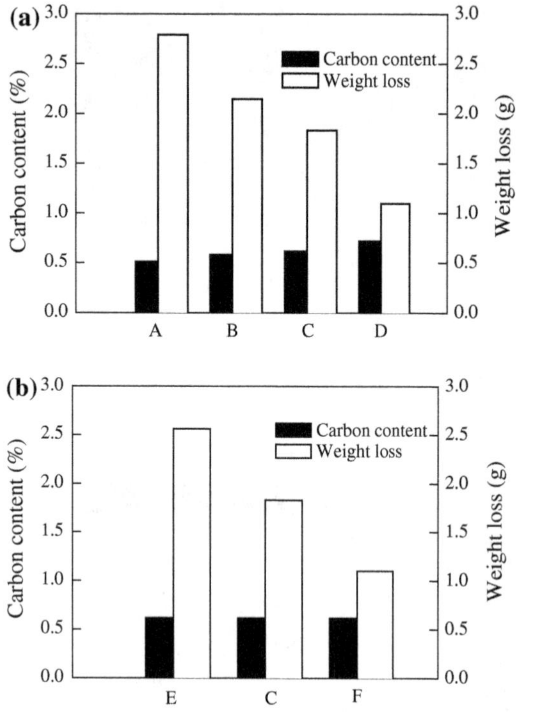

Fig. 7 The weight loss of six specimens. **a** For different materials. **b** For different rotational speeds

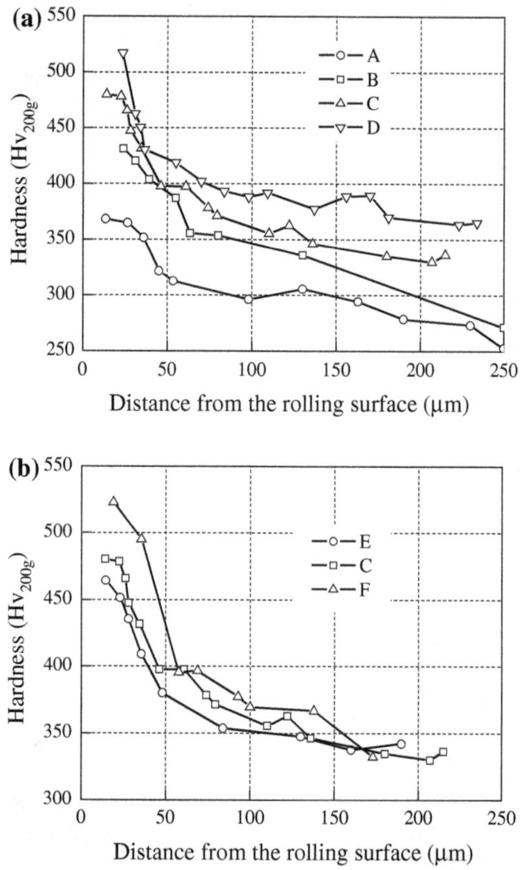

Fig. 8 The variation of hardness as a function of depth from the rolling–sliding surface

To some extent, strain accumulation (ratcheting) and wear are competitive mechanisms, with one reducing the influence the other has on the material. Ratcheting can cause subsurface hardening which can slow down the wear rate directly (higher hardness means greater wear resistance), and therefore ratcheting can reduce wear. Conversely, wear reduces the influence of ratcheting via material attrition. No matter which, if one process becomes dominant, then severe damage, such as severe wear and/or fatigue fractures, is the result. Therefore, the best relationship to maintain between rail and wheel is a balance of ratcheting and wear in order to achieve the state of steady wear [16], thereby facilitating maximum service life.

There also exists a competitive and restrictive coupling mechanism between wear and plastic deformation, which is described by Zhong [14]. The plastic deformation and fatigue cracks formed by the ratcheting effect [17] can improve the wear resistance of the material, but the increase in hardness of the surface and subsurface can reduce the fatigue resistance.

Analysis of the rolling surface of these materials shows the occurrence of flaking due to ratcheting and the adhesive wear mechanism for all specimens, but uneven wear gives evidence that the wear mechanism varies in different regions of the surface. The relationship between flaking due to ratcheting and the adhesive wear mechanism could impact the morphology of these scars.

The differing carbon content of the wheels leads to the different damage forms of their surfaces. As the carbon content increases, there is a gradual transition from a wear-dominant to a fatigue-dominant mechanism due to the competitive and restrictive coupling between wear and fatigue, with the stronger mechanism becoming dominant.

The material presents a wear-dominant mechanism when the carbon content is lower and a fatigue-dominant mechanism when the carbon content is higher. The wear mechanism is more apparent than other damage forms in specimens A and B, with specimen A showing more wear mechanism than specimen B. With an increase in the carbon content, the length of fatigue cracks increase, showing more evidence of the fatigue damage mechanism on the surface of specimens. Consequently, specimens C and D present a more dominant fatigue damage mechanism than specimens A and B.

In Fig. 7, the weight loss of specimens A and B is greater than specimens C and D, which means there is more wear damage on specimens A and B. When we look at the evidence given by the combined Figs. 9 and 11, we see that when wear is the dominant damage form (A, B), the plastic deformation is slighter and almost no fatigue cracks exist on the surface layer.

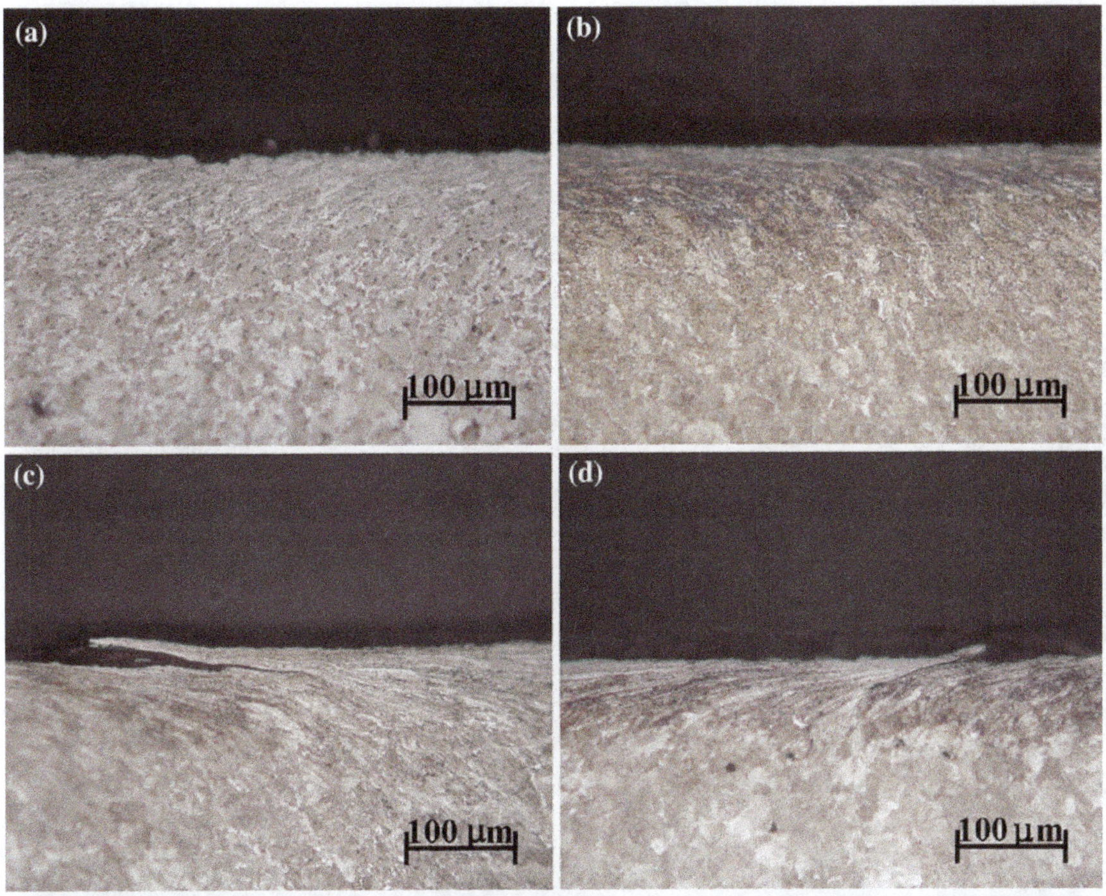

Fig. 9 Plastic deformation of specimens A to D. **a** Specimen A. **b** Specimen B. **c** Specimen C. **d** Specimen D (Etching: Nital 3 %)

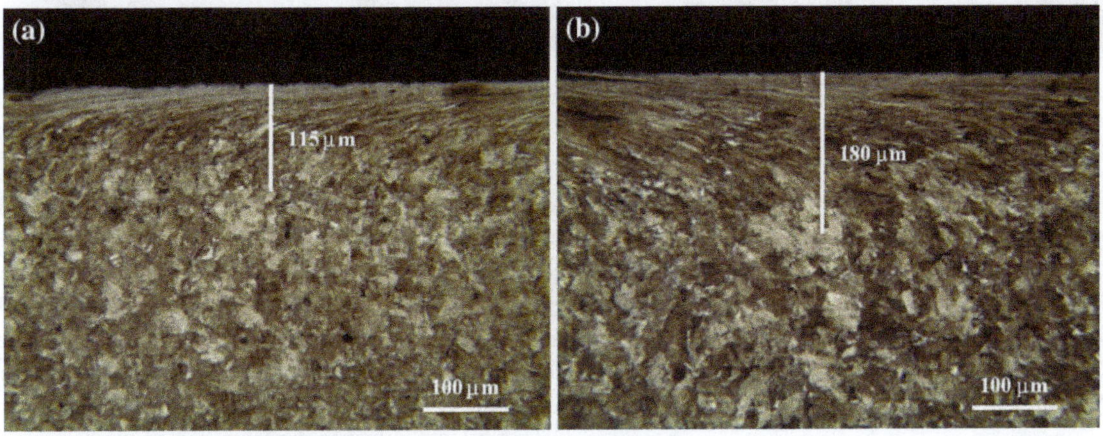

Fig. 10 Plastic deformation of specimens E and F. **a** Specimen E. **b** Specimen F (Etching: Nital 3 %)

The hardness increments of lower carbon content materials is smaller, as shown in Fig. 8a, and hardness has a direct influence on wear resistance. There are fewer surface cracks on specimens A, B, and C compared with specimen D in Fig. 5, which means that specimen D exhibits more of the fatigue damage mechanism than do the other specimens. A smaller hardness increment makes less of a contribution toward improving the wear resistance, so the weight loss of lower carbon content materials is much greater than in the others. This indicates that for these specimens the wear mechanism is the dominant damage form, and the SEM observations of worn surfaces in Fig. 5 corroborate this.

Changing the test rotation speed can also lead to a change of the damage mechanism on the wheel surface.

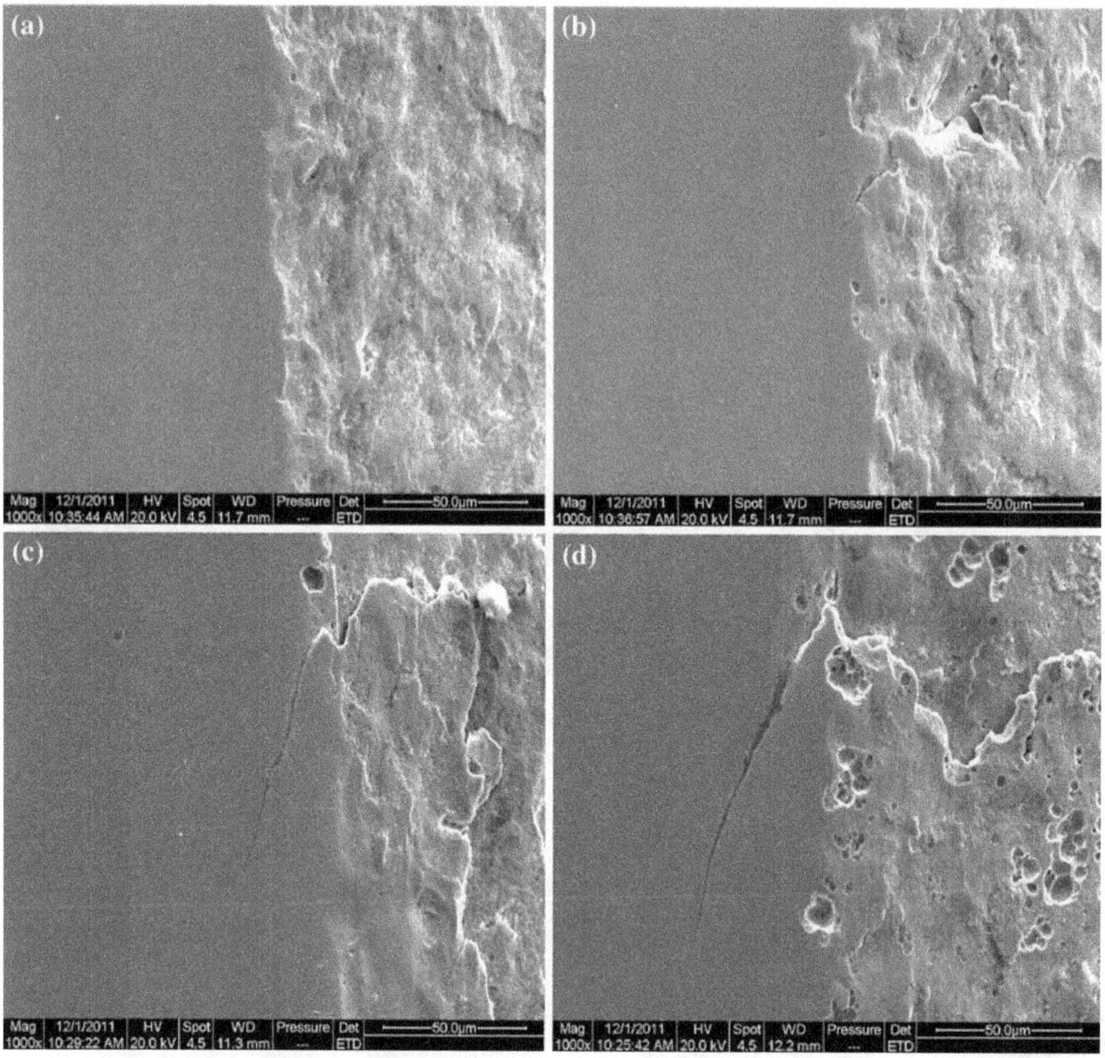

Fig. 11 The SEM of four specimens for different materials, the *left side* is cross section and the *right side* is contact surface. **a** A. **b** B. **c** C. **d** D

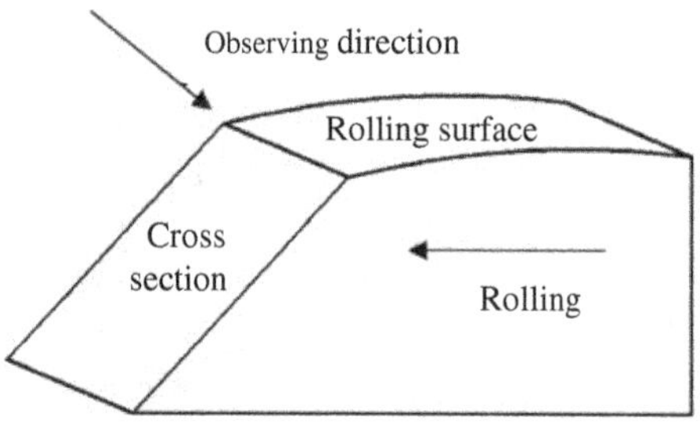

Fig. 12 Scheme of SEM observing

The specimen surfaces exhibit a wear-dominant mechanism when the test rotation speed is low and transition into a fatigue-dominant mechanism when the test rotation speed increases. Analyzing specimens E and F, we see that the cracks in specimen E are smaller and more numerous (Fig. 13a), and the weight loss of specimen F is much greater and the cracks are larger (Fig. 13b). This means that the wear mechanism is more dominant in specimen E than in F, and the cracks in specimen E were ground off before propagating. From the SEM observations of the plastic deformation (Fig. 10), we also find that the thickness of the plastic deformation layer in specimen E is smaller than F, which means that the fatigue damage mechanism is secondary and the wear mechanism is dominant.

Two effects influence the transition of the damage form mechanism upon changing the test rotation speed. On one hand, the increasing test rotation speed reduces the wear loss of the wheel roller so that the influence of the wear mechanism in specimen F is less. On the other hand, increasing the test rotation speed can increase the vibration of the simulated rollers, which can increase the dynamic load coefficient of the contact load between the simulated rollers. This effect can increase the fatigue damage and hardening of the contact surface, and the results in Figs. 8b,

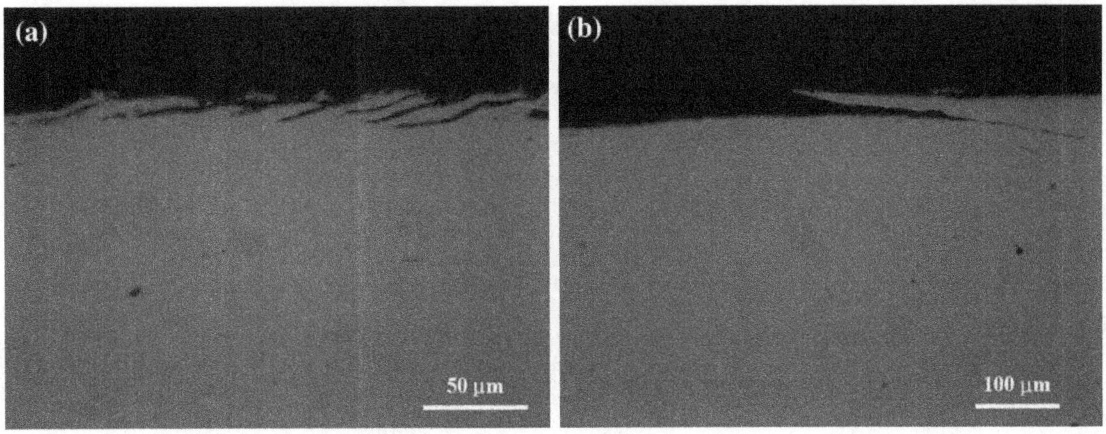

Fig. 13 Optical microscope image of three specimens on different rotational speed. **a** E. **b** F

10, and 13 all show evidence of this. Both of these effects transition the damage mechanism from wear-dominant into fatigue-dominant as the rotation speed increases.

According to the experimental results, under the same rolling conditions, the weight loss of a specimen reduces as the carbon content of the material increases. The surface hardness of all specimens is found to increase and the wear resistance thereby improves. The fatigue resistance of a material is also seen to be related to the carbon content.

Comparing the four materials, the wear resistance of the #4 material is better than the others, but because the wear of the wheel and the rail must both be taken into account, the wear resistance is not the only aspect to consider when choosing a material [1]. Although the specimens using the #1 material didn't exhibit any cracks in the surface cross section, these specimens experienced a much greater weight loss than specimens made from the other materials, making material #1 a poor choice of steel for actual wheels for economic reasons. Therefore, both #2 and #3 materials are proper choices for railway operation because material #2 has good fatigue resistance and material #3 has good wear resistance. The #2 material is suitable for high-speed railway operation and the #3 material is suitable for heavy axle operation.

5 Conclusions

In this paper, the fatigue and wear characteristics of four types of wheel materials are investigated using chemical composition analysis, rolling contact fatigue, and wear bench tests on a JD-1 apparatus, profile analysis and micro-morphology analysis. The conclusions from this study are as follows:

1. Both the wear resistance and the fatigue resistance are directly related to the carbon content of each material. The wear resistance has a positive correlation with the carbon content, while the fatigue resistance has a negative one. After 106 rotations during rolling tests in the JD-1 wheel/rail simulation facility, uneven wear occurred on all of the specimens.
2. The relationship between flaking due to ratcheting and the adhesive wear mechanism can influence the topography of scars.
3. As the rotation speed of the rolling test increases, the hardness increments, and the thickness of layers exhibiting a steep hardness gradient are seen to increase.
4. All cracks initiate at and propagate along the plastic deformation line, and the extent of the severity of fatigue damage, such as length of the fatigue crack, has a positive correlation with the carbon content of the material.

The analysis suggests that both #2 and #3 materials are proper choices for railway operation, because their resistance to wear and fatigue is moderate and they have the capacity to achieve a state of steady wear.

Acknowledgments The work is supported by National Natural Science Foundation of China (Nos. 51174282, U1134202), the Fundamental Research Funds for the Central Universities (No. SWJTU12CX037), and Project supported by the Innovative Research Teams in Universities (No. IRT1178).

References

1. Jin XS, Liu QY (2004) Tribology of wheel and rail. China Railway Press, Beijing
2. Robles Hernández FC, Kalay S, Stone D, Cummings S (2011) Properties and microstructure of high performance wheels. Wear 271:374–381

3. Mi GF, Liu YL, Zhang B et al (2009) Wear property of cast steel wheel material in rail truck. J Iron Steel Res Int 16(3):73–77
4. Cvetkovski K, Ahlström J, Karlsson B (2011) Monotonic and cyclic deformation of a high silicon pearlitic wheel steel. Wear 271:382–387
5. Mazzù A, Donzella G, Faccoli M et al (2011) Progressive damage assessment in the near-surface layer of railway wheel-rail couple under cyclic contact. Wear 271:408–416
6. Olofsson U, Telliskivi T (2003) Wear, plastic deformation and friction of two rail steels—a full-scale test and a laboratory study. Wear 254:80–93
7. Liu QY, Zhang B, Zhou ZR (2003) An experimental study of rail corrugation. Wear 255:1121–1126
8. Garnham JE, Davis CL (2011) Very early stage rolling contact fatigue crack growth in pearlitic rail steels. Wear 271:100–112
9. Kapoor A, Franklin FJ, Wong SK, Ishida M (2002) Surface roughness and plastic flow in rail wheel contact. Wear 253:257–264
10. Chen H, Ban T, Ishida M, Nakahara T (2002) Adhesion between rail/wheel under water lubricated contact. Wear 253:75–81
11. Sakamoto H, Toyama K, Hirakawa K (2000) Fracture toughness of medium-high carbon steel for railroad wheel. Mater Sci Eng A 285:288–292
12. Ringsberg JW, Lindbäck T (2003) Rolling contact fatigue analysis of rails including numerical simulations of the rail manufacturing process and repeated wheel-rail contact loads. Int J Fatigue 25:547–558
13. Wang WJ, Shen P, Song JH et al (2011) Experimental study on adhesion behavior of wheel/rail under dry and water conditions. Wear 271:2699–2705
14. Zhong W, Hu JJ, Li ZB et al (2011) A study of rolling contact fatigue crack growth in U75V and U71Mn rails. Wear 271:388–392
15. Zapata D, Jaramillo J, Toro A (2011) Rolling contact and adhesive wear of bainitic and pearlitic steels. Wear 271:393–399
16. Tyfour WR, Beynon JH, Kapoor A (1995) The steady state wear behaviour of pearlitic rail steel under dry rolling–sliding contact conditions. Wear 180:79–89
17. Beynon JH, Garnham JE, Sawley KJ (1996) Rolling contact fatigue of three pearlitic rail steels. Wear 192:94–111

Application of artificial neural networks for operating speed prediction at horizontal curves: a case study in Egypt

Ahmed Mohamed Semeida

Abstract Horizontal alignment greatly affects the speed of vehicles at rural roads. Therefore, it is necessary to analyze and predict vehicles speed on curve sections. Numerous studies took rural two-lane as research subjects and provided models for predicting operating speeds. However, less attention has been paid to multi-lane highways especially in Egypt. In this research, field operating speed data of both cars and trucks on 78 curve sections of four multi-lane highways is collected. With the data, correlation between operating speed (V_{85}) and alignment is analyzed. The paper includes two separate relevant analyses. The first analysis uses the regression models to investigate the relationships between V_{85} as dependent variable, and horizontal alignment and roadway factors as independent variables. This analysis proposes two predicting models for cars and trucks. The second analysis uses the artificial neural networks (ANNs) to explore the previous relationships. It is found that the ANN modeling gives the best prediction model. The most influential variable on V_{85} for cars is the radius of curve. Also, for V_{85} for trucks, the most influential variable is the median width. Finally, the derived models have statistics within the acceptable regions and they are conceptually reasonable.

Keywords Artificial neural networks · Horizontal curve · Multi-lane highways · Operating speed · Prediction models · Regression models · Roadway factors

A. M. Semeida (✉)
Civil Engineering Department, Specialization of Transportation and Traffic Engineering, Faculty of Engineering, Port Said University, Port Said, Port Fouad 42523, Egypt
e-mail: semeida77@hotmail.com

1 Introduction

Horizontal curves have long been recognized as having a significant effect on vehicle speeds. They have therefore been afforded a great deal of attention by researchers. Design features for multi-lane roadways, such as curvature and superelevation, are directly related to, and vary appreciably with, design speed. Other features, such as widths of lanes and shoulders are not directly related to design speed, but they also affect vehicle speeds. Therefore, wider lanes and shoulders should be considered for higher design speed [1]. In this work, a driver's speed under free-flow conditions avoids the effect of traffic flow on vehicle speed, as only the effect of horizontal curves and highway geometry on operating speed is considered, as executed by Hashim [2]. The 85th-percentile value of the distribution of observed speeds (V_{85}) is the most frequently used in characterizing measure of operating speed associated with a particular location or geometric feature.

Past researches did a lot of work for two-lane rural [3–7]. A large number of studies used radius as an explanatory variable for operating speed prediction while a few used length of curve. Previous research for speed prediction on two-lane rural highways indicates that there are several important elements in determining speed on horizontal curves. Curve radius, superelevation, deflection angle, degree of curvature, length of curve, and cross section are examples of variables that have been used in the regression equations to predict operating speeds on horizontal curves. Curve radius is considered to be the most important element in determining operating speed on horizontal curves; therefore, most researchers have used it as the dominant independent variable in their regression analyses [8].

Gong and Stamatiadis [9] studied 50 horizontal curves that are located in rural four-lane highways in Kentucky.

They derived two models for operating speed. The first one was for inside lane and the other was for outside lane, respectively. For the first one, they concluded that a surfaced shoulder and logarithm of horizontal curve length were positively correlated with V_{85}. For the other model, a surfaced shoulder, horizontal curve radius, and the ratio of the horizontal curve length to radius were positively correlated with V_{85}. In addition, the first model explained nearly 65 % of the variability in the 85th-percentile inside lane operating speeds. Also, the other model explained approximately 43 % of the variability in the 85th-percentile outside lane operating speeds.

Cheng et al. [10] studied 30 horizontal curves that are located in HU-Ning expressway, Jiangsu province, China. The correlation analysis results showed better correlations between V_{85} (operating speed) and alignment. This study proposed four speed predicting models (linear or curve fitting models), respectively, of cars and trucks. Results of modeling showed that deflection angle and radius of curve were the two most important parameters for predicting operating speeds of cars and trucks.

Prediction and estimation of speeds on multi-lane rural highways are of great significance to planners and designers; therefore, all proposed speed prediction models should be validated, and the accuracy of their results should be evaluated. In this paper, the first part in the analysis involves the prediction of operating speed for cars and trucks on horizontal curves using conventional regression models. The modeling of operating speed on curved roadway using ANN models is another aspect of this paper.

2 Study sites and field data and methodology

2.1 Study sites and field data

This work uses 78 horizontal curves from two categories of multi-lane highways in Egypt. These categories are as follows:

(1) Agricultural highways category which includes two roads as Cairo-Alexandria agricultural highway (CAA) and Tanta-Damietta agricultural highway (TDA).

(2) Desert highways category which includes two roads as Cairo-Alexandria desert highway (CAD) and Cairo-Ismailia desert highway (CID).

The collected data are divided into road geometric and spot speed data.

2.1.1 Road geometric data

This data presents the key independent variables in the analysis. Some of this data are collected directly from site investigation which includes lane width, right shoulder width, number of lanes in each direction, median width, and pavement width. The horizontal curves data are extracted from Abdalla [11] who worked with the survey team of General Authority of Roads, Bridges and Land Transport in Egypt (GARBLT) [12]. The horizontal curve properties include radius of curve, deflection angle, length of curve, and superelevation. All the previous variables, their symbols, and statistical analysis are provided in Table 1.

2.1.2 Spot speed data

Speed data presents the key dependent variables in analysis, which are divided into operating speed for cars and trucks separately. In this work, a driver's speed under free-flow conditions avoids the effect of traffic flow on vehicle speed, as only the effect of horizontal curves and highway geometry on operating speed is considered. Free-flow speeds are collected for passenger cars and trucks. The passenger cars include taxis, private cars, vans, and jeeps. While the term "truck" refers to any combination of single- or multi-unit vehicles having at least one axle with dual wheels. The trucks contain trucks, trucks with trailers, semi-trailers, and multiple trailer road trains. Spot speed data are collected using radar gun (version LASER 500

Table 1 Statistical analysis and symbols of independent variables

Variable	Variable symbol	Max.	Min.	Avg.	SD
Lane width (m)	L_W	3.65	3	3.6	0.1
Pavement width in one direction (m)	P_W	14.6	6	8.36	1.8
Right shoulder width (m)	S_W	3	1	1.17	0.44
Number of lanes in each direction in lanes	N_L	4	2	–	–
Median width (m)	M_W	8	2	5.2	1.8
Radius of curve (m)	R	780	40	402.5	231.7
Deflection angle (°)	D_A	70.55	14.25	27.89	16.68
Length of curve (m)	L_C	256.72	39.71	142	60.26
Superelevation (%)	e	12	2	4.7	3.5

with ±1 km/h accuracy) placed at many points along each horizontal curve in hidden places outside road so as not to be visible to drivers (see Fig. 1). Vehicles traveling in free-flow conditions are considered to have time headways of at least 5 s. Then, the main dependent variable is the average operating speed of a curve. The number of speeds collected at each horizontal curve ranges from 100 to 160 for each cars and trucks, which lead to nearly 18,000 spot speeds. Speeds are carried out in working days, during daylight hours. During all data collection periods, the weather is clear and the pavement is dry and in a good condition. Generally, to assure validity of Pearson correlation, there is a demand that each set of selected data should follow normal distribution. Using Kolmogorov–Smirnov test, it is found that the distribution of the data is normal and could not be rejected at the 95 % confidence level. Operating speed at each observation point is defined as the 85th percentiles of collected speed data (V_{85}). Then, the 85th percentiles of collected speed data of each observation point (V_{85}), respectively, of cars and trucks, are calculated for further correlation analysis and modeling as stated by Hashim [2]. The sample size requirements for V_{85} were determined by [1]:

$$N = \frac{\sigma^2 K^2 (2 + u^2)}{2E^2}, \tag{1}$$

where N is the least number of sample size, σ is the estimated sample standard deviation, K is the constant corresponding to the desired confidence level of 95 % (K=1.96), E is the permitted error in the average speed estimation (±2 km/h), and u is the constant corresponding to the V_{85} u=1.04.

The operating speed values for cars (V_{85C}) and trucks (V_{85T}) at each horizontal curve are provided in Table 2.

2.2 Methodology

The methodology of operating speed prediction in the present research includes two main methods: (1) regression models and (2) ANN models.

2.2.1 Regression models

There are nine independent variables (geometric variables) and two dependent variables (speed variables) as stated in the previous section. The present research proposes two speed predicting models, respectively, of cars and trucks. To obtain the best model for the prediction, multiple linear regression method is used in modeling.

First, the correlation between V_{85} and the selected independent variables is analyzed. The significant variables from the correlation analysis are chosen for the final prediction model. Second, stepwise regression analysis is used to select the most statistically significant independent variables with V_{85} in one model. Stepwise regression starts with no model terms. At each step, it adds the most statistically significant term (the one with lowest P value) until the addition of the next variable makes no significant difference. An important assumption behind the method is that some input variables in a multiple regression do not have an important explanatory effect on the response. Stepwise regression keeps only the statistically significant terms in the model. Finally, the adjusted R^2 and root mean square error (RMSE) values are calculated for each model.

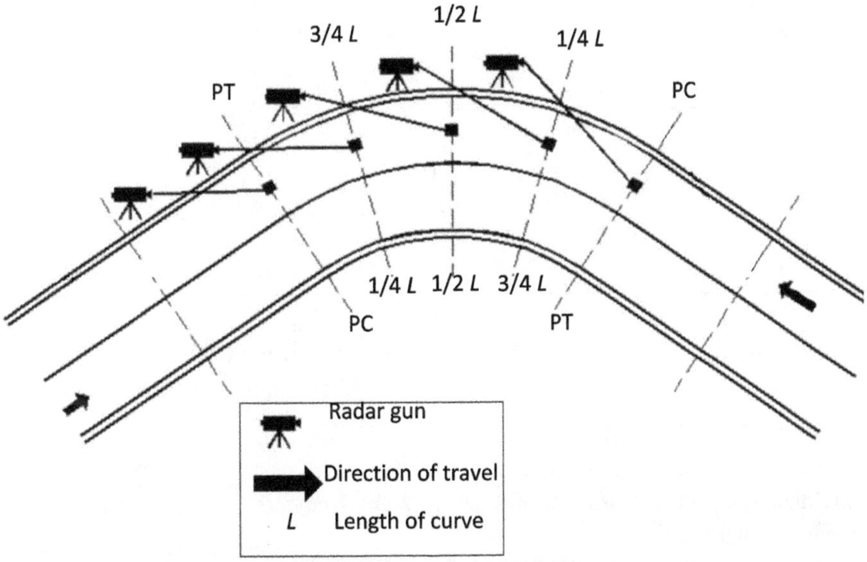

Fig. 1 Schematic location of radar meters on horizontal curves

Table 2 Operating speeds values for cars (V_{85C}) and trucks (V_{85T}) at all curves

Road name	Curve no.	V_{85C} (km/h)	V_{85T} (km/h)	Road name	Curve no.	V_{85C} (km/h)	V_{85T} (km/h)
CID	1	46.4	43.5	TDA	40	92.49	80.77
CID	2	39.1	30.8	TDA	41	69.47	60.67
CID	3	61	48.4	TDA	42	80.78	70.55
CID	4	56.1	47.8	TDA	43	86.89	75.88
CID	5	42.38	35.33	TDA	44	94.21	82.27
CAA	6	43.3	37.7	TDA	45	91.05	79.51
CAA	7	35.6	28.7	TDA	46	98.66	86.16
CAA	8	37.3	32.7	TDA	47	29.36	25.64
CAA	9	41	34.7	TDA	48	89.64	78.29
CAA	10	74.8	61.3	TDA	49	88.82	77.56
CAA	11	84.35	71.67	TDA	50	60.86	53.15
CAA	12	83.7	68.54	TDA	51	80.76	70.53
CAA	13	82	69.25	TDA	52	88.68	77.44
CAA	14	77.64	64.44	TDA	53	88.69	77.45
CAA	15	77.33	59.93	TDA	54	63.21	55.2
CAA	16	81	67.1	TDA	55	63.13	55.14
CAA	17	83.62	66.2	TDA	56	90.57	79.09
CAA	18	37.67	31.74	TDA	57	89.58	78.23
CAA	19	46.85	41.55	TDA	58	101.48	88.63
CAA	20	90.27	73.9	TDA	59	82.29	71.86
CAA	21	82.56	67.58	TDA	60	75.55	65.98
CAA	22	84.36	68.56	TDA	61	70.75	61.79
CAA	23	41.95	39.49	TDA	62	26.15	22.84
CAA	24	58.4	54.45	TDA	63	92.42	80.71
CAA	25	68.29	58.86	TDA	64	88.51	77.3
CAA	26	59.98	56.49	TDA	65	101.5	88.64
CAA	27	33.63	30.97	TDA	66	86.65	75.67
CAA	28	68.73	61.17	TDA	67	100.2	87.51
CAD	29	99.25	77.33	TDA	68	83.98	73.34
CAD	30	99.28	78.62	TDA	69	72.32	63.15
TDA	31	73.27	63.98	TDA	70	92.31	80.61
TDA	32	87.31	76.24	TDA	71	83.18	72.64
TDA	33	81.76	71.4	TDA	72	86.38	75.43
TDA	34	87.31	76.24	TDA	73	69.63	60.81
TDA	35	84.72	73.99	TDA	74	81.31	71.01
TDA	36	90.29	78.85	TDA	75	90.57	79.09
TDA	37	87.13	76.09	TDA	76	74.02	64.64
TDA	38	89.81	78.43	TDA	77	62.42	54.51
TDA	39	81.75	71.4	TDA	78	41.16	37.21

Several precautions are taken into consideration to ensure integrity of the model as follows [13]:

(1) The signs of the multiple linear regression coefficients should agree with the signs of the simple linear regression of the individual independent variables and agree with intuitive engineering judgment;

(2) There should be no multicollinearity among the final selected independent variables; and

(3) The model with the smallest number of independent variables, minimum RMSE, and highest R^2 value is selected.

2.2.2 *ANN models*

In general, ANNs consist of three layers, namely, the input, the hidden, and the output layers. In statistical terms, the input layer contains the independent variables and the output layer contains the dependent variables. ANNs typically start out with randomized weights for all their neurons. When a satisfactory level of performance is reached the training is ended and the network uses these weights to make a decision [14].

The experience in this field is extracted from Semeida [15, 16]. In his research, the multi-layer perceptron (MLP) neural network models give the best performance of all models. In addition, this network is usually preferred in engineering applications because many learning algorithm might be used in MLP. One of the commonly used learning algorithms in ANN applications is back propagation algorithm (BP) [17], which is also used in this work.

The overall dataset of 78 curve sections is divided into a training dataset and a testing dataset. As in [18], the training data set varies from 70 % to 90 % and the testing data set varies from 10 % to 30 %. Model performances are RMSE and R^2 for testing and training data set in one hand and for all data set in the other hand.

So many trials are done to reach the suitable percentage between training and testing data that gives the best performance for cars and trucks speed models. In addition, over fitting can be avoided by randomizing the 78 curves before training the network to reach the best performance for both training and testing data. The performance of testing data must be good as training data (R^2 must not be smaller than 0.7) [19].

Table 3 Correlations between V_{85C} and independent variables

	V_{85C}	N_L	L_W	P_W	S_W	M_W	L_C	D_A	e	R
V_{85C}										
PC	1	−0.063	0.197	−0.049	−0.298**	0.857**	0.639**	−0.89**	−0.562**	0.675**
Sig.		0.585	0.084	0.668	0.008	0.000	0.000	0.000	0.000	0.000
N_L										
PC	−0.063	1	−0.285*	0.992**	0.501**	−0.226*	0.197	0.019	−0.111	0.149
Sig.	0.585		0.011	0	0	0.047	0.084	0.866	0.334	0.194
L_W										
PC	0.197	−0.285*	1	−0.165	0.015	0.347**	−0.003	−0.162	−0.024	0.047
Sig.	0.084	0.011		0.149	0.897	0.002	0.982	0.158	0.838	0.684
P_W										
PC	−0.049	0.992**	−0.165	1	0.522**	−0.194	0.194	0.01	−0.112	0.15
Sig.	0.668	0	0.149		0	0.089	0.088	0.928	0.329	0.189
S_W										
PC	−0.298**	0.501**	0.015	0.522	1	−0.244*	−0.284*	0.361**	0.324**	−0.27**
Sig.	0.008	0	0.897	0		0.031	0.012	0.001	0.004	0.017
M_W										
PC	0.857**	−0.226*	0.347**	−0.194	−0.244*	1	0.483**	−0.109	−0.147	0.521**
Sig.	0	0.047	0.002	0.089	0.031		0	0.343	0.221	0
L_C										
PC	0.639**	0.197	−0.003	0.194	−0.284*	0.483**	1	−0.765**	−0.756**	0.965**
Sig.	0	0.084	0.982	0.088	0.012	0		0	0	0
D_A										
PC	−0.89**	0.019	−0.162	0.01	0.361**	−0.109	−0.765**	1	0.189	−0.811**
Sig.	0	0.866	0.158	0.928	0.001	0.343	0		0.094	0
e										
PC	−0.562**	−0.111	−0.024	−0.112	0.324**	−0.147	−0.756**	0.189	1	−0.744**
Sig.	0	0.334	0.838	0.329	0.004	0.221	0	0.094		0
R										
PC	0.675**	0.149	0.047	0.15	−0.27*	0.521**	0.965**	−0.811**	−0.744**	1
Sig.	0	0.194	0.684	0.189	0.017	0	0	0	0	

PC Pearson correlation coefficient, *Sig.* Sig. (two-tailed)

* Correlation is significant at the 0.05 level (two-tailed); ** correlation is significant at the 0.01 level (two-tailed)

Table 4 Correlations between V_{85T} and independent variables

	V_{85T}	N_L	L_W	P_W	S_W	M_W	L_C	D_A	e	R
V_{85T}										
PC	1	−0.146	0.192	−0.13	−0.355**	0.88**	0.625**	−0.891**	−0.557**	0.667**
Sig.		0.203	0.067	0.258	0.001	0	0	0	0	0
N_L										
PC	−0.146	1	−0.285*	0.992**	0.501**	−0.226*	0.197	0.019	−0.111	0.149
Sig.	0.203		0.011	0	0	0.047	0.084	0.866	0.334	0.194
L_W										
PC	0.231*	−0.285*	1	−0.165	0.015	0.347**	−0.003	−0.162	−0.024	0.047
Sig.	0.042	0.011		0.149	0.897	0.002	0.982	0.158	0.838	0.684
P_W										
PC	−0.13	0.992**	−0.165	1	0.522**	−0.194	0.194	0.01	−0.112	0.15
Sig.	0.258	0	0.149		0	0.089	0.088	0.928	0.329	0.189
S_W										
PC	−0.355**	0.501**	0.015	0.522**	1	−0.244*	−0.284*	0.361**	0.324**	−0.27*
Sig.	0.001	0	0.897	0		0.031	0.012	0.001	0.004	0.017
M_W										
PC	0.88**	−0.226*	0.347**	−0.194	−0.244*	1	0.483**	−0.111	−0.151	0.521**
Sig.	0	0.047	0.002	0.089	0.031		0	0.332	0.213	0
L_C										
PC	0.625**	0.197	−0.003	0.194	−0.284*	0.483**	1	−0.765**	−0.756**	0.965**
Sig.	0	0.084	0.982	0.088	0.012	0		0	0	0
D_A										
PC	−0.891**	0.019	−0.162	0.01	0.361**	−0.111	−0.765**	1	0.171	−0.811**
Sig.	0	0.866	0.158	0.928	0.001	0.332	0		0.124	0
e										
PC	−0.557**	−0.111	−0.024	−0.112	0.324**	−0.151	−0.756**	0.171	1	−0.744**
Sig.	0	0.334	0.838	0.329	0.004	0.213	0	0.124		0
R										
PC	0.667**	0.149	0.047	0.15	−0.27*	0.521**	0.965**	−0.811**	−0.744**	1
Sig.	0	0.194	0.684	0.189	0.017	0	0	0	0	

PC Pearson correlation coefficient, *Sig.* Sig. (two-tailed)

* Correlation is significant at the 0.05 level (two-tailed); ** correlation is significant at the 0.01 level (two-tailed)

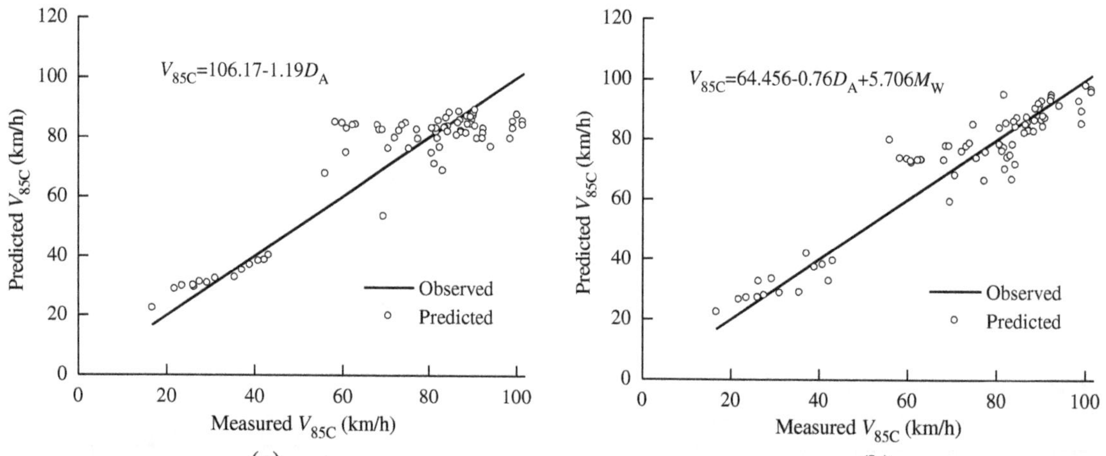

(a) (b)

Fig. 2 Measured and predicted V_{85C} for models 1 (**a**) and 2 (**b**)

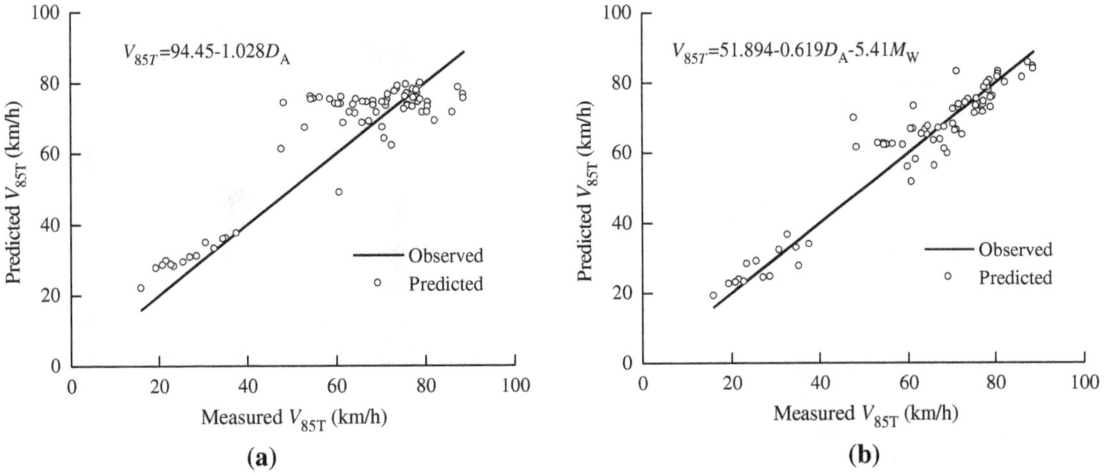

Fig. 3 Measured and predicted V_{85T} for models 1 (**a**) and 2 (**b**)

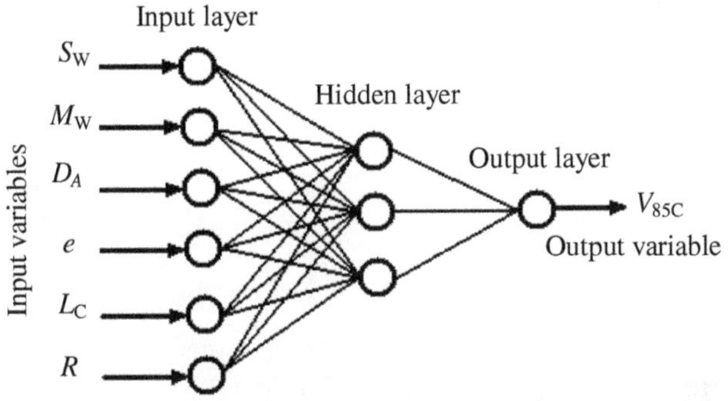

Fig. 4 MLP network architecture of V_{85C} model

Table 5 Performances for the best ANN model (cars and trucks)

Model type	Performance	Training (63 samples)	Testing (15 samples)	Overall model
Cars	R^2	0.935	0.901	0.932
	RMSE	5.84	5.51	5.77
Trucks	R^2	0.953	0.923	0.95
	RMSE	4.29	4.51	4.33

3 Data analysis, results, and final models

3.1 Correlation analysis

The correlations among V_{85C} and V_{85T} on curve sections and the nine independent variables are analyzed. As shown in Tables 3 and 4, respectively, Pearson correlation coefficient and the value of significant are calculated by SPSS. It can be seen from Table 3, there are relatively significant correlations among V_{85C} and six independent variables. These variables are S_W, M_W, L_C, D_A, e, and R. Also, Table 4 shows significant correlations among V_{85T} and these six independent variables.

Then, these variables are introduced into the multiple linear regression models. Consequently, stepwise regression analysis is used to select the most statistically significant independent variables with V_{85C} and V_{85T} in one model. Also, these variables are included in final ANN models.

3.2 Final models

3.2.1 Regression models

Car models There are two models that are statistically significant with V_{85C} after stepwise regression using SSPS Package. All the variables are significant at the 5 % significance level (95 % confidence level) for these two models. In other words, P value is less than 0.05 for all independent variables. Finally, many models are excluded due to poor significance with V_{85}. Therefore, the best models are shown in Eqs. (2) and (3), and in Fig. 2.

$$V_{85C} = 106.17 - 1.19 \times D_A, \tag{2}$$

whereas $R^2_{adj} = 0.79$, and RMSE $= 10.1$,

$$V_{85C} = 64.456 - 0.76 \times D_A + 5.706 \times M_W, \tag{3}$$

whereas $R^2_{adj} = 0.892$, and RMSE $= 7.2$.

Investigation of the previous results shows that:

- Model 2 is better than model 1 as it has better R^2_{adj}, and lower RMSE.
- The negative sign of the coefficient for D_A means that V_{85C} decreases with the increase of D_A. Then, the higher D_A for curves is more disturbing for drivers. This result is similar to Ref [20] and consistent with logic.
- The positive sign of the coefficient for M_W means that V_{85C} increases with the increase of M_W. In other word, the wider median width encourages the drivers to increase their speed on horizontal curve as the opposite

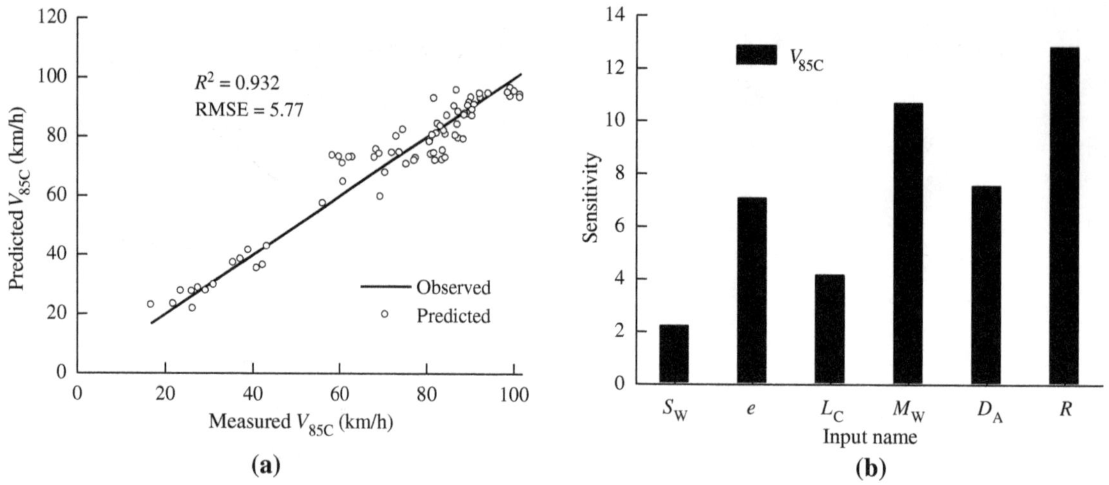

Fig. 5 Measured and predicted V_{85C} (**a**) and sensitivity analysis (**b**) for the best ANN model

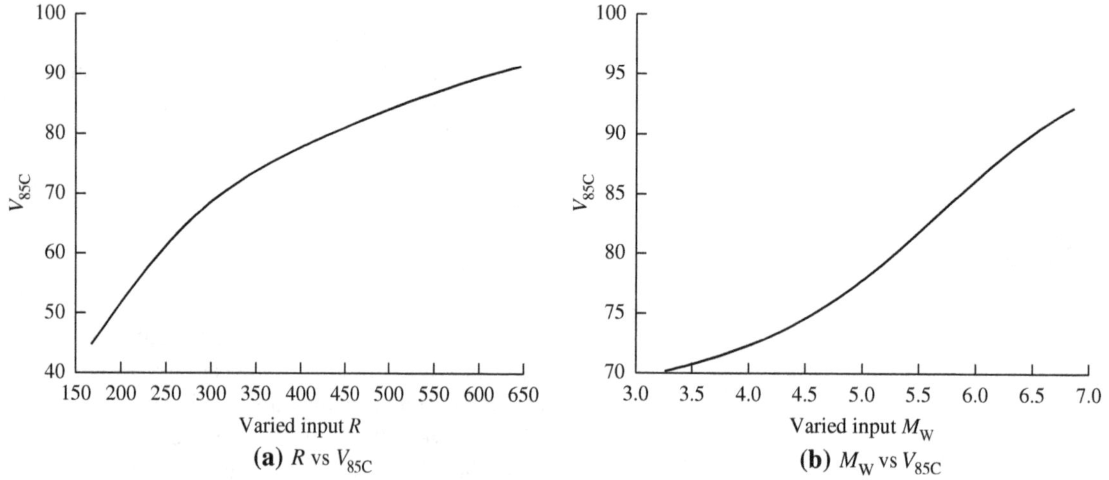

Fig. 6 Relations between the effective explanatory variables and V_{85C}

traffic is far from the field of vision. This result is similar to Refs [9, 21], and consistent with logic.

- Although S_W, R, e, and L_C have considerable effect on operating speed, but they are excluded from the statistical model, because there is multicollinearity among these independent variables. Therefore, the modeling with other technique is necessary to assure these results.

Truck models The same steps are used to reach the best truck models as car models. There are two models that are statistically significant with V_{85T}. These models are shown in Eqs. (4) and (5), and in Fig. 3.

$$V_{85T} = 94.45 - 1.028 \times D_A, \quad (4)$$

whereas $R^2_{adj} = 0.791$, and RMSE $= 9.19$,

$$V_{85T} = 51.894 - 0.619 \times D_A + 5.41 \times M_W, \quad (5)$$

whereas $R^2_{adj} = 0.915$, and RMSE $= 5.49$.

Investigation of the previous results shows that:

- Model 2 is the best model as it has the maximum R^2_{adj}, and the lowest RMSE.
- As car model, the coefficients of D_A and M_W have similar signs. Then, the same conclusions can be extracted.

3.2.2 ANN models

Car model As a result of correlation analysis, there are six independent variables that are highly correlated with V_{85C}. These variables are in input layer. One hidden layer is used, and one desired variable (V_{85C}) is in output layer with 78 observations used. The architecture of the ANN model is shown in Fig. 4. The curves are divided into training data set that has 63 curves (80 % of all observations), and

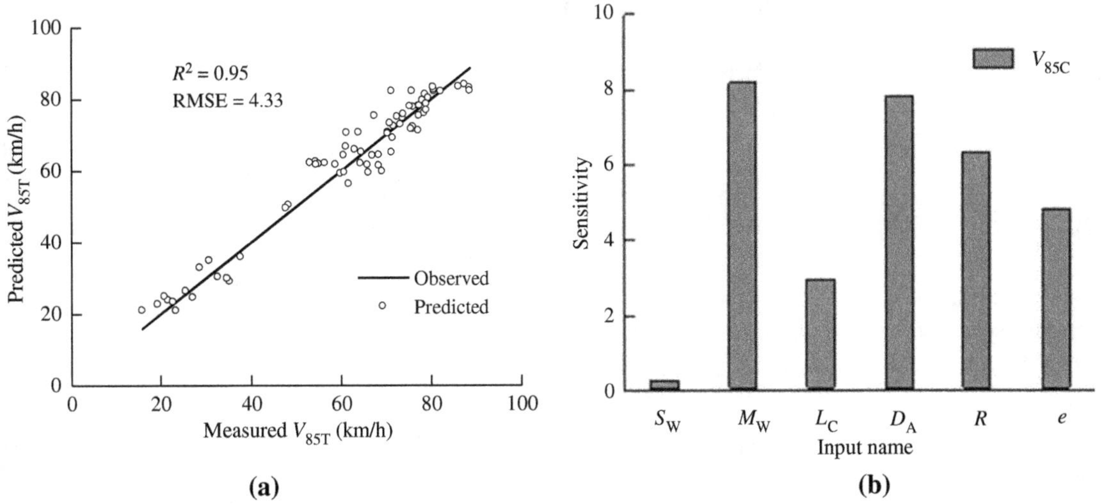

Fig. 7 Measured and predicted V_{85T} (**a**) and sensitivity analysis (**b**) for the best ANN model

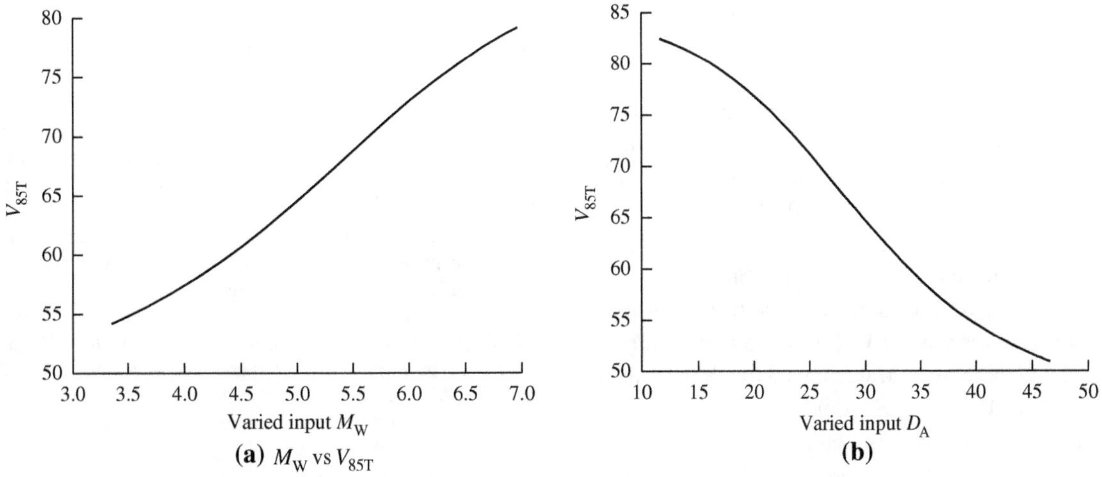

Fig. 8 Relations between the effective explanatory variables and V_{85T}

testing data set that has 15 curves (20 % of all sites). Many trials are done to reach this percentage between training and testing data which gives the best model performance in the present case.

The number of neurons in hidden layer is about half of the total number of neurons at the input and output layers (thee neurons), which is set based on generally accepted knowledge in this field. Using the learning rule of momentum, the suitable number of iterations is 5,000. The previous conditions are suitable for quick convergence of the problem [15, 16]. As a result of training and testing processing, the performances of the best model for training (63 samples) and testing (15 samples) data set are presented in Table 5.

The observed versus predicted values are shown in Fig. 5. It is clear that the ANN models give better and most confident results than the regression models. In order to measure the importance of each explanatory variable, general influence (sensitivity about the mean or standard deviation) is computed based on the trained weights of ANN. For the specified independent variable, if this value (sensitivity about the mean) is higher than other variables, this indicates that the effect of this variable on dependent variable (V_{85C}) is higher than other variables. Also, Fig. 5 shows the sensitivity of each explanatory variable in the selected model. It is found that the most influential variable on V_{85C} is R, followed by M_W. The relationships between each effective input variable and $V_{85\ C}$ are shown in Fig. 6. It is concluded that V_{85C} increases with the increase of R. In addition, V_{85C} decreases with the increase of D_A. These results are more accurate than the regression models and rational.

Truck model The same steps are executed as car models. Also, training of 63 samples and testing of 15 samples give the best model performances. These results are presented in Table 5. The predicted versus observed values and the sensitivity analysis are shown in Fig. 7. It is found that the most influential variable on V_{85T} is M_W, followed by D_A.

The relationships between each effective input variable and V_{85T} are shown in Fig. 8. It is concluded that V_{85T} increases with the increase of M_W. In addition, V_{85T} decreases with the increase of D_A. These results are more accurate than the regression models and consistent with logic.

4 Conclusions

The current paper presents new modeling techniques for predicting operating speed on horizontal curves at multi-lane highways in Egypt. The findings of this paper are summarized as:

(1) The ANN models give better, more confident, and logic results than the regression models in terms of predicting V_{85} for both cars and trucks.
(2) For cars, the best ANN model gives R^2 and RMSE equal to 0.932 and 5.77, respectively, for overall data set compared with the best regression model which gives R^2 and RMSE equal to 0.892 and 7.2, respectively, for all data set.
(3) For trucks, the best ANN model gives R^2 and RMSE equal to 0.95 and 4.33, respectively, for overall data set compared with the best regression model which gives R^2 and RMSE equal to 0.919 and 5.43, respectively, for all data set.
(4) For ANN model, the most influential variable on V_{85C} is R, followed by M_W. The increase of R from 168 to 642 m leads to an increase of V_{85C} from 45 to 92 km/h. In addition, the increase of M_W from 3.3 to 6.8 m leads to an increase in V_{85C} from 69 to 90 km/h.
(5) For ANN model, the most influential variable on V_{85T} is M_W, followed by D_A. The increase of M_W from 3.4 to 6.9 m leads to an increase in V_{85T} from 54 to 80 km/h. Also, the increase of D_A from 12° to 46° leads to a decrease of V_{85T} from 82 to 51 km/h.

The previous results are useful for controlling V_{85} on horizontal curves for multi-lane rural highways. V_{85} can be controlled by targeting curve factors to improve the safety performance of the curved sections of highways. This is so beneficial for road authorities in Egypt.

Finally, future research should be conducted to add two-lane rural roads and sloping sections to the present sites in order to explore the impact of them on operating speed.

Acknowledgments The author acknowledges Eng. Nasser Abdalla, Department of Civil Engineering, Faculty of Engineering, Al-Azhar University for his assistance with the acquisition of spot speed and geometric curves data at sites. Also, the author acknowledges the support of the General Authority of Roads, Bridges, and Land Transport (GARBLT) for their assistance with the acquisition of sites data.

References

1. Abdul-Mawjoud AA, Sofia GG (2008) Development of models for predicting speed on horizontal curves for two-lane rural highways. Arab J Sci Eng 33(2B):365–377
2. Hashim IH (2011) Analysis of speed characteristics for rural two-lane roads: a field study from Minoufiya Governorate, Egypt. Ain Shams Eng J 2:43–52.
3. Jessen DR, Schurr KS, McCoy PT et al (2001) Operating speed prediction on crest vertical curves of rural two-lane highways in Nebraska. Transp Res Rec 1751:67–75
4. Islam M, Seneviratne PN (1994) Evaluation of design consistency on two-lane highways. ITEJ 64(2):28–31
5. Ottesen JL, Krammes RA (2000) Speed-profile model for a design-consistency evaluation procedure in the United States. Transp Res Rec 1701:76–85
6. Polus A, Fitzpatrick K, Fambro DB (2000) Predicting operating speed on tangent sections of two-lane rural highways. Transp Res Rec 1737:50–57
7. McFadden J, Elefteriadou L (1997) Formulation and validation of operating speed-based design consistency models by bootstrapping. Transp Res Rec 1579:97–103
8. Bennett CR (1994) A speed prediction model for rural two-lane highways. PhD Dissertation, University of Auckland, New Zealand
9. Gong H, Stamatiadis N (2008) Operating speed prediction models for horizontal curves on rural four-lane highways. Transp Res Rec 2075:1–7
10. Cheng Y, Chen F, Huang X, Wang F, Liu M (2011) Predicting operating speed on curve sections of eight-lane expressway in plain area. In: 4th international symposium proceedings of highway geometric design, 2011
11. Abdalla NM (2010) Relationship between traffic flow characteristics and pavement conditions on roads in Egypt. MSc Dissertation, University of Al-Azhar, Egypt
12. General Authority of Roads, Bridges and Land Transport "GARBLT" (2009) System of Road Geometric Data, El-Nasr St., Cairo, Egypt. http://www.garblt.gov.eg/. Accessed 12 Feb 2009
13. Simpson AL, Rauhut JB, Jordahl PR et al (1994) Sensitivity analyses for selected pavement distresses. SHP report no. SHP-P-393
14. Singh D, Zaman M, White L (2011) Modeling of 85th percentile speed for rural highways for enhanced traffic safety. No. FHWA 2211. Oklahoma Department of Transportation, Oklahoma
15. Semeida AM (2013) Impact of highway geometry and posted speed on operating speed at multilane highways in Egypt. J Adv Res 4(6):515–523.
16. Semeida AM (2013) New models to evaluate the level of service and capacity for rural multi-lane highways in Egypt. Alex Eng J 52(3):455–466.
17. User's Manual "NeuroSolutions 7" (2010) NeuroDimension, Inc. Gainesville
18. Voudris AV (2006) Analysis and forecast of capsize bulk carriers market using artificial neural networks. MSc Dissertation, Massachusetts Institute of Technology, USA
19. Tarefder RA, White L, Zaman M (2005) Neural network model for asphalt concrete permeability. J Mater Civil Eng 17:19–27
20. Hashim IH (2006) Exploring the relationship between safety and the consistency of geometry and speed on rural single carriageways. J UTSG 2(A1):1–12
21. Ali A, Flannery A, Venigalla M (2007) Prediction models for free flow speed on urban streets. In: 86th annual meeting of the Transportation Research Board, Washington, DC, 2007

4

The time and space characteristics of magnetomotive force in the cascaded linear induction motor

Dajing Zhou · Jiaqing Ma · Lifeng Zhao · Xiao Wan · Yong Zhang · Yong Zhao

Abstract To choose a reasonable mode of three-phase winding for the improvement of the operating efficiency of cascaded linear induction motor, the time and space characteristics of magnetomotive force were investigated. The ideal model of the cascaded linear induction motor was built, in which the B and C-phase windings are respectively separated from the A-phase winding by a distance of d and e slots pitch and not overlapped. By changing the values of d and e from 1 to 5, we can obtain 20 different modes of three-phase winding with the different combinations of d and e. Then, the air-gap magnetomotive forces of A-, B-, and C-phase windings were calculated by the magnetomotive force theory. According to the transient superposition of magnetomotive forces of A-, B-, and C-phase windings, the theoretical and simulated synthetic fundamental magnetomotive forces under 20 different arrangement modes were obtained. The results show that the synthetic magnetomotive force with $d = 2$ and $e = 4$ is close to forward sinusoidal traveling wave and the synthetic magnetomotive force with $d = 4$ and $e = 2$ is close to backward sinusoidal traveling wave, and their amplitudes and wave velocities are approximately constant and equal. In both cases, the motor could work normally with a high efficiency, but under other 18 arrangement modes ($d = 1$, $e = 2$; $d = 1$, $e = 3$; $d = 1$, $e = 4$;...), the synthetic magnetomotive force presents obvious pulse vibration and moves with variable velocity, which means that the motor did not work normally and had high energy loss.

Keywords Linear induction motor · Three-phase winding · Magnetomotive force

D. Zhou (✉) · J. Ma · L. Zhao · X. Wan · Y. Zhang · Y. Zhao
Key Laboratory of Magnetic Levitation Technologies and Maglev Trains (Ministry of Education of China), Southwest Jiaotong University, Chengdu 610031, Sichuan, China
e-mail: zdj007008009@163.com

D. Zhou · J. Ma · L. Zhao · X. Wan · Y. Zhang · Y. Zhao
Superconductivity and New Energy R&D Center, Southwest Jiaotong University, Chengdu 610031, Sichuan, China

Y. Zhao
School of Materials Science and Engineering, University of New South Wales, Sydney, NSW 2052, Australia

1 Introduction

Linear motor is a kind of electrical equipment which can directly convert the electrical energy to linear movement. Compared with traditional rotating machine, the drive system of linear motor works without the intermediate gearing, which simplifies the driver system, and makes its linear velocity unlimited and moving process without mechanical touch. In addition, the noise level of linear motor is very low. For all these characteristics, linear motor is widely applied to high-speed ground transportation [1–4].

When the pole number of a linear motor is not less than six, the values of negative-sequence current and zero-sequence current are much small compared with the positive sequence current; thus their influence on the cascaded linear motor can be ignored [5, 6]. In this case, the asymmetry of three-phase current will not be considered. With three-phase symmetrical current flowing into three-phase winding, linear motor will produce a traveling magnetomotive force (MMF) wave in the air-gap between primary and secondary windings. Since the energy exchange in mechanic-electronics is achieved by the air-gap magnetic field of linear motor [7], the time and space characteristics of the MMF wave directly affect the operating efficiency and energy consumption in linear motor.

The motor can work efficiently with the waveform of the MMF close to a sinusoidal traveling wave [8].

The finite element method (FEM) is an effective and accurate method to investigate the linear motor characteristics. Lu et al. [9] analyzed the features of air-gap magnetic field in large air-gap linear induction motor. Selcuk and Kurum [10] built a simplified FEM model of an actual short primary linear induction motor and solved it for air-gap magnetic field distribution. Lu et al. [11] analyzed the two-dimensional transient air-gap magnetic field of long primary induction motor with the help of FEM. Li et al. [12] studied the characteristics of temperature field for tubular linear motor with the FEM. In this paper, we will solve the single-phase winding MMF and three-phase winding-synthesized MMFs in single-side linear induction motor (SLIM) by the classical theory of MMF [13] and adopt the FEM to validate the theoretical results. The work conditions and efficiency of SLIM are then optimized by analyzing MMF characteristics with different three-phase winding arrangement.

2 Theoretical analysis and simulation model for SLIM

2.1 The MMF of A-phase winding

The established theoretical model of SLIM is shown in Fig. 1, in which the leak flux in this model is ignored. In this model, it is assumed that primary and secondary are infinite long, primary iron yoke is not in magnetic saturation, the magnetoconductivity of primary and secondary is infinite, and magnetic induction intensity only contains the component in the y axis direction. It is also assumed that the current flows along the z axis direction, the equivalent air-gap δ between primary and secondary is distributed evenly along the x axis direction, the magnetic potential difference of three-phase winding is distributed evenly along the air-gap, the number of pole pairs is p and pole pitch is three slots pitch long, and the three-phase wingding is a bi-layered full-pitch winding.

Without loss of generality, the sinusoidal current as shown in Eq. (1) is assumed flowing into A-phase wingding, i.e.,

$$i_A = \sqrt{2} I_A \cos \omega t. \tag{1}$$

With reference to Fig. 1, according to the Ampere's circuital Law, A-phase winding MMF in air-gap can be expressed as [14]

$$f_A = H_y \cdot \delta = N \cdot i_A \quad \text{when } -\frac{\pi}{2} \pm 2m\pi \le \theta \le \frac{\pi}{2} \pm 2m\pi, \tag{2a}$$

$$f_A = -H_y \cdot \delta = -N \cdot i_A \quad \text{when } \frac{\pi}{2} \pm 2m\pi \le \theta \le \frac{3\pi}{2} \pm 2m\pi, \tag{2b}$$

where $m = 1, 2, 3, \ldots$, electrical angle $\theta = \pi x/\tau$, the number of turns in series winding $N = 2pN_c$, and N_c is the number of turns in a single coil.

Since the quantity $f_A(\theta)$ represents a periodic square wave along the air-gap, it can be represented by the Fourier series as

$$f_A(\theta) = f_{A1} \cos \theta + f_{A3} \cos 3\theta + f_{A5} \cos 5\theta + \cdots + f_{An} \cos n\theta, \tag{3}$$

where f_{An} is

$$f_{An} = \frac{4}{T} \int_0^{\frac{T}{2}} f_A(\theta) \cos(n\omega_0\theta) d\theta = \frac{1}{n} \cdot \frac{4}{\pi} \cdot N i_A \sin n\frac{\pi}{2}, \tag{4}$$

$n = 1, 2, 3, \ldots$

Substituting Eqs. (1) and (4) into Eq. (3), we have the instantaneous value of A-phase wingding MMF:

$$f_A(\theta, t) = \frac{4\sqrt{2}}{\pi} N I_A \left[\cos \theta - \frac{1}{3} \cos 3\theta + \frac{1}{5} \cos 5\theta - \cdots + \frac{1}{n} \sin(n\frac{\pi}{2}) \cos n\theta \right] \cos \omega t. \tag{5}$$

From Eq. (5), we can find that A-phase winding MMF after the Fourier series transformation can be decomposed

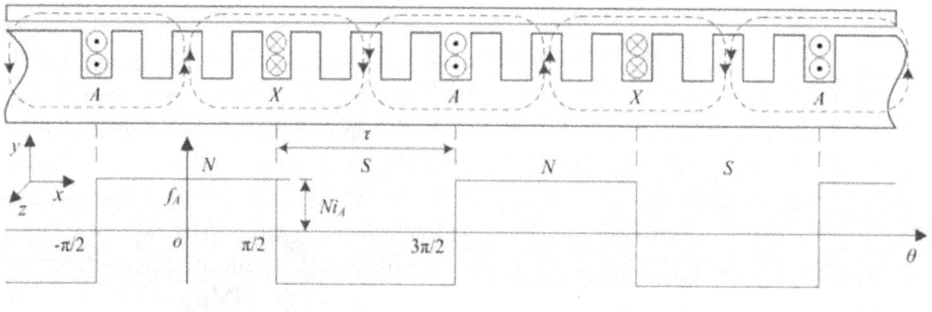

Fig. 1 The theoretical model of SLIM and A-phase winding MMF distribution diagram

into fundamental wave and a series of higher harmonics, and the amplitudes of fundamental wave and higher harmonics pulse over time with the current frequency. Because fundamental wave determines the energy conversion of linear motor and its main performance, it is most important and fundamental to analyze the fundamental MMF [15].

2.2 The MMFs of three-phase winding

As shown in Fig. 2, when the B and C-phase windings are respectively d and e slots pitch away from the A-phase winding, we can obtain the MMF of B-phase winding and C-phase winding respectively by moving the MMF of A-phase winding $\pi d/3$ and $\pi e/3$ along the positive direction of θ axis, respectively.

Let three-phase current flow into three-phase winding, through above analysis we can obtain the expression of the fundamental MMF for A, B, C phase windings, respectively,

$$F_A(\theta,t) = \frac{4\sqrt{2}}{\pi} NI_A \cos(\theta)\cos(\omega t), \tag{6a}$$

$$F_B(\theta,t,d) = \frac{4\sqrt{2}}{\pi} NI_A \cos\left(\theta - \frac{1}{3}\pi d\right)\cos\left(\omega t - \frac{2}{3}\pi\right), \tag{6b}$$

$$F_C(\theta,t,e) = \frac{4\sqrt{2}}{\pi} NI_A \cos\left(\theta - \frac{1}{3}\pi e\right)\cos\left(\omega t + \frac{2}{3}\pi\right). \tag{6c}$$

Adding the instantaneous values of fundamental MMF with A, B, C phase windings, we have the fundamental MMFs:

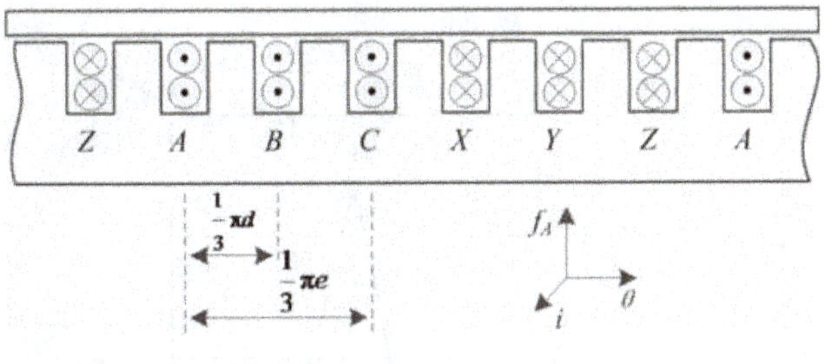

Fig. 2 Three-phase winding distribution diagram

$$F(\theta,t,d,e) = F_A(\theta,t) + F_B(\theta,t,d) + F_C(\theta,t,e). \tag{7}$$

According to the theoretical model of SLIM, the values of d and e can be taken from 1, 2, 3, 4 and 5 in a cycle of MMF, so that the number of arrangement modes of three-phase winding is 20 with the combination between d and e. The results of fundamental MMF_S with different arrangements of three-phase windings will be analyzed later.

2.3 The simulation model for SLIM

The simulation model of SLIM is established by Ansoft Maxwell as shown in Fig. 3. Table 1 presents the specific parameters related to this model. Different simulation data of MMFs will be obtained by changing the relative positions of three-phase winding.

3 Results and analysis

3.1 The MMFs with $d = 1$ and $e = 2$

Referring to Fig. 2, we move B-phase winding and C-phase winding from the position of A-phase winding 1 slot pitch and 2 slots pitch, respectively. Then, substitute $d = 1$ and $e = 2$ into Eq. (7) and the expression of fundamental MMFs is calculated by trigonometric formula [16]:

$$F(\theta,t,1,2) = \left[\frac{2\sqrt{2}}{\pi} NI_A [2\cos(\omega t + \theta) + \cos(\omega t - \theta) + \cos(\omega t + \theta - \pi) + \cos\left(\omega t - \theta - \frac{1}{3}\pi\right) + \cos\left(\omega t - \theta + \frac{4}{3}\pi\right)\right] \tag{8}$$

The MMF is composed of three forward traveling waves and two back traveling waves referring to Eq. (8), and the results are shown in Fig. 4. As can be seen in Fig. 4a, when the spatial MMF waveform is at the transient time of $\omega t = 0, \pi/2, \pi, 3\pi/2$, respectively, and with a scaling factor $\beta = 2\sqrt{2}NI_A/\pi$, the MMF pushes to the left in space and exists obvious vibration. The movement with variable velocity is shown in Fig. 4c, d. These features will reduce

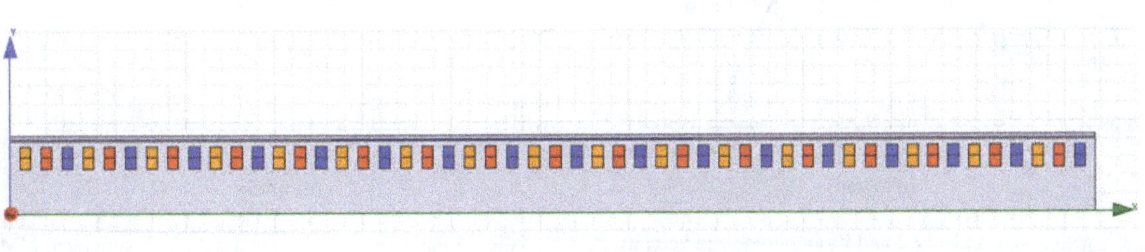

Fig. 3 The simulation model of SLIM

the efficiency of linear motor thrust in the horizontal direction, and at the same time also bring high energy loss.

According to the simulation model for SLIM, when arranging three-phase winding to be $d = 1$ and $e = 2$, the simulation transient waveform of the MMF is shown in Fig. 4b, which further validates the characteristics of MMF with $d = 1$ and $e = 2$.

3.2 The MMFs with $d = 2$, $e = 4$, and $d = 4$, $e = 2$

In order to keep the amplitude and wave velocity of MMF invariant over time and avoid the pulse vibration and movement with variable velocity of MMF, we solved the MMFs under different arrangements of three-phase winding. The result indicates that the MMFs in the condition of $d = 2$, $e = 4$ or $d = 4$, $e = 2$ are a traveling wave with its amplitude and velocity being constant as shown in Fig. 5a, c, which are expressed, respectively, by Eqs. (9a) and (9b):

Table 1 SLIM parameters for simulation model

Parameter	Value	Parameter	Value
Virtual current I_A	8 A	Air-gap δ	0.5 mm
Power frequency f	50 Hz	Slots per pole per phase q	1
Pole pitch τ	36 mm	Turns per coil N_c	60
Pole pairs p	8		

$$F(\theta, t, 2, 4) = \frac{6\sqrt{2}}{\pi} N I_A \cos(\omega t - \theta), \tag{9a}$$

$$F(\theta, t, 4, 2) = \frac{6\sqrt{2}}{\pi} N I_A \cos(\omega t + \theta), \tag{9b}$$

where $\theta = k \cdot x = \frac{\pi}{\tau} \cdot x$.

From Fig. 5a, c, we can see that $F(\theta, t, 2, 4)$ is a forward traveling wave and $F(\theta, t, 4, 2)$ is a backward traveling wave. Their amplitude and wave velocity can be expressed as

$$F_m(\theta, t, 2, 4) = F_m(\theta, t, 4, 2) = \frac{6\sqrt{2}}{\pi} N I_A, \tag{10a}$$

$$v(\theta, t, 2, 4) = v(\theta, t, 4, 2) = \frac{\omega}{k} = \frac{\tau \cdot \omega}{\pi}. \tag{10b}$$

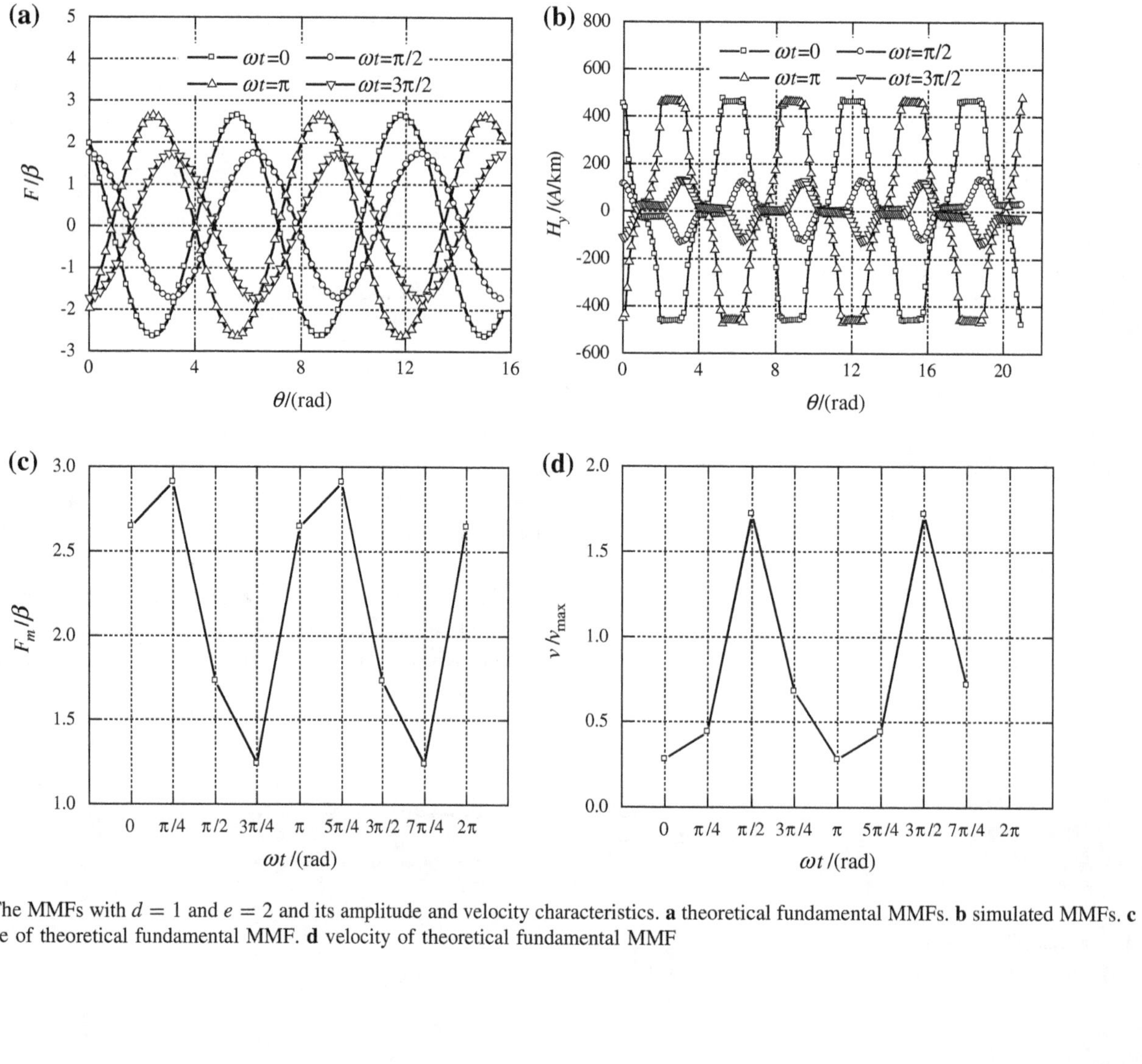

Fig. 4 The MMFs with $d = 1$ and $e = 2$ and its amplitude and velocity characteristics. **a** theoretical fundamental MMFs. **b** simulated MMFs. **c** amplitude of theoretical fundamental MMF. **d** velocity of theoretical fundamental MMF

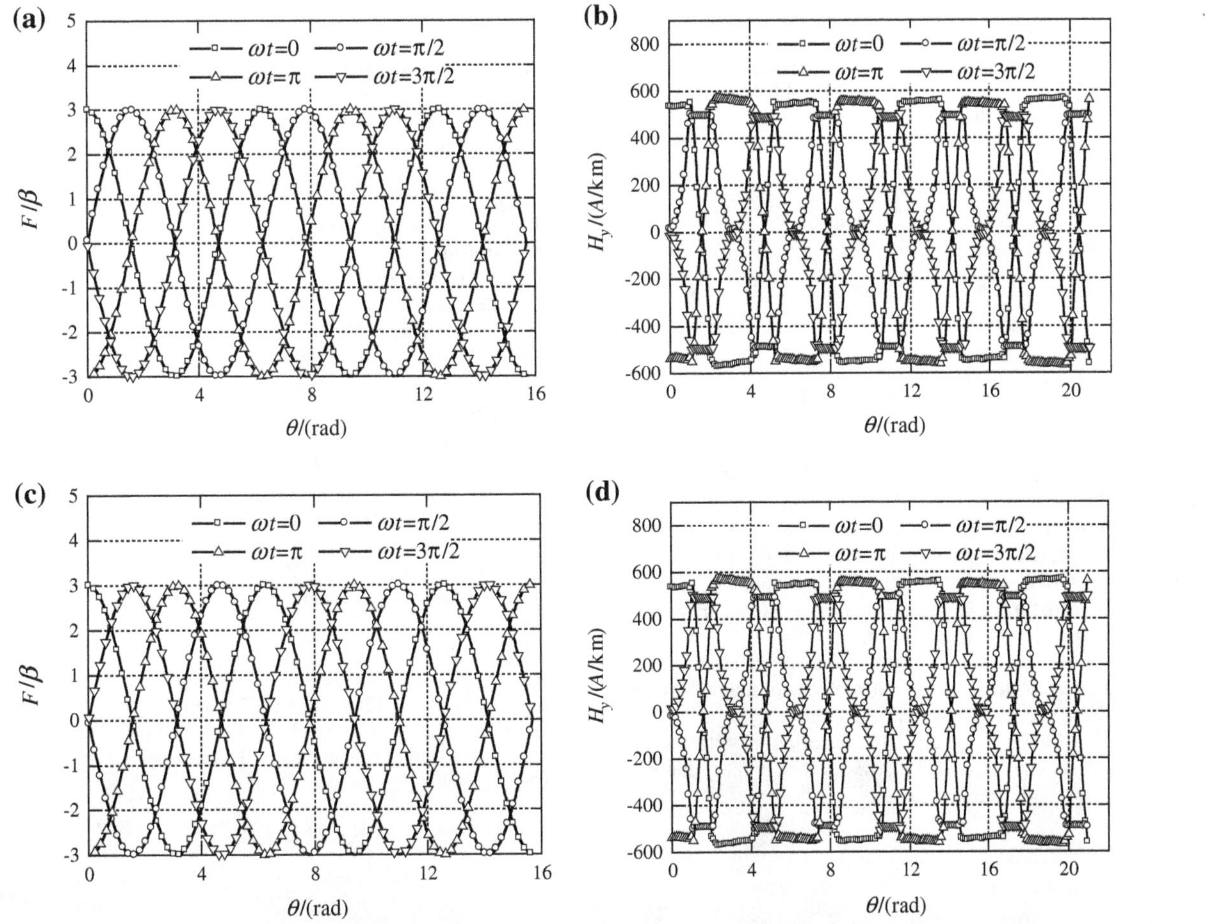

Fig. 5 The MMFs with $d = 2$, $e = 4$ and $d = 4$, $e = 2$. **a** theoretical fundamental MMFs with $d = 2$ and $e = 4$. **b** simulated MMFs with $d = 2$ and $e = 4$. **c** theoretical fundamental MMFs with $d = 4$ and $e = 2$. **d** simulated MMFs with $d = 4$ and $e = 2$

Based on the simulation model for SLIM, we also changed the arrangement mode of three-phase winding to $d = 2$, $e = 4$ and $d = 4$, $e = 2$, respectively, and obtained the MMFs simulated by Ansoft Maxwell. The results are shown in Fig. 5b, d. Since the simulation model for SLIM is not fully ideal and its MMFs contain other higher harmonics, the MMFs are close to sinusoidal traveling wave with constant amplitude and velocity.

4 Conclusion

Based on the classical MMF theory of rotating machine, we established the theoretical model and simulation model for SLIM, in which three-phase winding has 20 arrangement modes in a cycle of MMF, and the arrangement mode determines the time and space characteristics of MMFs. Through the calculation with electromagnetic theory and finite element software simulation, we solved and discussed the MMFs under 20 arrangement modes of three-phase winding, respectively. On the basis of the above analysis results, the conclusions are drawn as follows:

(1) The MMFs with $d = 2$ and $e = 4$ are close to sinusoidal wave, and it travels toward the positive direction with the constant amplitude and velocity. In this case, the motor can work normally with a high efficiency.
(2) The MMFs with $d = 4$ and $e = 2$ are close to sinusoidal wave, and travel toward the negative direction with the constant amplitude and velocity. Also the motor can work normally with a high efficiency.
(3) In other 18 conditions, the MMFs show obvious pulse vibration and movement with variable velocity, such as the MMFs under $d = 1$, $e = 2$. This means that the motor does not work normally, and instead it has high energy loss.

Acknowledgments This work was supported by the National Magnetic Confinement Fusion Science Program 2011GB112001,

Program of International S&T Cooperation S2013ZR0595, the financial support of the National Natural Science Foundation of China (No. 51271155), the Fundamental Research Funds for the Central Universities (SWJTU11ZT16, SWJTU11ZT31), the Science Foundation of Sichuan Province 2011JY0031, 2011JY0130.

References

1. Ye YY (2000) Principle and application of linear motor. China Machine Press, Beijin (in Chinese)
2. Wang TC (1971) Linear induction motor for high-speed ground transportation. IEEE Trans Ind Gen Appl 7(5):632–642
3. Deng JM, Chen TF, Tang JX et al (2013) Optimum slip frequency control of Maglev single-sided linear induction motors to maximum dynamic thrust. Proc CSEE 33(12):123–130 (in Chinese)
4. Wang K, Shi LM, He JW et al (2009) A decoupling control of normal-and-thrust forces in single-sided linear induction motor. Proc CSEE 29(6):100–104 (in Chinese)
5. Lu JY, Ma WM, Sun ZL et al (2009) Research on static longitudinal end effect of linear induction motor with multi-segment primary. Proc CSEE 29(33):95–101 (in Chinese)
6. Sun ZL, Ma WM, Lu JY et al (2010) Research of static longitudinal end effect and impedance matrix for long primary double-sided linear induction motors. Proc CSEE 30(18):72–77 (in Chinese)
7. Long LX (2006) Theory of linear induction motor and its electromagnetic design method. Science Press, Beijing (in Chinese)
8. Fan SD (1996) Determination for magnetic density wave in the gap of the permanent magnet DC motor. Small Spec Electr Mach 3:2–5 (in Chinese)
9. Lu QF, Fang YT, Ye YY (2005) A study on force characteristic of large air gap linear induction motor. Proc CSEE 25(21):132–136 (in Chinese)
10. Selcuk AH, Kurum H (2008) Investigation of end effects in linear induction motors by using the finite-element method. IEEE Trans Magn 44(7):1791–1795
11. Lu JY, Ma WM, Xu J (2008) Modeling and simulation of high speed long primary double-sided linear induction motor. Proc CSEE 28(27):89–94 (in Chinese)
12. Li LY, Huang XZ, Kou BQ et al (2013) Numerical calculation of temperature field for tubular linear motor based on finite element method. Trans China Electrotech Soc 28(2):132–138 (in Chinese)
13. Lipo TA (2004) Introduction to AC machine design. University of Wisconsin Press, Chicago
14. Liu CY, Wang H, Zhang ZJ et al (2011) Research on thrust characteristics in permanent magnet linear synchronous motor based on analysis of nonlinear inductance. Proc CSEE 31(30):69–76 (in Chinese)
15. Xie MT, Zhang GY (2004) Electromechanics. University of Chongqing Press, Chongqing (in Chinese)
16. Zhou SL (2003) The analysis of stator magnetic potential in the three-phase asynchronous motor. J Tongling Coll 2(2):73–74 (in Chinese)

Optimal control system of the intermittent bus-only approach

Jie Jiang · Yu-lin Chang

Abstract To guarantee bus priority with a minimum impact on car traffic at intersections, an optimal control system of the intermittent bus-only approach (IBA) was proposed. The problems of the existing system are first solved through optimization: the judgment time of the IBA system was advanced to allow a bus to jump car queues if the bus was detected to arrive at the intersection, and the instant that the IBA lane became available to cars was controlled dynamically to increase the capacity of the IBA lane. The total car delay in one cycle was then analyzed quantitatively when implementing the optimal control system. The results show that in comparison with the existing system of the IBA, the car delay is greatly reduced and the probability of a car stopping twice is low after optimizing the IBA system.

Keywords Intermittent bus-only approach · Bus priority · Optimal control · Car delay · Bus detection

1 Introduction

Since intersections are bottlenecks of urban traffic, a public transport priority strategy can only be truly implemented by guaranteeing bus priority at intersections. There are two aspects of providing priority to buses at intersections: time prioritizing and space prioritizing. The assigning of a dedicated bus lane (DBL) removes one lane from car use to provide buses with space priority, while transit signal prioritization (TSP) provides buses with time priority through adjustment of the signal phases of traffic lights. The benefits of these solutions have been highlighted by a handful of studies [1–3]. Unfortunately, there are two major problems associated with bus prioritization. (1) When the frequency of buses arriving at an intersection is low, the assignment of a DBL may be a waste of road resources and may reduce traffic capacity. (2) TSP is less effective for heavy traffic, since the signals have to accommodate not only the bus but also the car traffic in which the bus is embedded.[1]

To overcome these drawbacks, Viegas and Lu proposed the concept of an intermittent bus lane (IBL) that is intermittently open to buses exclusively and then all vehicles when not being used by buses [4–6]. This system restricts cars from changing into the bus lane ahead of a bus, but does not request those cars already in the bus lane to leave the lane. Therefore, there are both buses and cars in the IBL at an intersection. To ensure bus priority, TSP is often included to flush the queues in arterial streets and clear the way for the bus. Nevertheless, these signal adjustments may decrease the amount of green-phase time allocated to side streets, thus reducing their capacity and increasing delay. Eichler and Daganzo [7] studied a bus lane with intermittent priority (BLIP), which is a variant of the IBL. In the case of the BLIP, cars are forced out of the lane reserved for the bus with variable-message signs (VMSs), and buses can jump car queues at intersections. Therefore, the BLIP does not require changes to the settings of signals. In this paper, the author employs kinematic

J. Jiang · Y. Chang (✉)
School of Automobile and Traffic Engineering, Jiangsu University, Zhenjiang 212013, China
e-mail: ylchang@ujs.edu.cn

[1] To leave side streets as unaffected as possible, a reduction in the duration of their green phase due to the passage of a bus is made up in subsequent cycles by an offsetting increase of the same magnitude. Thus, the arterial red-phase time will increase in the headway following the passage of a bus. This will increase the car delay which is more than offset by the benefit to bus when car demand is high.

wave theory to evaluate the BLIP roughly. The control system of the BLIP is considered to be a black box with no explicit commands or steps.

Xie et al. [8] introduced a control system of an intermittent bus-only approach (IBA) for a single intersection, which focuses on bus spatial prioritization at an intersection with no change in signal timing. This system sets the start of the red phase as the initial time of one cycle, and then, in the period of the system reaction time before the red phase starts, judges whether there is a bus arriving at the intersection within a given period of the next signal cycle, where the period is the queuing and dissipating time of cars when the IBA lane is available to cars. If there is a bus arriving, the IBA lane is reserved for the bus and cars are restricted from entering the lane in this given period; if not, the IBA lane is available to cars in the next cycle. Unfortunately, this control system does not consider that some cars that enter the IBA toward the end of the green phase in this cycle may fail to pass the intersection, and they queue at the stop line ahead of the bus in the next cycle, thus delaying the bus. Moreover, the period in which cars are restricted from using the IBA lane may reduce the road capacity when car traffic is heavy.

For the case of an intersection with low bus frequency, this paper optimizes the existing system of the IBA to solve the above issues and then quantitatively analyzes car delay to determine the efficiency of the optimal IBA system.

2 Optimal control system of the IBA

The following assumptions and simplifications are made.

(1) Car arrival and dissipation rates at intersections are constant.
(2) No residual car queue persists for more than one cycle even after the implementation of the IBA.
(3) When there is more than one approach in the same direction, cars choose the approach for which the queue is shorter. Approaches having the same queue length have the same car inflow.
(4) The bus frequency is low. There is no more than one bus arriving at the intersection within one cycle.
(5) The delay associated with the acceleration and deceleration of cars and buses is not taken into account.

To guarantee a bus spatial priority, it is necessary to ensure the bus reaches the stop line of intersection without a car ahead of it during the red phase, or the bus passes through the intersection without interference from dissipation of the downstream car queue during the green phase. Therefore, the IBA system should make a judgment at an earlier instant time before the red phase of the next cycle starts to clear the IBA lane in the case when a bus is arriving. The judgement time can be advanced quantitatively to match the time cars take to travel from the location of a VMS to the stop line. Furthermore, after cars are restricted from entering the IBA lane, when the IBA lane will again be available to cars is controlled dynamically depending on the bus arrival time. If the bus is arriving at the stop line during the red phase of the next cycle, once it has been detected passing the VMS, cars following the bus are allowed to enter the IBA lane thereafter. If the bus is arriving during the green phase, cars reaching the stop line during the green phase are allowed to enter the IBA lane, because these cars can pass through the intersection without creating a queue that would disrupt the bus. While considering that it takes time for cars to travel from the location of the VMS to the stop line, the IBA lane will be open to cars prior to the start of the green phase.

We next specify how the optimal control system of the IBA works. Figure 1 depicts the optimal IBA layout. The figure shows a VMS set on a section of road. The VMS informs drivers of the status of the IBA lane so that they can choose an appropriate approach in advance. The distance between the VMS and stop line is l.[2]

The optimal control system of the IBA comprises the following steps.

(1) At the time instant that is $(\Delta t + l/v_c)$ earlier than the start of the red phase in the next cycle, the queuing and dissipation time t_m of cars when the IBA is open to cars is calculated, where Δt is the reaction time of the system and v_c is the average speed of cars.
(2) The initial researching radius R' is calculated. $R'_d < R' < R'_u$, where R'_u and R'_d are the upper and lower limits of the initial researching radius, respectively. $R'_u = (t_m + \Delta t + l/v_c)v_b$ and $R'_d = (\Delta t + l/v_c)v_b$, where v_b is the average speed of buses.
(3) The number of bus stops i and j that are within the upper and lower limits of the researching radius, respectively, are detected.
(4) The researching radius R is determined; $R_d < R < R_u$. $R_u = (t_m + \Delta t + l/v_c - it_s)v_b$ and $R_d = (\Delta t + l/v_c - jt_s)v_b$, where t_s is the average dwell time at one bus stop.
(5) The distance L between the current position of the bus and the stop line is calculated.
(6) If $L \in \{R \mid R_d < R < R_u\}$, cars are restricted from entering the IBA lane immediately, and the system

[2] The positioning of the VMS at the start of the lane-changing section of road is slightly unrealistic. To ensure that cars can change lanes appropriately, the VMS should be moved far from the intersection. We do not consider the suitability of length l in this paper, because it would require a study of driver lane-changing maneuvers affected by many factors.

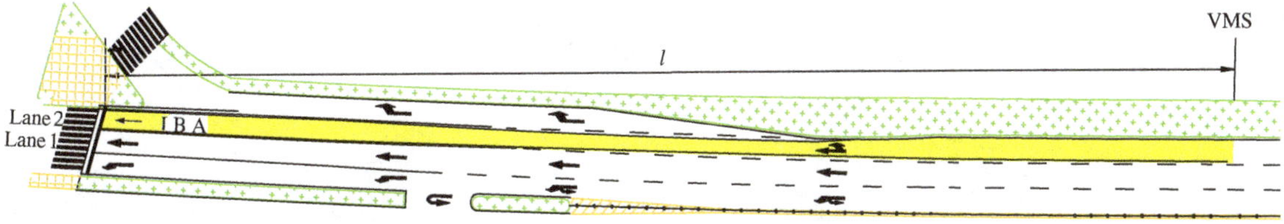

Fig. 1 Optimal layout of the IBA

advances to step (7). If not, cars are allowed to enter the IBA lane in the next cycle.

(7) If the bus is detected to pass the IBA before the instant that is ($\Delta t + l/v_c$) earlier than the start of the green phase, the IBA lane is made available to cars as soon as the bus is detected. If not, the IBA will be made available to cars at the instant ($\Delta t + l/v_c$) earlier than the start of the green phase.

This system uses the Global Positioning System to get information of the current bus position, provide basic data to calculate L, and to detect whether the bus has passed the VMS.

3 Analysis of car delay

Since the bus can jump car queues and travel unhindered by cars at the intersection, bus spatial priority is guaranteed. There is no change in the bus delay before and after the IBA control system is optimized. The paper, therefore, focuses on analyzing the car delay when implementing the optimal control system of the IBA.

As illustrated by Fig. 1, we assume that the bus at the signalized intersection is traveling straight ahead and that there are two lanes in which vehicles can travel straight through the intersection: lane 1, which is open to cars, and lane 2, which is the IBA lane. We then set the instant that the red phase starts as the initial time of the signal cycle and denote the cycle length as c, the duration of the red phase as r, the duration of the green phase as $g = c-r$ (the duration of the yellow phase and the time taken for the bus to depart are not considered), the car arrival rate as q, and the saturation flow rate of a single lane as s. It is noteworthy that q is usually less than s.[3]

If there is no bus arriving in the next cycle or the bus is arriving at the stop line in the interval (t_m, c], then L is out of the research radius and lane 2 is made available to cars in the next cycle. Figure 2 depicts the course of cars arriving and leaving when the IBA lane is made available to cars. The slope of the line AB is the car arrival rate q, and

[3] This is reasonable in that we note that the car demand is bounded by the approach capacity of the signalized intersection, namely $q < 2sg/c$, where the green ratio, g/c is often less than 1/2.

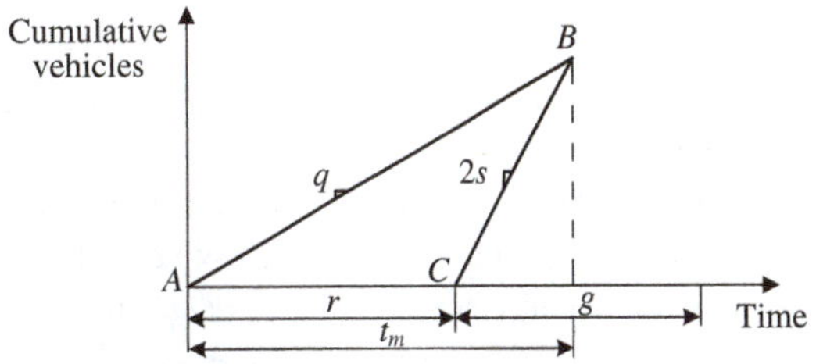

Fig. 2 Course of cars arriving and leaving when the IBA lane is available to cars

the slope of the line CB is the car dissipation rate $2s$. The queuing and dissipating time is $t_m = 2sr/(2s-q)$ and the total car delay is $D = qsr^2/(2s-q)$.

If there is a bus arriving at the stop line in the interval (0, t_m], L is in the research radius and cars are prevented from entering lane 2. The time at which lane 2 is again made available to cars depends on the bus arrival time, which affects the IBA utilization rate, thus resulting in a difference in the car delay. Hence, according to the dissipation time of lane 2, (0, t_m] can be divided into three parts separated by t_1 and r. When the bus arrives in the interval (0, t_1], the car queues fully dissipate in the two lanes at the same time; when the bus arrives in the interval (t_1, r], queued cars dissipate more quickly in lane 2 than in lane 1; when the bus arrives in the interval (r, t_m), cars pass through the green signal without queuing in lane 2.

Let t be the instant that the bus arrives at the stop line. Figure 3 shows the course of cars arriving and leaving when $t \in (0, t_1]$. In the figure, the lines $ABCD$ and $GFCD$ are the traffic arrival curves of lanes 1 and 2, respectively. Cars are allowed to drive only in lane 1 owing to the restriction of lane 2 in the interval $(0, t]$, which means that the slope of the line segment AB is q. Lane 2 then becomes available to cars at t. According to existing criteria of passenger car equivalents [9], the bus arriving at t is equal to two cars and can be represented by the line GF, which has length of 2. Since the queue in lane 2 is shorter than that in lane 1, cars following the bus will choose to enter lane 2 instead of lane 1 (recalling assumption (3)), and the slope of line FC is thus q and that of line BC is zero. After the car queues in the two lanes become the same length, arriving cars divide into the two flows with the same flow rate, and the slope of CD is thus $q/2$. The traffic dissipation

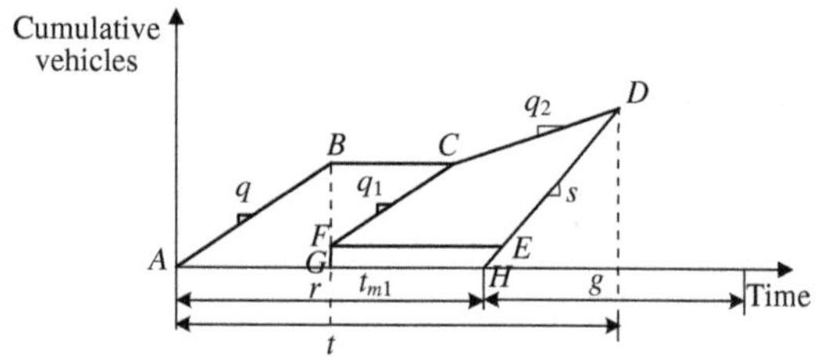

Fig. 3 Course of cars arriving and leaving when $t \in (0, t_1]$

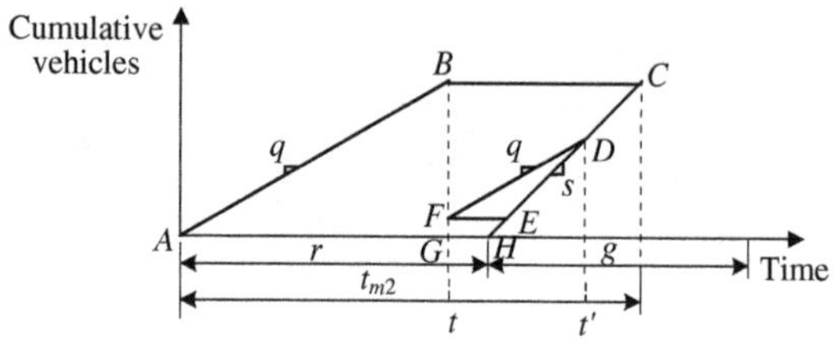

Fig. 4 Course of cars arriving and leaving when $t \in (t_1, r]$

curves of lanes 1 and 2 overlap as represented by the line segment *HD*, with slope *s*.

In Fig. 3, the areas of polygons *ABCDH* and *FCDE* are, respectively, the car delays in lanes 1 and 2. We denote the total car delay in one cycle when $t \in (0, t_1]$ as D_{op1}, the area of *ABCDH* as S_{ABCDH}, and the area of *FCDE* as S_{FCDE}. It follows from geometry that

$$D_{op1} = S_{ABCDH} + S_{FCDE} = \frac{qr^2s^2 + 2qrs - 2s + 2q}{s(2s-q)} \qquad (0 < t \le t_1). \tag{1}$$

The queuing and dissipation time t_{op1} shown in Fig. 3 is given by

$$t_{op1} = (2 + 2sr)/(2s - q) \qquad (0 < t \le t_1). \tag{2}$$

We note that D_{op1} and t_{op1} are independent of *t*. From a macroscopic view, cars arrive at a flow rate *q* and discharge from the two lanes at a saturation flow rate 2*s*. Consequently, D_{op1} and t_{op1} are constant and approximate *D* and t_m. Furthermore, we find the differences between D_{op1} and *D* and between t_{op1} and t_m are caused by the bus in front of the queue of cars.

When the bus arrives at critical time t_1, the queue length of lane 2 is exactly equal to that of lane 1 at the end moment of dissipation. We then have

$$t_1 = \begin{cases} (sqr + 2s)/(2qs - q^2) & s > q^2r/(qr-2) \\ r & q < s \le q^2r/(qr-2) \end{cases}. \tag{3}$$

The course of cars arriving and leaving when $t \in (t_1, r]$ is shown in Fig. 4. The lines *ABC* and GFD are, respectively, the traffic arrival curves of lanes 1 and 2. Again, cars are only allowed to enter lane 1 in the interval $(0, t]$, and the slope of the line *AB* is *q*. Lane 2 is then made available to cars at instant *t* (represented by the line GF). Before the car queue dissipates completely from lane 1, the queue length of lane 2 never exceeds that of lane 1. Therefore, no car will choose to enter lane 1 after *t*, the slope of *BC* is zero, and the slope of *FD* is *q*. The line segments *HC* and *HD* are the traffic dissipation curves of lanes 1 and 2, with the same slope *s*. Note that *HD* is shorter than *HC*, meaning that the queue in lane 2 will fully dissipate at an early instant time t'.

In Fig. 4, the areas of polygons *ABCH* and *FDE* are, respectively, the car delays in lanes 1 and 2. The total car delay in a single cycle when $t \in (t_1, r]$ is denoted as D_{op2}, the area of *ABCH* as S_{ABCH}, and the area of FDE as S_{FDE}. We then have

$$D_{op2} = S_{ABCH} + S_{FDE} = qrt - \frac{qt^2}{2} + \frac{q^2t^2}{2s} + \frac{qs(r - t + 2/s)^2}{2(s-q)} \qquad (t_1 < t \le r). \tag{4}$$

Let t_{op2} denote the queuing and dissipation time, such that

$$t_{op2} = qt/s + r \qquad (t_1 < t \le r). \tag{5}$$

From Eqs. (4) and (5), we find that there is a quadratic functional relation between D_{op2} and *t* and a linear functional relation between t_{op2} and *t* when $t \in (t_1, r]$.

Figure 5 depicts the course of cars arriving and leaving when $t \in (r, t_m]$. The line *ABC* is the traffic arrival curve of lane 1. Lane 1 is open to cars, whereas cars are restricted from entering lane 2 during the red phase, and the slope of the line *AB* is thus *q*. Since cars arriving at the stop line during the green phase can pass through the intersection without stopping, no car will choose to enter lane 1 after *r* until the previous queue has discharged, and the slope of the line *BC* is thus zero. The line segment *DC* is the traffic dissipation curve of lane 1, having slope *s*.

In Fig. 5, the area of polygon *ABCD* is the car delay in lane 1. Because cars can pass through the green signal from lane 2 without delay, the total car delay is the area of *ABCD*. We denote the total car delay in a single cycle when $t \in (r, t_m]$ as D_{op3}, and the area of *ABCD* as S_{ABCD}. We then have

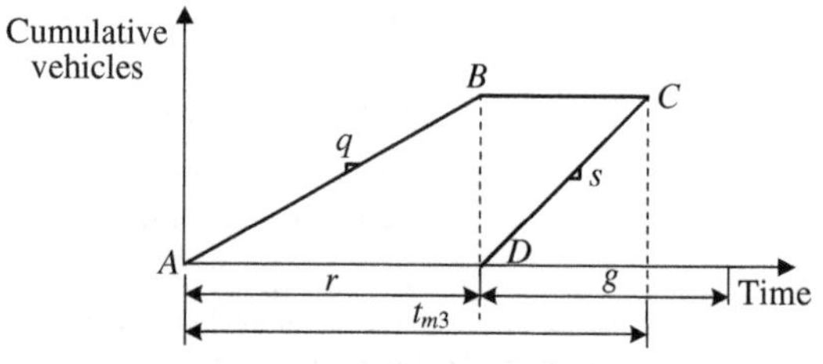

Fig. 5 Course of cars arriving and leaving when $t \in (r, t_m]$

$$D_{op3} = S_{ABCD} = \frac{q^2r^2}{2s} + \frac{qr^2}{2} \qquad (r < t \leq t_m). \tag{6}$$

The queuing and dissipation time t_{op3} shown in Fig. 5 is given by

$$t_{op3} = qr/s + r \qquad (r < t \leq t_m). \tag{7}$$

As we shall see, D_{op3} and t_{op3} do not change with t, because the timing of the activation of the IBA (i.e., when the IBA lane is made available to cars) is independent of t.

Let $\bar{D}_{op}$ be the average total car delay when the IBA is activated in a single cycle. $\bar{D}_{op}$ can then be obtained as

$$\bar{D}_{op} = \frac{\int_0^{t_1} D_{op1}d(t) + \int_{t_1}^{r} D_{op2}d(t) + \int_r^{t_m} D_{op3}d(t)}{t_m}. \tag{8}$$

4 Analysis of examples

For an intersection without an IBA lane, when $t \in (0, t_m]$, the bus will mix with a queue of cars. The average total car delay is then $\bar{D}_{op} = (D_{op1} + D)/2$, and the queuing and dissipation time is $t_{no} = t_{op1}$.[4] For an intersection with an IBA lane and the existing control system, when $t \in (0, t_m]$, cars are restricted from entering the IBA lane in a given period $(0, t_m]$. The total car delay is then $D_{ex} = qrt_m - qt_m^2/2 - q^2t_m^2/(2s)$, and the queuing and dissipation time is $t_{ex} = qt_m/s + r$; this has been analyzed by Xie et al. [8].

We take as fixed parameters s = 1,600 vehicles/h and r = 55s and conduct an analysis for different car flows of q = 600, 1,000, and 1,400 vehicles/h.

Table 1 compares $\bar{D}_{no}, D_{ex}$ and $\bar{D}_{op}$ for the different car flows. It is seen that $\bar{D}_{op}$ is obviously lower than $\bar{D}_{ex}$. This means that the car delay is greatly reduced by optimizing the existing control system of the IBA, and these delay savings tend to be greater as q increases. Furthermore, $\bar{D}_{op}$ increases modestly compared with $\bar{D}_{no}$ as the car flow increases; thus, the optimized IBA system does not remarkably delay cars in the interest of giving buses priority.

$\bar{D}_{op}$ is the average total car delay when the optimal IBA system is implemented. In fact, the car delay varies with the bus arrival time t. These variations are depicted specifically in Fig. 6. Overall, the car delays for different time quanta differ slightly. When $t \in (t_1, r]$, we find D_{op2} appears to have a weakly increasing trend. Because the number of queued cars in the IBA lane decreases as t increases, the queued cars will fully dissipate at an earlier time from the IBA lane than from the other lane. The IBA capacity therefore reduces, leading to greater car delay. Meanwhile, we find that the critical time t_1 approaches r as q increases, as shown in Fig. 6. Thus, this added car delay can be negligible if q approximates s.

Note also from Fig. 6 that the critical time r is a discontinuity on the car delay curve. There are two possible scenarios when a bus arrives at this instant. One is that the bus stops exactly at the end of the red phase, and the time taken for the bus to depart interrupts the upstream car traffic. The other is that the bus passes through the intersection at the start of the green phase without stopping, and thus no following car is impeded. Similar to the two conditions of critical time r, since the queued cars do not dissipate until the bus has departed, the bus delays the upstream cars when $t \in (0, r]$. Since the bus can pass through the green signal without stopping, no car discharging from the IBA lane will be delayed when $t \in (r, t_m]$. Therefore, despite the fact that the IBA capacity reduces to q during the green phase, the car delay does not increase as much as expected when $t \in (r, t_m]$. As can be seen in Fig. 6, when q is relatively high, the car delay saved by the bus posing no impediment is even more than that added by the capacity reduction of the IBA lane.

Recall that the car delay discussed above is based on our assumption that no car will stop twice at the intersection. This assumption holds if the cycle length is no less than the queuing and dissipation time of cars.[5]

Figure 7 presents the curve of car queuing and dissipation time after the control system of the IBA is optimized. The figure shows that the value of t_{op3} is greatest for $t \in (r, t_m]$. Thus, whenever a bus arrives, the residual car queue fully clears in one cycle if the cycle length is no less than t_{op3}. Furthermore, Table 2 compares c_{no} (equal to t_{no}), c_{ex} (equal to t_{ex}), and c_{op} (equal to t_{op3}), the minimum cycle lengths that can guarantee that no residual queue forms at

Table 1 Car delays under three flow conditions: without the IBA and with the IBA before and after optimizing the control system

q (veh/h)	$\bar{D}_{no}$ (s)	D_{ex} (s)	$\bar{D}_{op}$ (s)
600	322	382	342
1,000	636	889	674
1,400	1,088	1,859	1,119

[4] If a bus arrives at the start of the red phase, the bus in front of the car queue will strongly interfere with upstream cars, and the car delay is thus D_{op1}. Meanwhile, if a bus arrives at t_m, the bus causes no interference at the rear of the car queue, and the car delay is thus D. Therefore, we can regard the average of the range from D to D_{op1} as the car delay when a bus queues with cars. Additionally, the position of the bus in the car queue has little effect on the queuing and dissipation time, and $t_{no} = t_{op1}$ is thus reasonable.

[5] In actuality, since it takes time for cars at the end of a queue to reach the stop line, the cycle length should be greater. Fortunately, implementation of the optimal IBA system does not increase the length of the car queue greatly. Thus, for simplicity, we neglect the factor of the car queue length.

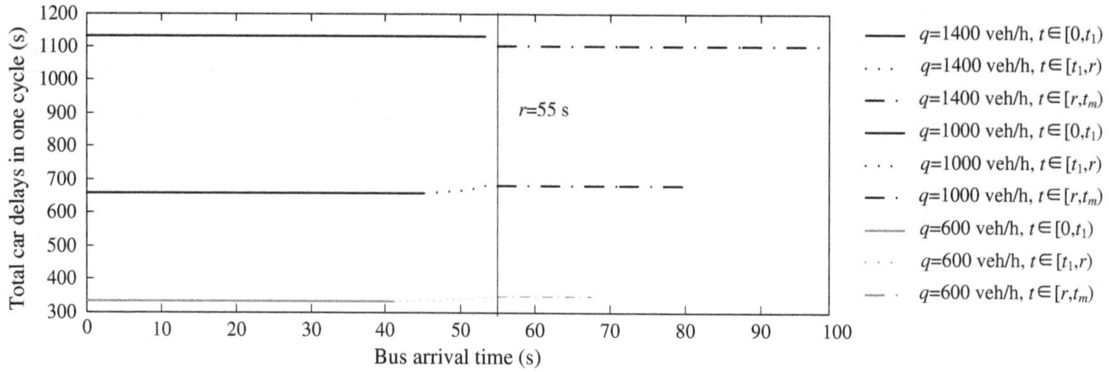

Fig. 6 Car delay curves when implementing the optimal control system of the IBA

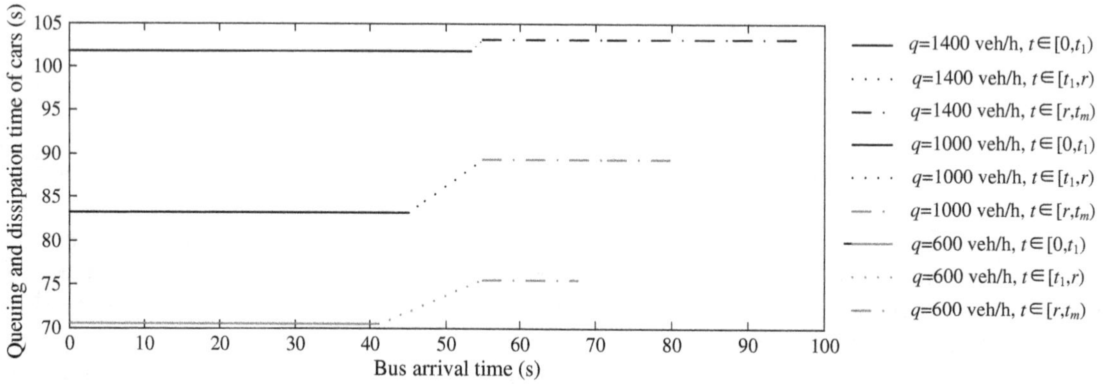

Fig. 7 Car queuing and dissipation time curves when implementing the optimal control system of the IBA

Table 2 Minimum cycle lengths under three conditions: without the IBA and with the IBA before and after optimization of the control system

q veh/h	c_{no} s	c_{ex} s	c_{op}
600	70	80	76
1,000	83	105	89
1,400	102	141	103

the end of one cycle. As shown in the table, because the minimum green ratio $(c_{ex}-r)/c_{ex}$ is much greater than the actual green ratio (usually a little greater than $(c_{no}-r)/c_{no}$) without setting the IBA, more cars will stop twice as q increases when the existing IBA system is implemented. Meanwhile, when the optimal IBA system is implemented, we find that c_{op} is just slightly larger than c_{no}. Thus, even a residual car queue exists. It can dissipate within the next cycle without imposing long-term delays. Moreover, as shown in Fig. 7, if a bus arrives during this time quantum (0, t_1], the optimal IBA system does not create a residual car queue at the intersection because $t_{op1} = c_{no}$. Hence, compared with the existing system of IBA, the probability of a car stopping twice in one cycle is notably lower after the IBA system is optimized.

5 Conclusions

This paper proposed an optimal control system of the IBA and analyzed the car delay according to the bus arrival time. Then, by presenting an example, we compared the total car delays in one cycle under three conditions: without the IBA and with the IBA system before and after optimization. The results show that the optimal IBA system does not significantly delay cars, and the delay is much less than that induced by the existing system. While we analyzed the car delay on the basis that the green ratio is sufficient to clear the car queue in one cycle, in actuality, with no change in the signal timing, a residual car queue might form at the intersection after implementing the IBA, thus imposing additional car delays on following cycles. Fortunately, by analyzing the car queuing and dissipation time, we find the probability of a car stopping twice is low when implementing the optimal IBA system.

The assumptions made in this paper allow for the quantitative analysis of car delay. In particular, ordinary car traffic is considered to arrive at a constant rate, which is not true for a real intersection. Hence, the total car delay is, of course, only approximate. Additionally, we neglect the lane-changing maneuvers as drivers choose their lanes. It will be necessary to set up a micro-simulation in investigating a more complex case in future work. Furthermore, we will study the position of the VMS to guarantee that all cars change the lane smoothly with no interference of downstream queues.

References

1. Balke K et al (2000) Development and evaluation of intelligent bus priority concept. Transp Res Rec J Transp Res Board 1727(1): 12–19
2. Janos M, Furth P (2002) Bus priority with highly interruptible traffic signal control: simulation of San Juan's Avenida Ponce de Leon. Transp Res Rec J Transp Res Board 1181:157–165
3. Yin BC, Yang XG (2005) Study on the bus priority signal control theory of single intersection. J Highw Transp Res Dev 22(12): 123–126
4. Viegas J, Lu B (1999) Bus priority with intermittent bus lane. In: Proceedings of Seminar D,European Transportation conference, Cambridge, UK, 27–29 Sept
5. Viegas J, Lu B (2001) Widening the scope for bus priority with intermittent bus lane. Transp Plan Technol 24(2):87–110
6. Viegas J, Lu B (2004) The intermittent bus lane signals setting within an area. Transp Res C 12(6):453–469
7. Eichler M, Daganzo C (2006) Bus lanes with intermittent priority: strategy formulae and an evaluation. Transp Res B 40(9):731–744
8. Xie QF, Li WQ, Jia XH et al (2012) Research on traffic flow conditions for set intermittent bus-only approach. J Transp Eng Inf 10(2):117–124
9. TRB (2000) Highway capacity manual (HCM). National research council, Washington, DC

Parametric analysis of wheel wear in high-speed vehicles

Na Wu · Jing Zeng

Abstract In order to reduce the wheel profile wear of high-speed trains and extend the service life of wheels, a dynamic model for a high-speed vehicle was set up, in which the wheelset was regarded as flexible body, and the actual measured track irregularities and line conditions were considered. The wear depth of the wheel profile was calculated by the well-known Archard wear law. Through this model, the influence of the wheel profile, primary suspension stiffness, track gage, and rail cant on the wear of wheel profile were studied through multiple iterative calculations. Numerical simulation results show that the type XP55 wheel profile has the smallest cumulative wear depth, and the type LM wheel profile has the largest wear depth. To reduce the wear of the wheel profile, the equivalent conicity of the wheel should not be too large or too small. On the other hand, a small primary vertical stiffness, a track gage around 1,435–1,438 mm, and a rail cant around 1:35–1:40 are beneficial for dynamic performance improvement and wheel wear alleviation.

Keywords Parametric analysis · Wheel profile wear · Flexible wheelset · High-speed railway · Vehicle dynamic model · Finite element method

1 Introduction

With the rapid development of high-speed railways, study on wheel profile wear has become increasingly important [1, 2]. Wheel and rail wear is a fundamental problem in railways; the change of profile shape affects the dynamic characteristics of railway vehicles such as stability and passenger comfort and, in the worst case, can cause derailment [3]. Therefore, it is very important to establish a reliable vehicle dynamic model and wheel/rail wear model to analyze the influence of vehicle parameters on the wear of the wheel profile. However, it is difficult to predict wheel and rail wear simultaneously using state-of-the-art numerical techniques [4]; so we focus on predicting the wear of railway wheels in this work.

To date, many papers on wheel/rail wear prediction have been published. The existing research work of wheel profile wear prediction mainly focus on three aspects: 1) to establish a prediction model based on the vehicle dynamics model, wheel–rail rolling contact model, and wheel material wear model; 2) to confirm the maximum limit value while updating wheel profiles; and 3) to analyze the influence of vehicle track parameters on the wear.

For the wheel profile wear prediction model and maximum limit value as the interval for the wheel profile updating in the repeated dynamic analysis of the vehicle, some scholars carried out studies in different ways. Fries et al. [5] compared four existing wear models, predicting the wear of a freight wagon wheel profile when travel ing in straight lines. The results showed that there was no significant difference between the four wear models. Pearce et al. [6] proposed a wear model for a simple wheel profile by calculating the global contact forces and creepage acting on the contact patch. The amount of material removed was calculated through a wear index (later called the "Derby wear index"), and the wear process was analyzed on a combined straight line and S-curve route. They established that a distance of 1,100 km could be traveled before the wear surface needed upgrading. Li et al. [7]

N. Wu (✉) · J. Zeng
State Key Laboratory of Traction Power, Southwest Jiaotong University, Chengdu 610031, China
e-mail: wunapp@126.com

J. Zeng
e-mail: zeng@swjtu.edu.cn

adopted SIMPACK software to simulate vehicle dynamics. They analyzed the wheel–rail contact with the non-Hertzia multi-point and conformal contact model based on CONTACT, and the wear depth of 0.1 mm is considered as the interval for the wheel profile updating in the repeated dynamic analysis of the vehicle. Jendel et al. [8] developed total simulation conditions using discrete and grouped different curve radii, and analyzed the wheel–rail contact problem using the Hertz theory, Fastsim method, and the Archard model for wear calculations. The update of the wheel profile wear was established when the maximum wear value reached 0.1 mm or the operation distance reached 1,500 or 2,500 km.

The wheel–rail wear is influenced by many factors, and governed by a complex mechanism. Some researchers addressed this problem by analyzing the effect of vehicle and track parameters on the wheel wear. Luo et al. [9] analyzed the influence of the vehicle parameters on the wheel profile wear with a frictional work model. Ignesti et al. [10] developed a mathematical model for wheel–rail wear evaluation in complex railway lines and compared the performance provided by different wheel profiles in terms of resistance to wear and running stability. Pombo [11] used a computational tool to simulate the dynamic performance of an integrated railway system and predict the wear evolution of wheel profiles, taking into account the influence of track condition. Agostinacchio et al. [12] evaluated the influence of the geometrical and mechanical parameters of the superstructure on the dynamic response of the railway. Fergusson et al. [13] presented an analysis of wheel wear as a function of the relationship between the lateral and longitudinal primary suspension stiffness and the coefficient of friction at the center plate between the wagon body and the bolster. Li et al. [14] studied the relationship between the rail cant and wheel–rail rolling contact behavior. The results showed that the rail cant had a great influence on the wheel–rail rolling contact behavior. Wang et al. [15] analyzed the rolling contact geometrical parameters and creepage of the JM3 wheelset and 60 kg/m rail track in static rolling contact under different structural parameters of the track such as rail cant and rail gage. Chen et al. [16] simulated and analyzed the influence on wheel/rail wear caused by vehicle speed, rail cant, super-elevation on curve, and rail lubrication.

Most of the above studies regarded wheel as a rigid body when carrying out the wheel wear prediction. However, when the vehicle passed through a small radius curve, the influence of wheel profiles on the wheel–rail normal force, the contact patch size, position of the contact point, adhesion area, and the distribution of the slide area were different for a flexible wheelset and a rigid wheelset. Chang et al. [17] studied the wheel–rail wear by establishing a three-dimensional dynamic finite element model. Baeza et al. [18] built a model that coupled rotating flexible wheelset and a flexible track model for simulating vehicle–track interaction at high frequencies when investigating growth in rail roughness. Due to the increase of the vehicle speed and the presence of roughness, contact geometry perturbations induce a variation of forces in the vertical and tangential direction, and the torsional vibration of the wheelset axle may, therefore, be excited at high frequency. These vibrations directly affect the contact dynamic action of wheel/rail, and then influence on the wheel profile wear. Therefore, in the prediction of wheel profile wear, wheelset should be considered as flexible body.

In addition, when analyzing the impact of rail and vehicle parameters on wheel profile wear, the above studies completed the wheel profile wear prediction by a single iteration. However, wheel profile deformation caused by wear will change the tendency of these parameters' influence on the wheel profile wear. Therefore, when analyzing the influence of the rail and vehicle parameters on the wheel profile wear, wheel profile should be updated many times in calculation.

In the present work, in order to reduce wear of the wheel profile and extend the service life of wheels, a dynamic model for a high-speed vehicle was set up, in which the wheelset was regarded as flexible body, and the actual measured track irregularities and line conditions were considered. The wear depth of the wheel profile was calculated by the well-known Archard wear law [19]. Through this model, the influence of the wheel profile, primary suspension stiffness, track gage, and rail cant on the wear of the wheel profile were studied through multiple iterative calculations.

2 Model descriptions

2.1 Vehicle dynamic model

The rigid-flexible coupling dynamic model of a high-speed vehicle was established, and the vehicle system included a car-body, two bogie frames, four wheelsets, and eight axle boxes. To take into account the effect of wheel–rail high-frequency vibration on the wear, the wheelset was considered to be flexible, and the other bodies assumed to be rigid. The nonlinearities caused by wheel–rail interaction and suspension parameters were considered in the model. The vehicle system dynamic equations can be expressed in the following form:

$$\boldsymbol{M}\ddot{\boldsymbol{x}} + \boldsymbol{F}(\boldsymbol{x},\dot{\boldsymbol{x}}) = \boldsymbol{P}(\boldsymbol{x},\dot{\boldsymbol{x}},t), \tag{1}$$

where $\boldsymbol{x}$ denotes the displacement vector, $\boldsymbol{M}$ indicates the system mass matrix, $\boldsymbol{F}$ is the nonlinear suspension forces, and $\boldsymbol{P}$ is an item related to the nonlinear wheel/rail forces and track inputs.

The wheelset finite element model was set up using ANSYS software, in which one axle, two wheels, and three brake discs were included. The eight-node hexahedral 3D solid element mesh division was adopted for the modeling, and the whole unit had 70,592 elements with 83,616 nodes. The wheelset finite element model is shown in Fig. 1.

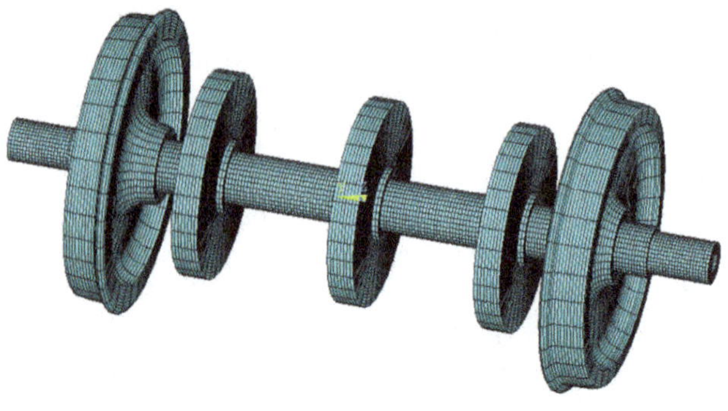

Fig. 1 Wheelset finite element model

Using the Guyan reduction method and maintaining the overall shape of the structure, a freedom set with a uniform distribution was selected. Through modal analysis using the finite element model and without imposing any constraints, the first 30 modes were obtained and imported to the SIMPACK dynamic analysis software. The rigid-flexible coupling dynamic model of the vehicle system was then built. The mode shapes of the flexible wheelset are shown in Fig. 2. The flexible wheelset had many mode shapes which might affect the wear of the wheel profile, and thus the flexibility of the wheelset could not be ignored.

2.2 Wear model

Archard's wear model is a function of the sliding distance, normal force, and hardness of the material. The wear volume of the material worn away is proportional to the product of the sliding distance and the normal force, and

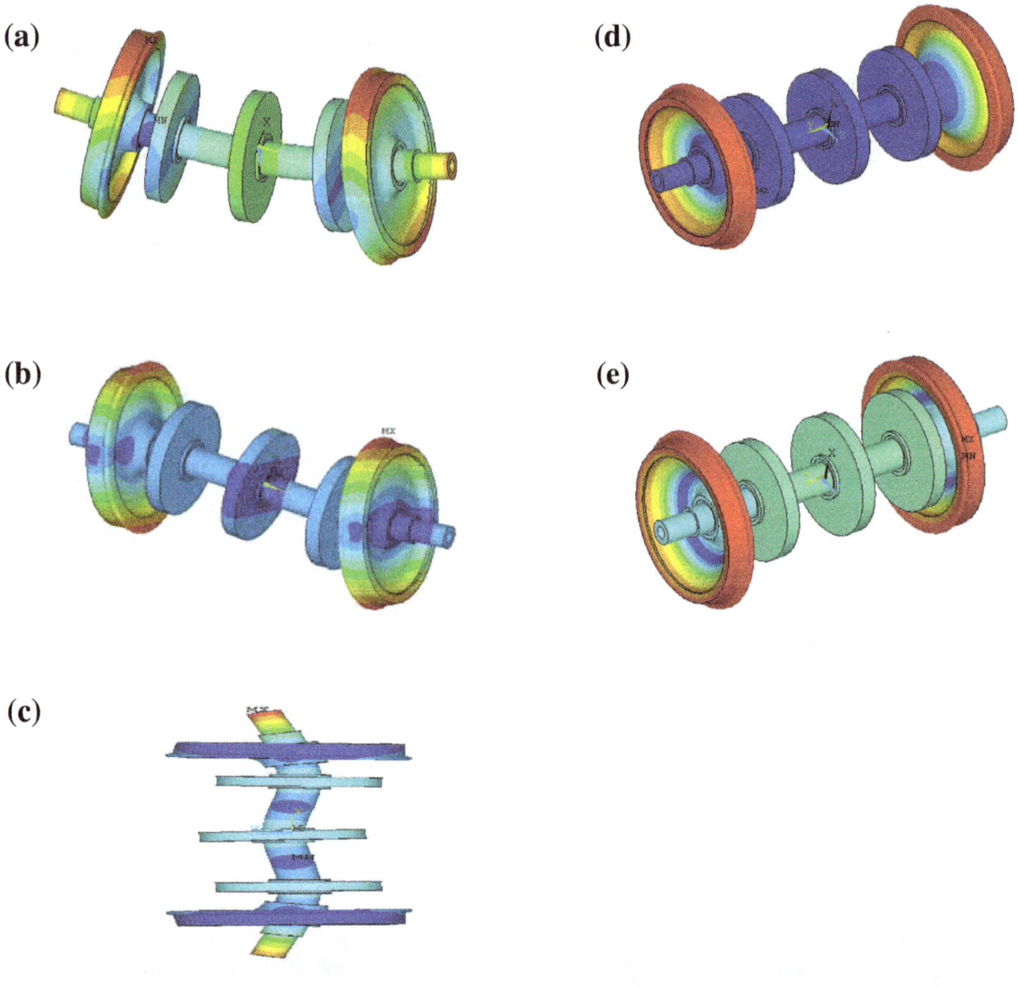

Fig. 2 Mode shapes of the flexible wheelset. **a** First vertical and horizontal bending modes (77 Hz). **b** Second vertical and horizontal bending modes (133 Hz). **c** Third vertical and horizontal bending modes (575 Hz). **d** First umbrella mode (225 Hz). **e** Second umbrella mode (282 Hz)

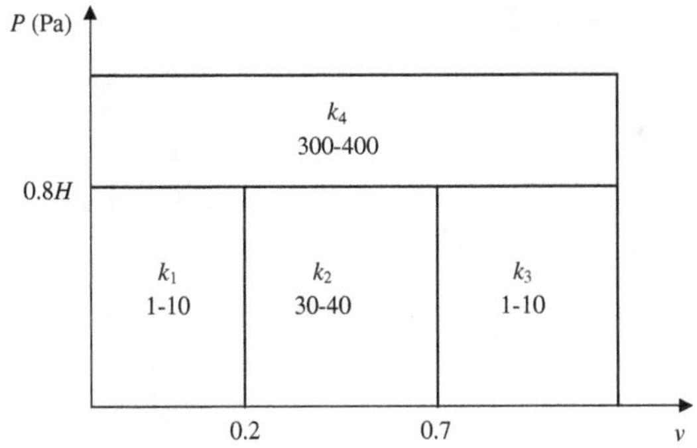

Fig. 3 Coefficient k

inversely proportional to the hardness of the worn material. It can be described by Eq. (2):

$$\boldsymbol{V}_{\text{wear}} = k\frac{\boldsymbol{Nd}}{H}, \tag{2}$$

where $\boldsymbol{V}_{\text{wear}}$ is the wear volume matrix, $\boldsymbol{d}$ the sliding distance vector, $\boldsymbol{N}$ the normal force matrix, H is the hardness of the worn material, and k is the wear coefficient.

The wear coefficient k can be determined by laboratory tests or by performing extensive field measurements. It is generally a function of the sliding velocity, contact pressure, temperature, and contact environment. The wear coefficient used in the present calculation is described in the wear chart from Ref. [7]. It can be expressed in Fig. 3, in which the horizontal and vertical axes are the sliding velocity and contact pressure, respectively. This figure has been derived under dry contact conditions. The tread contact occurs in region k_2, and the flange contact occurs in the regions k_1, k_2, k_3, and k_4. In this study, k is taken as the middle value in each region.

The wheel–rail contact model was set up using the simplified Kalker's algorithm Fastsim in which the wheel/rail contact ellipse is divided into many elements. In each element, the normal contact pressure P, sliding distance d, and wear depth z are expressed, respectively, by Eqs. (3)–(5):

$$P(x,y) = \frac{3N}{2\pi ab}\sqrt{1-\left(\frac{x}{a}\right)^2-\left(\frac{y}{b}\right)^2}, \tag{3}$$

$$d(x,y) = \Delta x\sqrt{(\xi-\phi y)^2+(\eta+\phi x)^2}, \tag{4}$$

$$z(x,y) = \frac{3Nk\Delta x}{2\pi abH}\sqrt{1-\left(\frac{x}{a}\right)^2-\left(\frac{y}{b}\right)^2}\sqrt{(\xi-\varphi y)^2+(\eta+\varphi x)^2}, \tag{5}$$

where x denotes the longitudinal direction of the contact plane; y is the transversal direction of the contact plane, and the element center point (x, y) are the Cartesian coordinates of the contact patch; ξ, η, and ϕ denote the longitudinal creepage, lateral creepage, and spin creepage, respectively; a and b denotes the long axis and short axis of contact patch, respectively.

2.3 Process of wear prediction

The vehicle parameters, wheel–rail initial profile, mode shapes of wheelset, track random inputs, and track line conditions were taken into account in the vehicle system dynamic model. The contact patch location, size, creepage, and normal stress distribution were then calculated. Subsequently, the amount of wear for the wheel was calculated using the wear model. Finally, the wheel wear distribution was obtained, and the wheel profile was updated using the smoothing method of cubic spline interpolation. The wheel–rail contact patch was divided into 50×50 elements. The wear model predicted the change in the wear of the wheel profile through multiple iterations. The integrated simulation process for wheel wear is shown in Fig. 4.

To accelerate the wear prediction, the following hypotheses for calculating wheel wear were developed:

(1) During one integrated simulation of wear prediction, the profile of the wheel remained unchanged, and the tread was updated when the wear depth was 0.1 mm or the vehicle had traveled through 1,500 km according to previous studies [4, 8]. On the basis of the field analysis of the measured data for a high-speed vehicle, a running distance of 1,000 km was taken as the step length for updating the wheel profile in this study.
(2) The vehicle structure was symmetrical, and the left and right rails on the curved track were arranged

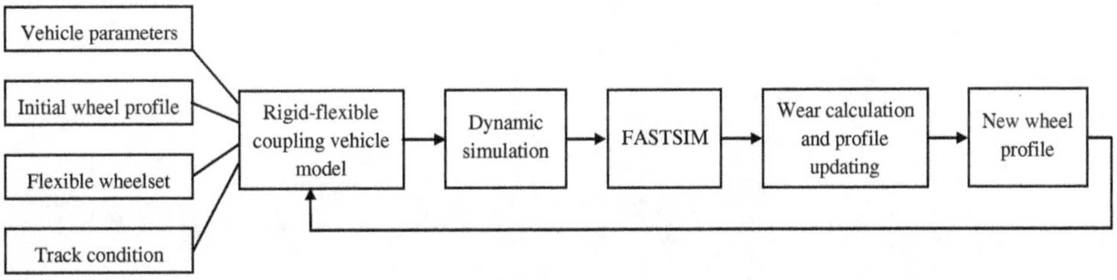

Fig. 4 Integrated simulation process

Table 1 Typical scenarios

Track radius (m)	Vehicle speed (km/h)	Percentage (%)
2,200	160	4
2,800	180	4
4,000	200	4
Straight	200	88

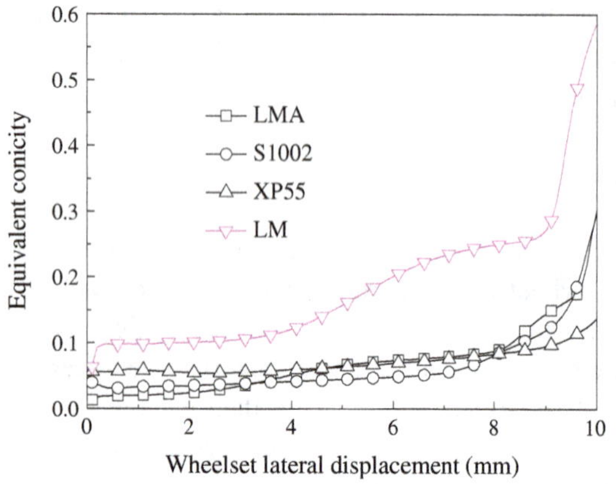

Fig. 5 Equivalent conicities for different types of wheel profiles

symmetrically. The vehicle always traveled forward and backward on the line, thus the wheel wear for wheelsets one and four was the same, and for wheelsets two and three was also the same.

(3) The track excitation was random, and the impact of rail wear on wheel wear was not considered.

For the wheel wear calculation, the vehicle was assumed to pass through a prescribed track consisting of three curved tracks and one straight track [20], which is shown in Table 1.

3 Parametric analysis of wheel wear

It is essential to acquire a better understanding on how the operation conditions influence wear evolution of the wheel profile. Therefore, the following analysis focuses on vehicle/track parameters influence of wheel profile on wear.

3.1 Influence of wheel profile on wear

The wear depth and distribution of four types of wheel profiles (LM, LMA, XP55, and S1002) were compared on the prescribed line conditions for the same operating mileage. Figure 5 illustrates the equivalent conicities of the

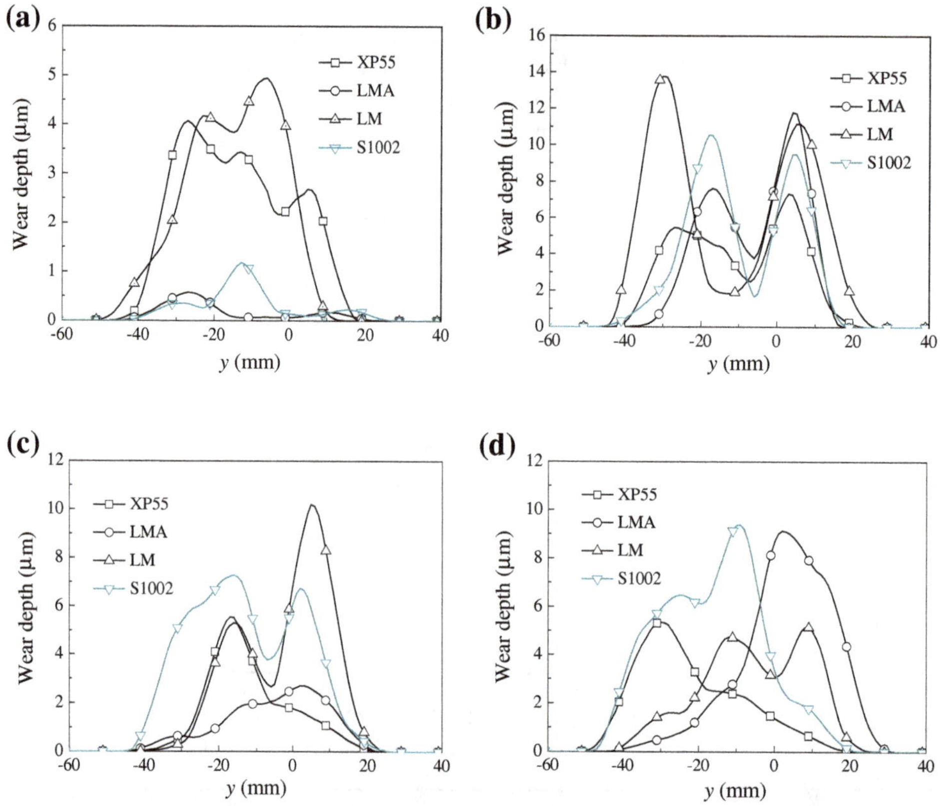

Fig. 6 Wear comparison for different types of wheel profiles. **a** Wear stage 1. **b** Wear stage 2. **c** Wear stage 3. **d** Wear stage 4

Table 2 Wear distribution zone for different wheel profiles (mm)

Wear stage	Wheel profiles			
	XP55	LMA	LM	S1002
1	−47–+18	−44–+ 26	−52–+ 16	−46–+29
2	−44–+24	−43–+24	−45–+26	−51–+24
3	−44–+26	−45–+26	−44–+25	−47–+27
4	−51–+26	−45–+32	−43–+31	−51–+31

four wheel profiles, and Fig. 6 shows the wear depth changes for the four wheel profiles. Figure 6 includes four wear stages for the wheel wear prediction, labeled (a), (b), (c), and (d), respectively, and each stage covered a distance of 1,000 km. After each stage, the wheel profile was updated to enable the calculations for the next stage. In the figures, the abscissa y is the horizontal axis of the wheel profile, and the origin is at the wheel nominal rolling circle position.

As known, the wear depth initially increased rapidly and then slowed with the increase in running mileage, and the wear range slowly broadened.

Table 2 shows the distribution of different types of wheel profiles for the four wear stages. In stage 1, because the type LM profile had the largest equivalent conicity, the wheel wear was close to the flange; the S1002 profile had the smallest equivalent conicity, and the wheel wear was near the outside of the wheel. For the later wear stages, the wear volume for LM was the greatest and the wear ranges the widest; and the wear volume for XP55 the smallest. Therefore, the selection of an appropriate wheel profile and equivalent conicity is very important for the actual wheel wear. If the equivalent conicity was too large, then the large contact angle would cause the spin creep to increase. At the same time, the wheel rail contact point was closer to the flange, and the wheel wear would be more serious. On the other hand, if the difference in the rolling radii between the left and right wheels increased, then the deviation from the center position of the wheelset caused greater longitudinal creep and increased the wear depth. If the equivalent conicity was too small, then the lateral motion of the wheelset would be greater to widen the wear range because of the weak centering ability. Therefore, a too large or too small equivalent conicity will intensify the wheel wear.

3.2 Influence of primary vertical stiffness on wear

To compare the influence of primary vertical stiffness on the wear depth and range, the primary vertical stiffness with values 0.8, 1.0, 1.2, and 1.4 MN/m was adopted for the calculation. Figure 7 shows that the primary vertical stiffness has little influence on the wear range. In the wear

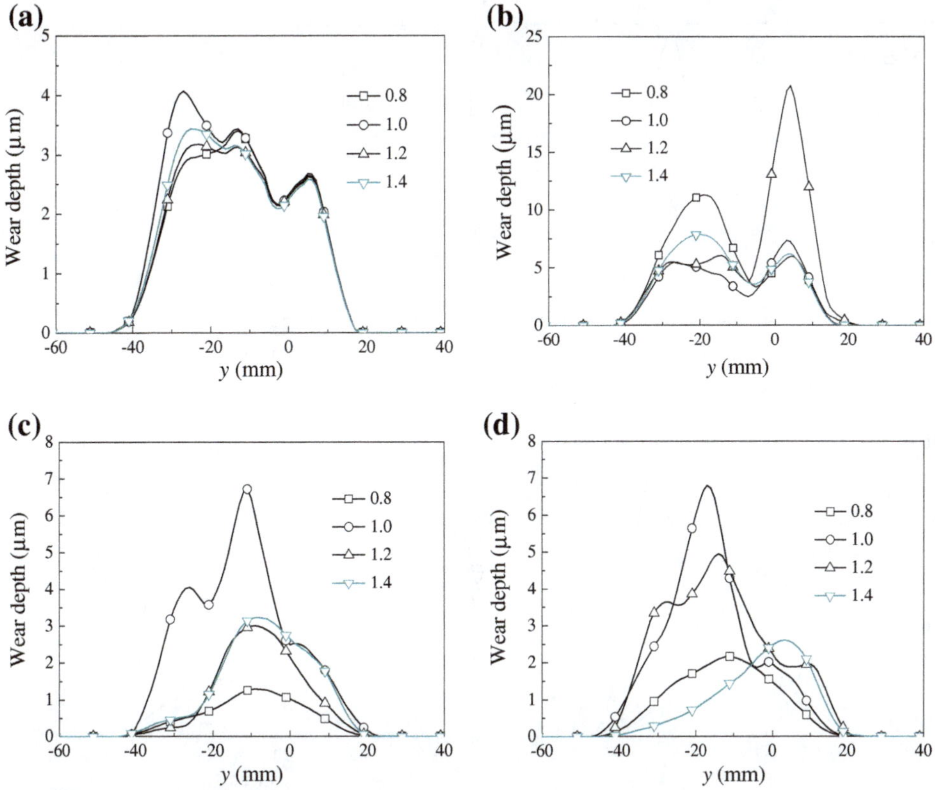

Fig. 7 Wear comparison for different primary vertical stiffnesses. **a** Wear stage 1. **b** Wear stage 2. **c** Wear stage 3. **d** Wear stage 4

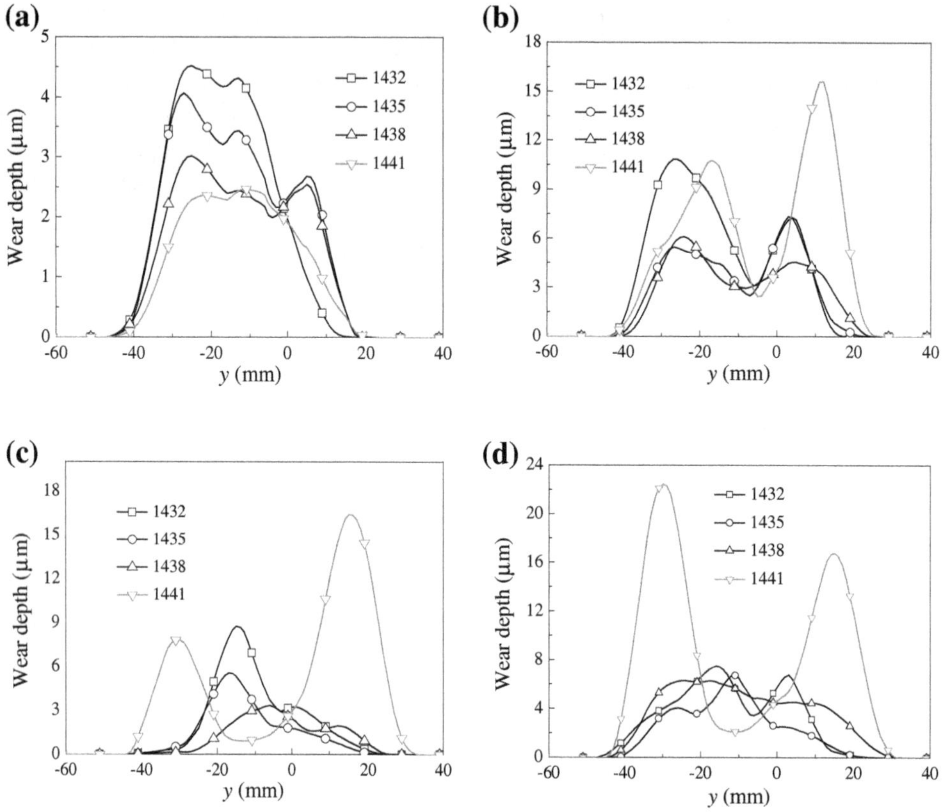

Fig. 8 Wear comparison for different track gages. **a** Wear stage 1. **b** Wear stage 2. **c** Wear stage 3. **d** Wear stage 4

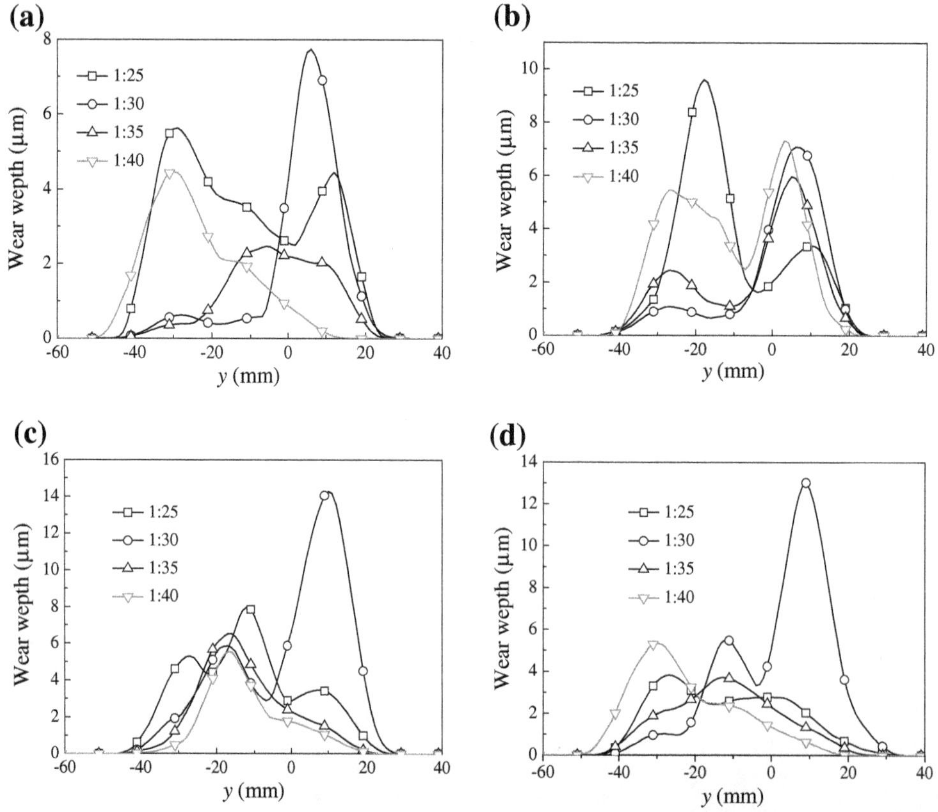

Fig. 9 Wear comparison for different rail cants. **a** Wear stage 1. **b** Wear stage 2. **c** Wear stage 3. **d** Wear stage 4

stage 3, the primary vertical stiffness has little effect on the wear depth; in the further wear stages, the stiffness 1.0 MN/m will cause greater wear than the others, and the stiffness 0.8 MN/m has the least wear. With the increasing of running mileage, the wear depths increase rapidly in the early stages and slowly in the later stages.

3.3 Influence of track gage on wear

The computed track gages were 1,432, 1,435, 1,438, and 1,441 mm, and their influences on the wear depth and distribution of the wheel profiles were compared in Fig. 8. We can see that wear depth reduces and wear range gradually moves away from the flange with the increase in track gage in the first stage. In the later stages, track gages 1,435 and 1,438 mm have the least wheel wear depth. Thus, slightly widening the track gage is advantageous for reducing wheel wear.

3.4 Influence of rail cant on wear

Rail cants of 1:25, 1:30, 1:35, and 1:40 were selected for the wheel wear calculation. As shown in Fig. 9, the wear depth was small when the rail cant angles were 1:35 and 1:40, and the wear distribution was near the flange. Thus, the rail cant should be between 1:35 to 1:40 to reduce wheel wear.

4 Conclusions

To study wheel profile wear, a vehicle dynamic model and a wheel profile wear model were established. The influence of wheel profile, primary vertical stiffness, track gage, and rail cants on wheel profile wear was investigated through numerical simulations, and the following conclusions were reached:

(1) The shape of the wheel profile has a significant influence on wheel wear depth and range. Among the four types of wheel profiles, the type XP55 wheel had the smallest cumulative wear depth, and LM had the largest wear. To reduce the wear of the wheel profile, an appropriate wheel equivalent conicity needs to be designed, and must not be too large or too small.
(2) Using a small primary vertical stiffness can have a better dynamic performance and reduce wheel wear.
(3) The track gage should be between 1,435 and 1,438 mm, a too large or too small gage will aggravate the wheel wear, and rail cant should be between 1:35 to 1:40.

Acknowledgments The authors would like to acknowledge the support of the National Natural Science Foundation of China (No. 51005189) and the National Key Technology R&D Program of China (2009BAG12A01).

References

1. Jin XS, Liu QY (2004) Tribology of wheel and rail. China Railway Press, Beijing
2. Zhou L, Shen ZY (2011) Progress in high-speed train technology around the world. J Mod Transp 19(1):1–6
3. Braghin F (2006) A mathematical model to predict railway wheel profile evolution due to wear. Wear 261:1253–1264
4. Li X, Jin XS, Wen ZF et al (2011) A new integrated model to predict wheel profile evolution due to wear. Wear 271:227–237
5. Fries RH, Dávila CG (1988) Analytical methods for wheel and rail wear prediction. Proceedings 10th IAVSD Symposium, Swets and Zeitlinger: 112–125
6. Pearce TG, Sherratt ND (1991) Prediction of wheel profile wear. Wear 144:343–351
7. Li ZL, Kalker JJ, Wiersma PK et al. (1998) Non-Herztian wheel-rail wear simulation in vehicle dynamical systems. Proceedings 4th International Conference on Railway Bogies and Running Gears. Budapest: 187–196
8. Jendel T (2002) Prediction of wheel profile wear-comparisons with field measurements. Wear 253:89–99
9. Luo R, Zeng J, Wu PB et al (2009) The influence of high speed trains between wheel and rail parameter on wheel profile wear. Traffic Transp Eng 9(6):47–63
10. Ignesti M, Innocenti A, Marini L, Meli E, Rindi A (2013) Development of a wear model for the wheel profile optimisation on railway vehicles. Veh Syst Dyn 51(9):1363–1402
11. Pombo J, Ambrosio J, Pereira M, Verardi R, Ariaudo C, Kuka N (2011) Influence of track conditions and wheel wear state on the loads impose on the infrastructure by railway vehicles. Comput Struct 89:1882–1894
12. Agostinacchio M, Ciampa D, Diomedi M, Olita S (2013) Parametrical analysis of the railways dynamic response at high speed moving loads. J Mod Transp 21(3):169–181
13. Fergusson SN, Frohling RD, Klopper H (2008) Minimising wheel wear by optimising the primary suspension stiffness and centre plate friction of self-steering bogies. Veh Syst Dyn 46(1):457–468
14. Li X, Wen ZF, Zhang J, Jin XS (2009) Effect of rail cant on wheel/rail rolling contact behavior. J Mech Strength 31(3):475–480
15. Wang WJ, Guo J, Liu QY (2010) Effects of track parameters on rolling contact behavior of wheel-rail. J Sichuan Univ (Eng Sci Ed). 42(6):213–218, 226
16. Chen P, Gao L, Hao JF (2007) Simulation study on parameters influencing wheel/rail wear in railway curve. China Railw Sci 28(5):19–23
17. Chang CY, Wang CG, Jin Y (2008) Numerical analysis of wheel/rail wear based on 3D dynamic finite. China Railw Sci 29(4):89–95
18. Baeza L, Vila P, Xie G, Simon Iwnicki D (2011) Prediction of rail corrugation using a rotating flexible wheelset coupled with a flexible track model and an on-Hertzian/non-steady contact model. J Sound Vib 330:4493–4507
19. Telliskivi T (2003) Wheel-rail interaction analysis. Doctoral Thesis, Department of Machine Design, KTH
20. He HW (2003) Speed up technique of 200 Km/h on Chinese previous railway. China Railway Publishing House, Beijin

7

Detecting land subsidence near metro lines in the Baoshan district of Shanghai with multi-temporal interferometric synthetic aperture radar

Tao Li · Guoxiang Liu · Hui Lin · Rui Zhang · Hongguo Jia · Bing Yu

Abstract Land subsidence is a major factor that affects metro line (ML) stability. In this study, an improved multi-temporal interferometric synthetic aperture radar (InSAR) (MTI) method to detect land subsidence near MLs is presented. In particular, our multi-temporal InSAR method provides surface subsidence measurements with high observation density. The MTI method tracks both point-like targets and distributed targets with temporal radar backscattering steadiness. First, subsidence rates at the point targets with low-amplitude dispersion index (ADI) values are extracted by applying a least-squared estimator on an optimized freely connected network. Second, to reduce error propagation, the pixels with high-ADI values are classified into several groups according to ADI intervals and processed using a Pearson correlation coefficient and hierarchical analysis strategy to obtain the distributed targets. Then, nonlinear subsidence components at all point-like and distributed targets are estimated using phase unwrapping and spatiotemporal filtering on the phase residuals. The proposed MTI method was applied to detect land subsidence near MLs of No. 1 and 3 in the Baoshan district of Shanghai using 18 TerraSAR-X images acquired between April 21, 2008 and October 30, 2010. The results show that the mean subsidence rates of the stations distributed along the two MLs are −12.9 and −14.0 mm/year. Furthermore, three subsidence funnels near the MLs are discovered through the hierarchical analysis. The testing results demonstrate the satisfactory capacity of the proposed MTI method in providing detailed subsidence information near MLs.

T. Li · G. Liu (✉) · R. Zhang · H. Jia · B. Yu
Dept. of Remote Sensing and Geospatial Information Engineering, Southwest Jiaotong University, Chengdu 610031, Sichuan, China
e-mail: rsgxliu@swjtu.edu.cn

T. Li · H. Lin · R. Zhang
Institute of Space and Earth Information Science, Chinese University of Hong Kong, Hong Kong, China

Keywords Multi-temporal InSAR · Subsidence · Baoshan district · Shanghai · Metro lines

1 Introduction

Due to the potential effects of land subsidence geo-hazards on metro lines (MLs) in terms of human lives and economic losses, great effort has been put forth to develop sustainable solutions for these hazards. To date, several methods have been investigated, including leveling [1], global positioning systems [2, 3], interferometric synthetic aperture radar (InSAR) [4], and multi-temporal InSAR (MTI) [5]. The MTI is a newly developed technique and provides a sound tool to assess land subsidence on Earth's surface. It is capable of providing millimetric-precision subsidence rates. Moreover, the method offers a synoptic view of subsidence funnels. Therefore, many methods using MTI to detect slow-moving land subsidence have been presented.

The most frequently used methods are PSInSAR^{TM} [6], SqueeSAR^{TM} [7], Small Baseline Subset [8], the Stanford Method for Persistent Scatterers [9], Interferometric Point Target Analysis [10], the Quasi-PS technique [11], PS Networking Analysis [12], and Temporarily Coherent Point InSAR [13]. Most methods focus on point-like target (PTs). Moreover, SqueeSAR^{TM} indicate that distributed targets (DTs) maintain temporal radar backscattering steadiness. However, combined PT and DT analyses are rarely jointly considered to provide land subsidence information.

To increase the observation density of surface subsidence measurements, an improved MTI method is proposed to jointly process both PTs and DTs. The MTI method uses both low-amplitude dispersion index (ADI) [6, 14] and high Pearson correlation coefficient [15] to select all valid pixels corresponding to PTs and DTs from the image series. Furthermore, to control error propagation, a hierarchical analysis strategy is presented. The method is first introduced at a high level. It consists of three major steps. They are briefly described below.

First, the subset of potential pixels (SPP) of the PTs is selected by considering pixels with low-ADI values. The SPP is processed using a least-squared estimator on an optimized freely connected network to obtain the PTs. Second, the SPP of the DTs is selected by considering pixels with high-ADI values, which are classified according to the ADI intervals. To reduce error propagation, the DTs and their subsidence rates are detected using the hierarchical analysis strategy and quality assessments on each SPP subgroup. Finally, the nonlinear subsidence components for all valid pixels are estimated using phase unwrapping and spatiotemporally filtering of the phase residuals.

The proposed MTI method is used to detect land subsidence near MLs in the Baoshan district of Shanghai using 18 TerraSAR-X images acquired between April 21, 2008 and October 30, 2010. Due to the high-spatial resolution (approximately 2 m in both ground range and azimuth directions), the TerraSAR-X images are capable of providing detailed subsidence information when used in the newly modified MTI method.

The remainder of the paper is organized as follows. In Sect. 2, the basic MTI method theories are explained. In Sect. 3, the data and experimental results are discussed. Finally, several conclusions are shown in Sect. 4.

2 Method

In this section, the core models of the three major steps in the MTI method are provided. Unlike the aforementioned MTI methods [6–13], both temporal and spatial phase data correlations are jointly considered using ADI and Pearson correlation coefficient. First, the equations for the subsidence rate extraction at the PTs with low-ADI values are discussed. Then, to extract the subsidence rates at the DTs, the pixels with high-ADI values are processed using the hierarchical analysis strategy. Finally, the nonlinear subsidence components are obtained through spatiotemporally filtering all valid pixels of the phase residuals.

2.1 Estimating subsidence rates at the PTs

The SPP with ADI values <0.4 [9] is selected to estimate the subsidence rates at the PTs. The SPP is connected using an optimized freely connected network [12], which provides basic observations for estimating linear

subsidence rates. A least-squared estimator [13, 16] is then applied on each arc of the network to calculate the relative subsidence parameters (i.e., the subsidence rate and DEM error).

It is assumed that N interferograms are obtained from $N + 1$ images. For each interferogram k, the differential model is applied to each arc of the freely connected network:

$$\Delta\varphi^k_{(x_i,y_i;x_j,y_j)} = -\frac{4\pi}{\lambda}\Delta v_{(x_i,y_i;x_j,y_j)}T^k - \frac{B^k_\perp}{R\sin\theta}\Delta\varepsilon_{(x_i,y_i;x_j,y_j)} + \Delta\varphi^k_{RE(x_i,y_i;x_j,y_j)}, \tag{1}$$

where $\Delta\varphi^k_{(x_i,y_i;x_j,y_j)}$ is the differential phase between two pixels, λ is the wavelength of the sensor, $\Delta v_{(x_i,y_i;x_j,y_j)}$ is the relative subsidence rate between two pixels, T^k is the temporal baseline, $B^k_\perp$ is the local perpendicular baseline, R is the slant range distance from the reference pixel to the sensor in the master image, θ is the local look angle, $\Delta\varepsilon_{(x_i,y_i;x_j,y_j)}$ is the relative DEM error, and $\Delta\varphi^k_{RE_{(x_i,y_i;x_j,y_j)}}$ is the differential phase residual.

We assume that there are no phase ambiguities in the arcs. The assumption allows applying the least-squared estimation method. And the phase residuals calculated from least-squared estimation can be used to locate the arcs affected by phase ambiguities. The problematic arcs determined should be discarded for further analysis. The least-squared estimation is used by jointly considering N interferograms [13, 16]

$$\Delta\boldsymbol{\Phi} = \boldsymbol{A}\begin{pmatrix}\Delta v\\ \Delta\varepsilon\end{pmatrix} + \boldsymbol{W}, \tag{2}$$

where

$$\begin{aligned}\Delta\boldsymbol{\Phi} &= \left(\Delta\varphi^1_{(x_i,y_i;x_j,y_j)}\ \Delta\varphi^2_{(x_i,y_i;x_j,y_j)}\cdots\Delta\varphi^N_{(x_i,y_i;x_j,y_j)}\right)^{\mathrm{T}}\\ \underset{N\times 2}{\boldsymbol{A}} &= -\begin{pmatrix}\frac{4\pi}{\lambda}T^1 & \frac{4\pi}{\lambda}T^2 & \cdots & \frac{4\pi}{\lambda}T^N\\ \frac{B^1_\perp}{R\sin\theta} & \frac{B^2_\perp}{R\sin\theta} & \cdots & \frac{B^N_\perp}{R\sin\theta}\end{pmatrix}^{\mathrm{T}}\\ \underset{N\times 1}{\boldsymbol{W}} &= \left(\Delta\varphi^1_{RE(x_i,y_i;x_j,y_j)}\ \Delta\varphi^2_{RE(x_i,y_i;x_j,y_j)}\cdots\Delta\varphi^N_{RE(x_i,y_i;x_j,y_j)}\right)^{\mathrm{T}}.\end{aligned} \tag{3}$$

The most probable values of Δv and $\Delta\varepsilon$ are retrieved according to

$$\begin{pmatrix}\Delta\hat{v}\\ \Delta\hat{\varepsilon}\end{pmatrix} = (\boldsymbol{A}^{\mathrm{T}}\boldsymbol{A})^{-1}\boldsymbol{A}^{\mathrm{T}}\Delta\boldsymbol{\Phi}, \tag{4}$$

where $\hat{\cdot}$ denotes the estimated value. The vector of estimated residual phases can be represented by

$$\hat{W} = \left(\Delta\,\hat{\varphi}^{1}_{RE(x_i,y_i;x_j,y_j)}\,\Delta\,\hat{\varphi}^{2}_{RE(x_i,y_i;x_j,y_j)}\cdots\Delta\,\hat{\varphi}^{N}_{RE(x_i,y_i;x_j,y_j)}\right)^{\mathrm{T}} = \Delta\boldsymbol{\Phi} - \boldsymbol{A}(\boldsymbol{A}^{\mathrm{T}}\boldsymbol{A})^{-1}\boldsymbol{A}^{\mathrm{T}}\Delta\,\boldsymbol{\Phi}. \tag{5}$$

After the least-squared estimation, the spatiotemporal phase data correlation between two adjacent pixels connected by the freely connected network is measured using the modeling coherence factor $\left(\gamma_{(x_i,y_i;x_j,y_j)}\right)$ [6]:

$$\gamma_{(x_i,y_i;x_j,y_j)} = \left|\frac{1}{N}\sum_{k=1}^{N}\exp\left[\Delta\hat{\varphi}^{k}_{RE(x_i,y_i;x_j,y_j)}\right]\right|. \tag{6}$$

If $\gamma_{(x_i,y_i;x_j,y_j)}$ is <0.8 [14], the arc is discarded from the freely connected network, which reasonably assures that low-quality pixels are eliminated.

Besides, an easy and efficient assessment used to detect outlier was applied by Zhang et al. [16] to discard the arcs affected by phase ambiguities:

$$Max\left(\left|\Delta\,\hat{\varphi}_{RE(x_i,y_i;x_j,y_j)}\right|\right) > E\left(\Delta\,\hat{\varphi}_{RE(x_i,y_i;x_j,y_j)}\right) + 2\delta\left(\Delta\,\hat{\varphi}_{RE(x_i,y_i;x_j,y_j)}\right), \tag{7}$$

where $Max(\cdot)$ means the maximum value in a vector or matrix. If the threshold value in Eq. (7) is reached, the tested arc is considered to contain an outlier at 95 % confidence level [16].

The subsequent processing estimates the absolute subsidence parameters for all PTs using the relative linking parameters in the freely connected network. The most probable estimates of the two subsidence parameters are calculated using the following formulas provided by Liu et al. [12]:

$$\underset{Q\times M}{B}\;\underset{M\times 1}{X} = \underset{Q\times 1}{L} + \underset{Q\times 1}{R}$$

$$P = \begin{pmatrix} \gamma_1^2 & 0 & 0 & 0 \\ 0 & \gamma_2^2 & 0 & 0 \\ \vdots & \vdots & \vdots & \vdots \\ 0 & 0 & 0 & \gamma_Q^2 \end{pmatrix}, \tag{8}$$

$$X = (B^{\mathrm{T}}PB)^{-1}B^{\mathrm{T}}PL$$

where B is the design matrix for weighted LS estimation; the elements of B are either 1 or -1. Moreover, L is the vector of $\Delta\,\hat{v}$, R is the vector of residuals, X is the vector of unknown PT linear subsidence rates (or DEM errors), Q is the number of arcs, and M is the number of PTs.

2.2 Estimating subsidence rates at the DTs

To increase the observation density of the subsidence rate map, the pixels with high-ADI values are further treated on a group-by-group basis. All pixels with ADI values >0.4 are classified into G groups according to the ADI intervals, i.e., (0.4, 0.5), …, $(0.3 + 0.1 \times i,\ 0.4 + 0.1 \times i)$, …, $(0.3 + 0.1 \times G,\ 0.4 + 0.1 \times G)$, where $i = 1, 2, \ldots, G$. The PTs are treated as Group 0. After classification, the subsidence rates in the radar line-of-sight direction are hierarchically analyzed for each group.

Group i is used here as an example. The phase quality of each pixel in Group i is assessed through Pearson correlation coefficient [15] to determine whether the POI phase is consistent with any neighboring pixels already accepted in the previous i groups (i.e., Groups 0 to i–1). The Pearson correlation coefficient is described as follows:

$$\rho_{(x_i,y_i,x_j,y_j)} = \frac{\sum_{k=1}^{n}\left[\varphi^{k}_{(x_i,y_i)} - \overline{\varphi_{(x_i,y_i)}}\right)\left(\varphi^{k}_{(x_j,y_j)} - \overline{\varphi_{(x_j,y_j)}}\right]}{\sqrt{\sum_{k=1}^{n}\left[\varphi^{k}_{(x_i,y_i)} - \overline{\varphi_{(x_i,y_i)}}\right]^2\sum_{k=1}^{n}\left[\varphi^{k}_{(x_j,y_j)} - \overline{\varphi_{(x_j,y_j)}}\right]^2}}. \tag{9}$$

When the Pearson correlation coefficient between two vectors is >0.75, the two vectors are considered to be highly correlated [15]. Therefore, we introduce 0.75 as a threshold to discard the point pairs whose phase time series are uncorrelated. The relative subsidence parameters between these pixels are derived using Eqs. (1)–(6) and consequently added to those at the neighboring pixel to determine the absolute parameters.

Besides Pearson correlation coefficient, two other constraints are also introduced to assess the pixel quality when the subsidence parameters are obtained through the aforementioned processing. They are spatial autocorrelation of the two subsidence parameters for the given pixel. The spatial autocorrelation is denoted by the localized standard deviations. For example, if there are L valid pixels near a pixel (x_i, y_i) in a given window size (e.g., 50 × 50 pixels), the standard deviations of the two subsidence parameters are

$$\delta v = \sqrt{\frac{1}{L}\sum_{j=1}^{L}\left(v_{(x_i,y_i)} - v_{(x_j,y_j)} - \Delta\,\hat{v}_{(x_i,y_i;x_j,y_j)}\right)^2}, \quad \delta\varepsilon = \sqrt{\frac{1}{L}\sum_{j=1}^{L}\left(\varepsilon_{(x_i,y_i)} - \varepsilon_{(x_j,y_j)} - \Delta\,\hat{\varepsilon}_{(x_i,y_i;x_j,y_j)}\right)^2} \tag{10}$$

where (x_j, y_j) are the valid pixels that are maintained from the previous i groups. The two constraints ensure the quality of the pixels around the investigated pixel. If the pixels are totally stable in radar backscattering, and $\Delta\,\hat{v}_{(x_i,y_i;x_j,y_j)}$ can be estimated without error, the value of $\left(v_{(x_i,y_i)} - v_{(x_j,y_j)} - \Delta\hat{v}_{(x_i,y_i;x_j,y_j)}\right)$ will be zero. A high value of δv indicates a greater disturbance introduced by the

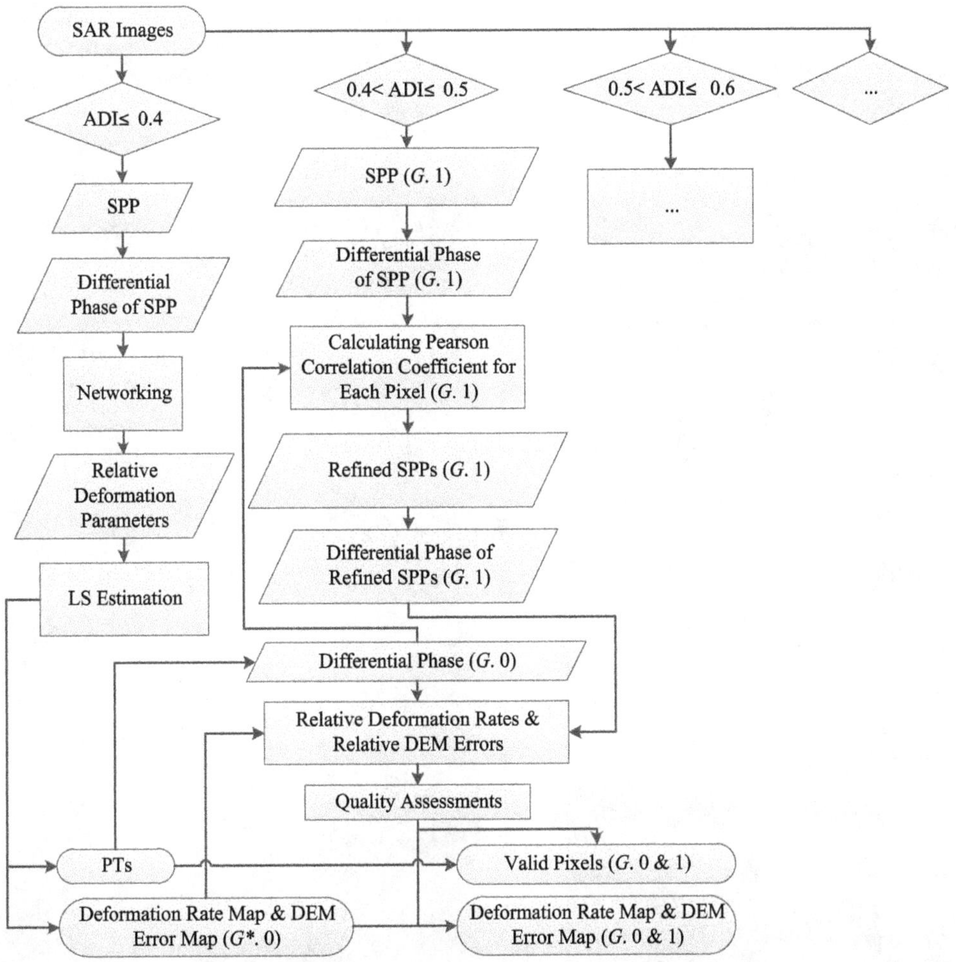

Fig. 1 Layout of linear subsidence rates and DEM errors estimation algorithm. "*G.*" denotes *Group*

noise. Meanwhile, $\delta\varepsilon$ is used on the same purpose. Empirically, δv and $\delta\varepsilon$ should not be greater than 5 mm/year and 10 m, respectively [12]. The pixel is discarded for further analysis if either constraint is not met. All pixels in Group i are processed in the same manner. The reserved pixels are used to analyze the subsequent groups. After all G groups are treated according to this group-by-group basis, all valid pixels are obtained. Moreover, the DTs are affirmed. A detailed layout of the algorithm is shown in Fig. 1 to clearly present the processing sequences for calculating the two subsidence parameters.

2.3 Estimating nonlinear subsidence components

To determine the complete subsidence evolution, the phase residuals on the valid pixels must be unwrapped and the nonlinear subsidence components must be extracted and subsequently added to the linear components. According to previous studies [9, 17, 18], the nonlinear subsidence components exhibit different characteristics from noise and the atmospheric phase screen in both space and time domains. Therefore, spatiotemporally filtering [9, 17, 18] on the unwrapped phase residuals is applied to estimate the nonlinear subsidence values.

The phase unwrapping technique is applied on the phase residuals embodied in Eq. (5). Because three phase data dimensions are provided, i.e., two in space and one in time, the three-dimensional phase unwrapping method described by Hooper and Zebker [19] is used. The unwrapping results can be summarized as

$$\varphi^k_{RE(x_i,y_i)} = \varphi^k_{nl(x_i,y_i)} + \varphi^k_{at(x_i,y_i)} + \varphi^k_{nl(x_i,y_i)}, \quad (11)$$

where the three terms on the right-hand side denote the unwrapped phase of the nonlinear subsidence, atmospheric phase screen, and decorrelation noise, respectively.

After phase unwrapping, spatial low-pass filtering is applied for each interferogram given the condition that the decorrelation noise is spatially uncorrelated, and the other two terms are spatially correlated [9]. The filtered phase data contain only the nonlinear subsidence components and atmospheric phase screen, i.e.,

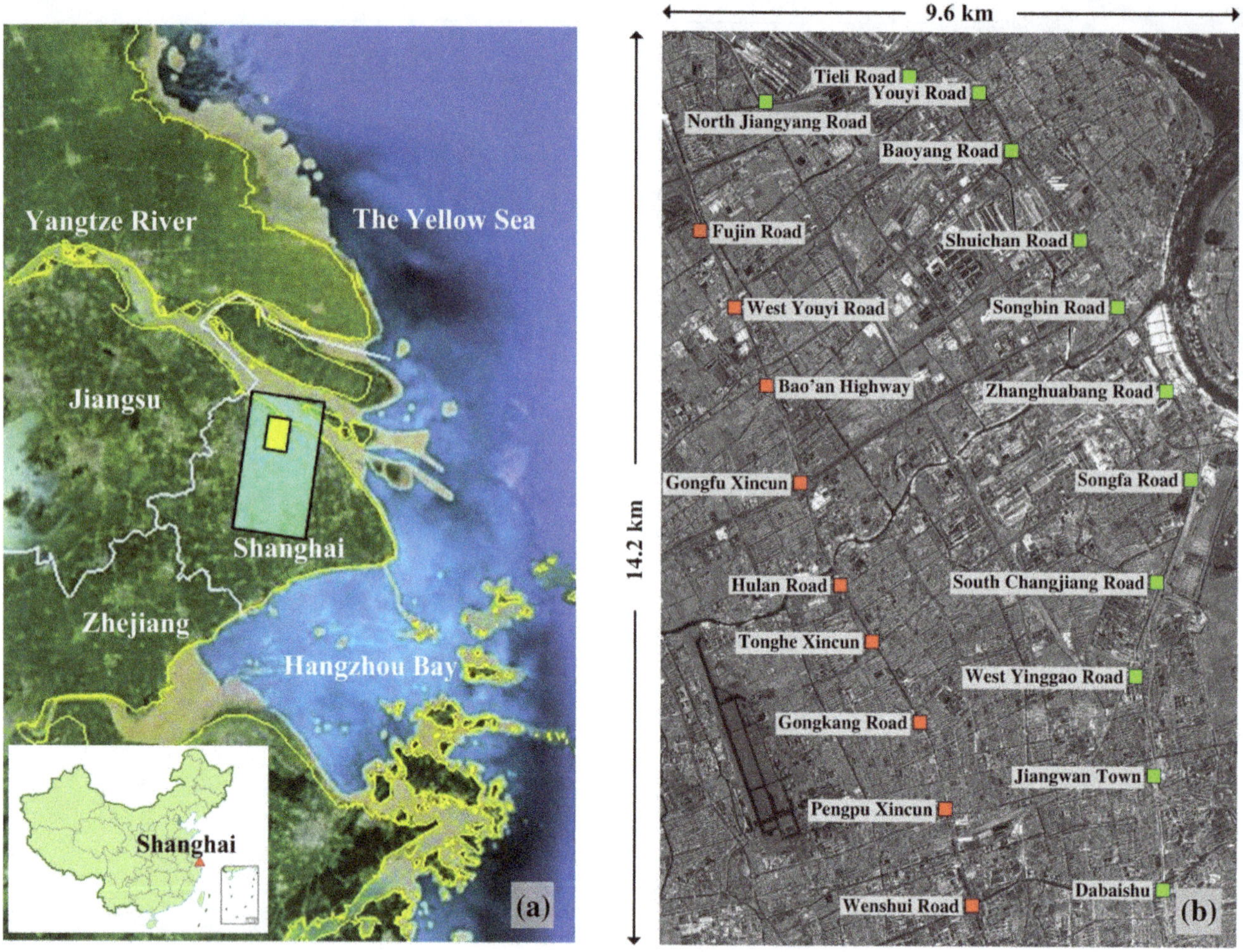

Fig. 2 Map and study area of the Baoshan district, Shanghai. The *larger and small rectangles* in **a** denote the full TerraSAR-X scene coverage and the experimental area, respectively. The amplitude image in **b** is averaged from 18 TerraSAR-X images. Nine metro stations along Line 1 and 12 metro stations along Line 3 are annotated using *red squares and green squares*, separately

Table 1 Eighteen TSX images and the interferometric parameters used in the experiment

Image index	Imaging dates (YYYYMMDD)	$B_\perp^k$ (m)	T^k (days)	Image index	Imaging dates (YYYYMMDD)	$B_\perp^k$ (m)	T^k (days)
1	20080421	53	−517	10	20091012	−18	22
2	20080820	−8	−396	11	20091023	−92	33
3	20090328	24	−176	12	20091114	45	55
4	20090408	103	−165	13	20091206	138	77
5	20090419	1	−154	14	20091217	148	88
6	20090511	23	−132	15	20091228	205	99
7	20090624	−62	−88	16	20100108	31	110
8	20090829	−158	−22	17	20100119	34	121
9	20090920	0	0	18	20100130	−67	132

$$\left(\varphi^k_{RE(x_i,y_i)}\right)^S_{LP} \approx \varphi^k_{nl(x_i,y_i)} + \varphi^k_{at(x_i,y_i)}, \tag{12}$$

where $(\cdot)^S_{\mathrm{LP}}$ is the spatial low-pass filter operator, typically the convolution with a two-dimensional Gaussian function [9]. The width of the Gaussian is as narrow as 50 m to include the entire useful signal except the noise.

To obtain the nonlinear subsidence components, which are expected to be temporally correlated, the results from

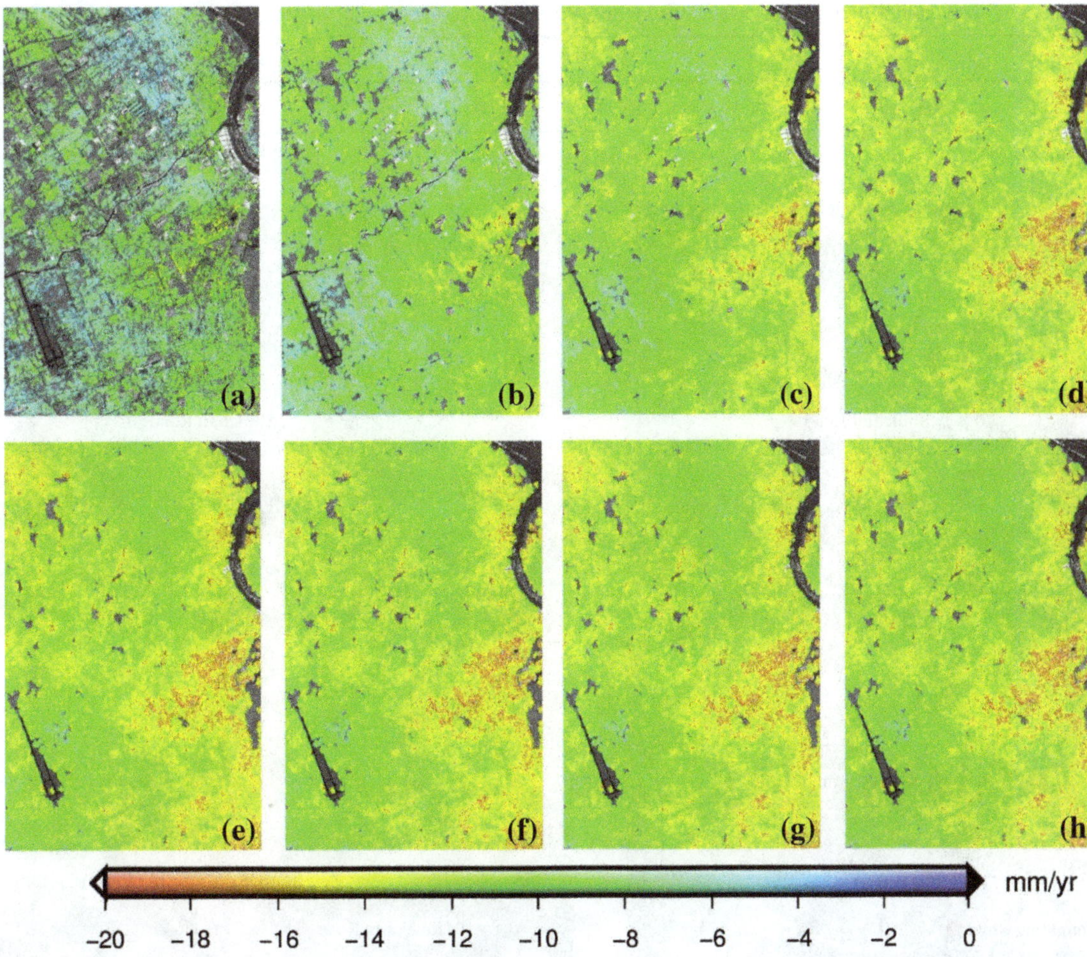

Fig. 3 Subsidence rate maps for the study area. **a–h** are the combined subsidence rates extracted from the previous and current groups. For example, **a** is extracted from Group 0; **h** is extracted from Groups 0 to 7

Eq. (12) are filtered using a temporal low-pass filter to obtain

$$\left(\left(\varphi^k_{RE(x_i,y_i)}\right)^S_{LP}\right)^T_{LP} \approx \varphi^k_{nl(x_i,y_i)}, \tag{13}$$

where $(\cdot)^T_{\mathrm{LP}}$ is the temporal low-pass filter operator, which is typically a Kaiser temporal filter [18]. The cut-off frequency of the filter is empirically set to 25 %.

Finally, for the given pixel (x_i, y_i) of the k-th image, the full resolution subsidence value $\left(S^k_{f(x_i,y_i)}\right)$ is the sum of the linear components $\left(S^k_{l(x_i,y_i)}\right)$ and nonlinear components $\left(S^k_{\mathrm{nl}(x_i,y_i)}\right)$:

$$\begin{aligned} S^k_{full(x_i,y_i)} &= S^k_{l(x_i,y_i)} + S^k_{nl(x_i,y_i)}, \\ S^k_{l(x_i,y_i)} &= v_{(x_i,y_i)}T^k, \\ S^k_{nl(x_i,y_i)} &= \frac{\lambda}{4\pi}\phi^k_{nl(x_i,y_i)}. \end{aligned} \tag{14}$$

3 Experimental results and discussion

3.1 Study area and data source

The experiment is conducted using images acquired over the Baoshan district, Shanghai, to detect land subsidence near the No. 1 and 3 MLs. As shown in Fig. 2a, Shanghai is located on the deltaic deposit of the Yangtze river. Consolidation of the soft layer causes land subsidence in Shanghai, where land subsidence was first observed in China [20]. It is reported that the largest accumulated subsidence value from 1921 to 1965 was 2.63 m [20]. More than 400 km^2 of land was affected by this geological hazard. However, since 1965, Shanghai has been controlled by restricting groundwater extraction and several subsidence funnels can still be observed [21, 22]. Land subsidence directly damages sewerage systems, roads, and buildings, etc. Moreover, land subsidence is a major factor affecting the stability of the Shanghai Metro.

Table 2 Number of valid pixels in the eight groups obtained using hierarchical analysis

Group No.	0	1	2	3	4	5	6	7
Number of valid pixels	303,219	694,854	165,945	57,414	18,387	4,493	1,001	222

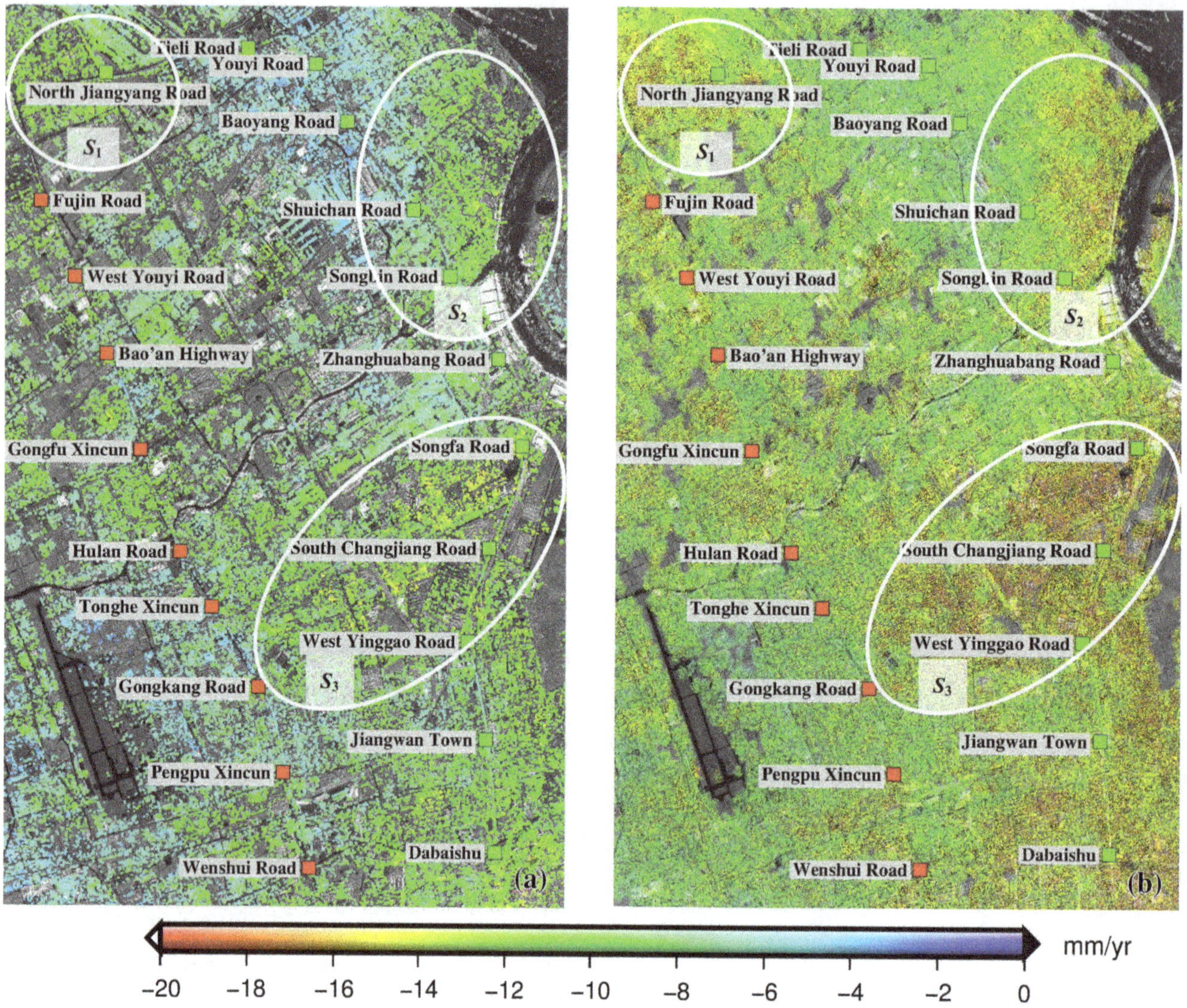

Fig. 4 Detailed subsidence rate map of the study area. **a** is close-up of Fig. 4a. **b** is close-up of Fig. 4h. The *red and green squares* are the metro stations for the No. 1 and 3 MLs, respectively. S_1, S_2, and S_3 represent the three subsidence funnels

The Shanghai Metro is one of the fastest-growing rapid transit systems in the world [23]. The first line was opened in 1993. On October 16, 2013, the operating route length reached 468 km. Moreover, the planned route length is 970 km, which would make the Shanghai Metro the world's longest metro system. Shanghai Metro has become one of the most popular means of travel. On March 9, 2013, the Shanghai Metro set a daily ridership record of 8.486 million passengers. The social position of the MLs is important for maintaining their sustainability and stability.

The proposed MTI method is applied to detect land subsidence near the No. 1 and 3 MLs in the Baoshan district of Shanghai. Eighteen TerraSAR-X images are used in the experiment (See Table 1). The TerraSAR-X images were provided by Infoterra GmbH. The observation period encompassed April 21, 2008, to October 30, 2010. All the images were collected with an incidence angle of 26 degrees in HH polarization mode. The original datasets were provided as single look complex images with slant range pixel spacing of 0.91 m (2.04 m in ground range) and azimuthal pixel spacing of 1.97 m.

The September 09, 2009 image is chosen as the unique master image. All other images are co-registered to the master image to produce 17 valid interferograms. Furthermore, a DEM provided by the Shuttle Radar Topography Mission is introduced to eliminate the topographic phase in each interferogram. The differential InSAR-related functions are all accomplished using the GAMMA DIFF module [24].

The amplitude image (4,688 × 7,208 pixels, equivalent to 9.6 × 14.2 km^2) of the test area is shown in Fig. 2b. The stations distributed along the No. 1 and 3 MLs are depicted with red squares and green squares, respectively. Line 1 originates at Fujin Road station; nine stations (red squares) along Line 1 are studied. Line 3 begins at North Jiangyang road station; 12 stations (green squares) along Line 3 are examined. The two MLs are distributed beneath the main roads in this district. Subsidence of the two MLs may directly damage roads and buildings in the surrounding areas.

3.2 Results and discussion

To obtain the subsidence fields in the test area, the three major steps of the aforementioned MTI method are followed. First, the SPP with ADI ≤ 0.4 are processed. The number of PT pixels extracted from the SPP is 303,219. Afterward, the pixels with ADI > 0.4 are analyzed on a group-by-group. Seven other groups are analyzed in this experiment. All eight groups are shown in Fig. 3. Each sub-figure *i* is plotted with the valid pixels extracted from all previous *i* groups (i.e., Groups 0 to $i - 1$). For example, the first figure is plotted with Group 0; the eighth figure is plotted with Groups 0 to 7. Table 2 shows the number of valid pixels for the eight groups. The number of useful pixels decreases as the ADI values increase. Groups 0 to 2 contribute 93 % of the useful information to the total valid pixels in the study area. Group 7 which contains ADI values greater than 1.0 provides very little useful information. Therefore, the hierarchical analysis strategy stops at Group 7. After hierarchical analysis, the total number of valid pixels is 1,245,526, which is 311 % more than provided by the initial group. This result indicates a large increase in the observation density of the subsidence rate map. The figure series shows that three subsidence funnels emerge when the hierarchical analysis is performed. The three areas are located in the northwestern, northeastern, and east-central parts of the test area. These subsidence funnels are clearly depicted in Fig. 3h. However, the subsidence funnels cannot be identified in Fig. 3a.

For further analysis of the subsidence field, close-ups of Fig. 3a, h are shown in Fig. 4a, b respectively. The metro stations along the No. 1 and 3 MLs are depicted with red and green squares. Moreover, S_1, S_2, and S_3 are the aforementioned three subsidence funnels. S_1 is located at the No. 3 ML origination point. North Jiangyang road is located in this area. S_2 is in a coastal region; compression of the silty marine clay, which is approximately 40 m thick and less than 70 m deep may be the primary reason for the subsidence [20]. S_3 is a subsidence funnel presented in many previous studies [22, 25]. The three subsidence funnels are not clearly depicted in Fig. 4a, which is plotted with valid pixels from Group 0. However, the subsidence funnels are clearly observed in Fig. 4b, which is plotted with valid pixels from Groups 0 to 7. Pixels extracted from Group 0 are primarily PTs, such as rocks, iron fences, and corner reflectors. However, pixels extracted from Groups 1 to 7 are mainly DTs, e.g., asphalt roads, building tops, and bare lands. The results indicate that the DTs are stable in radar backscattering time series. Therefore, the DTs can provide more subsidence details than the PTs.

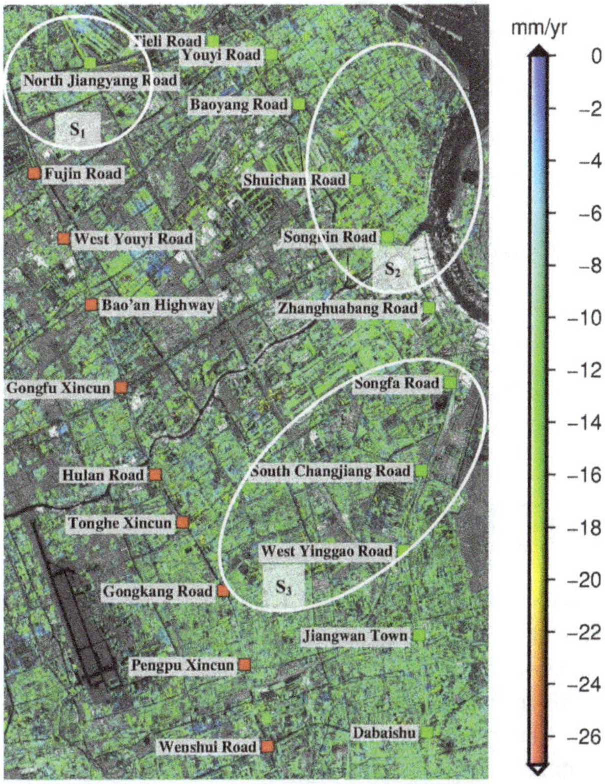

Fig. 5 Subsidence rate map extracted by considering the pixels with ADI < 1.1 without using hierarchical analysis strategy

If no hierarchical analysis strategy is applied, meaning that all of the points with ADI < 1.1 are processed simultaneously. 1,891,004 point candidates are selected and finally 1,557,626 points are maintained after using the method presented in Sect. 2.1. We display the result in Fig. 5. In this figure, the pixel number is greater than that of the hierarchical analysis result. That is because the point quality is too low to be controlled simply by considering

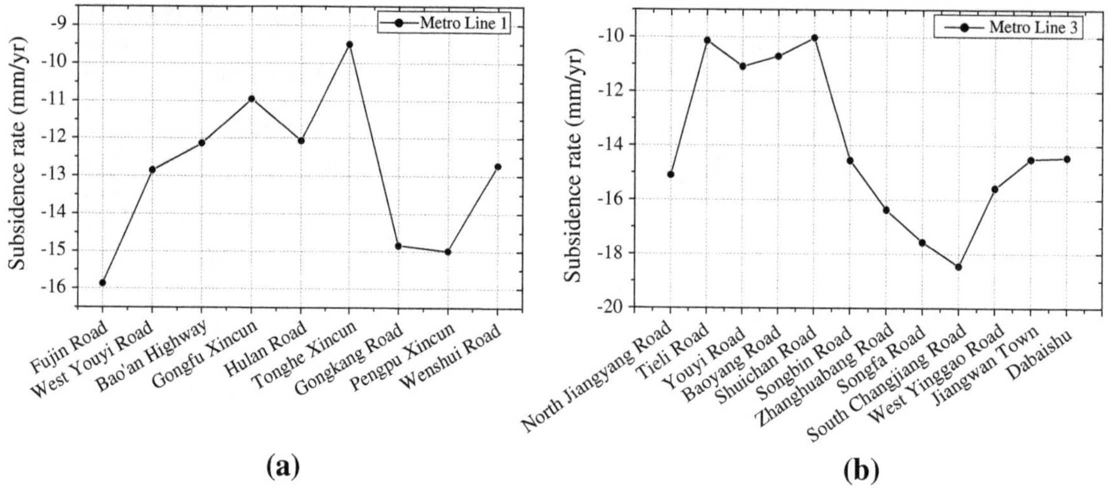

Fig. 6 Station subsidence rates along Metro Lines 1 and 3

the two constraints presented in Eqs. (6) and (7). Therefore, some invalid points are included in the final subsidence rate map. Meanwhile, all of the three subsidence areas detected in Fig. 4b are not exhibited in Fig. 5. It is concluded that the useful information on the subsidence funnels is totally overwhelmed by the noise information. The subsidence rate map in Fig. 5 might be biased or misleading. However, the in situ ground truth leveling data are needed for in-depth validation.

The subsidence rates on the stations are plotted in Fig. 6. Mean subsidence rates for the No. 1 and 3 MLs are −12.9 and −14.0 mm/year, respectively. Along Line 1, Fujin Road station has the maximum subsidence rate and Tonghe Xincun has the minimum subsidence rate. The differential subsidence rate between the two stations is approximately 6.4 mm/year. The subsidence rate vibration (σ_{dv}) along Line 1 is 2.0 mm/year, which indicates that homogeneous subsidence exists along Line 1. A similar result is obtained for Line 3, along which the maximum subsidence rate difference between the Shuichan Road and South Changjiang Road stations is 8.4 mm/year. σ_{dv} is 2.9 mm/year along this ML. The first station of Line 3 is at North Jiangyang Road, a metro station with one of the largest subsidence rates in S_1. The following four stations are also affected by small subsidence: Tieli Road, Youyi Road, Baoyang Road, and Shuichan Road. However, subsidence rates change slowly among these four stations. Specifically, σ_{dv} is 0.5 mm/year for the four stations, indicating that the station subsidence is uniform and that the stations are relatively stable. The last three stations also exhibit similar subsidence rates, i.e., σ_{dv} for the West Yinggao Road, Jiangwan Town, and Dabaishu stations is 0.6 mm/year.

After obtaining the linear subsidence rate map, the nonlinear components are calculated to provide the complete subsidence evolution for each pixel. For definitiveness and to avoid loss of generality, the nonlinear subsidence components are shown for three stations, i.e., Jiangwan Town, Fujin Road, and Hulan Road, in Fig. 7. Figure 7 shows that the first two SAR images (with acquisition dates of April 21, 2008, and August 20, 2008) are isolated from the image cluster. However, the two images provide useful information for the experiment. In Fig. 7, the nonlinear subsidence components indicate a slow vibration during the first three observation dates. Afterward, the nonlinear components suggest that the ground surface rises to a maximum in August and decreases to a minimum in January. The Shanghai climate data [26] show that precipitation might be the primary reason for the weather-related trend. The climate data from 1971 to 2000 show that the average precipitation amounts in July and December are 169.6 and 37.1 mm, respectively, which also correspond to the maximum and minimum rain capacities. The precipitation indicates that the pore-fluid pressure in the Shanghai soft clay is recharged in summer; the pore-fluid pressure provided by groundwater is transferred to the granular skeleton in winter.

4 Conclusions

An improved MTI method that includes linear subsidence rate extraction, hierarchical analysis strategy, and nonlinear subsidence extraction is presented to analyze the stations distributed along No. 1 and 3 MLs in the Baoshan district, Shanghai. The technique provides a method to derive useful information from a set of SAR images by tracking both PTs and DTs, therefore, increasing the subsidence information observation density. The linear subsidence rates are first obtained from the PTs. Thereafter, the hierarchical analysis strategy is introduced to

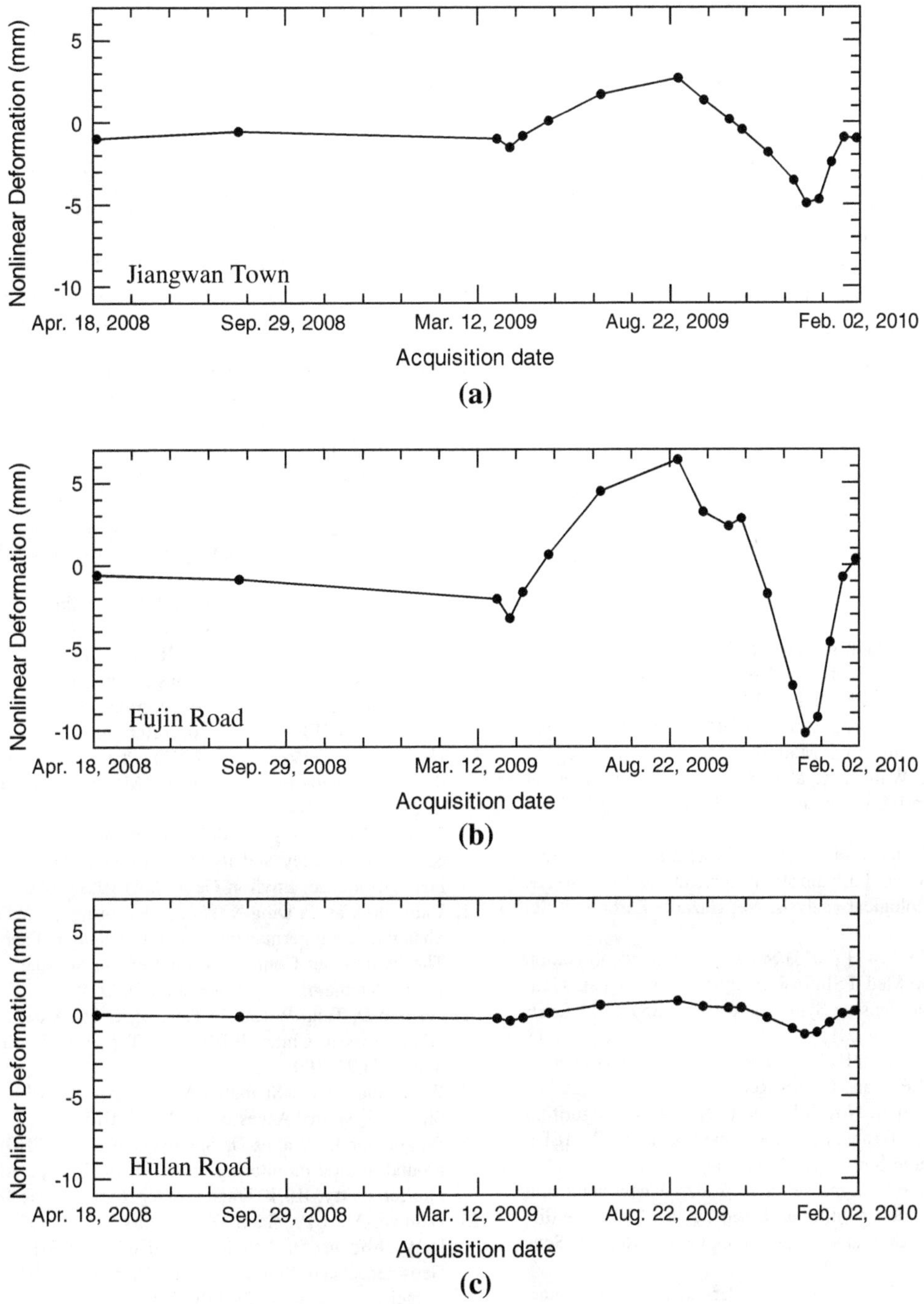

Fig. 7 Nonlinear subsidence components for three stations: Jiangwan town, Fujin road, and Hulan road

extract the linear subsidence rate from the DTs. The nonlinear subsidence values are estimated using 3D phase unwrapping and spatiotemporally filtering on the phase residuals.

Furthermore, 18 TerraSAR-X images are used to determine the subsidence rate map for the test site and stations along the two MLs. The number of valid pixels increases from 303,219 to 1,245,526 after performing the hierarchical analysis. Groups 0 to 2 contribute 93 % of the useful information to the total valid pixels in the study area. The number of valid pixels decreases as the ADI increases. Group 7 provides very little useful information. The valid pixels extracted from Group 0 are primarily PTs, and those extracted from the subsequent groups are mostly DTs. The experiment shows that DTs are largely capable of providing high observation density subsidence information. Furthermore, three subsidence funnels are identified after the hierarchical analysis, indicating that our MTI method is capable of providing detailed information for subsidence monitoring.

In-depth analysis is performed on the No. 1 and 3 MLs. The mean subsidence rates for these MLs are −12.9 and −14.0 mm/year, respectively. Moreover, vibrations of the relevant subsidence rate are 2.0 and 2.9 mm/year, respectively, indicating a homogenous subsidence pattern along the MLs. This finding may help in the management of metro line stations and in hazard prediction and prevention. Subsidence rate maps with high observation density are expected to facilitate metro line stability assessment for the transportation department in Shanghai.

References

1. Wang J, Li P (2012) Vault settlement discussions and precision analysis of Metro Line 3 of Qingdao tunnel monitoring work. Urban Geotech Investig Surv 1:122–125
2. Qin Z, Tao L (2011) The application of GPS technology in Wuxi urban rail transit engineering. Mod Surv Mapp 34(5):34–36
3. Yang J, Wang C, Wang J et al (2006) The establishment of Chengdu Metro No.1 Line control network. Surv Mapp Sichuan 29(1):42–44
4. Chen F, Lin H, Zhang Y et al (2012) Ground subsidence geo-hazards induced by rapid urbanization: implications from InSAR observation and geological analysis. Nat Hazards Earth Syst Sci 12(4):935–942
5. Wang Z, Perissin D, Lin H (2011) Subway tunnels identification through Cosmo-SkyMed PSInSAR analysis in Shanghai. Geoscience and Remote Sensing Symposium (IGARSS), 2011 IEEE International, 24-29 July 2011
6. Ferretti A, Prati C, Rocca F (2001) Permanent scatterers in SAR interferometry. IEEE Trans Geosci Remote Sens 39(1):8–20
7. Ferretti A, Fumagalli A, Novali F et al (2011) A new algorithm for processing interferometric data-stacks: squeeSAR. IEEE Trans Geosci Remote Sens 49(9):3460–3470
8. Berardino P, Fornaro G, Lanari R et al (2002) A new algorithm for surface deformation monitoring based on small baseline differential SAR interferograms. IEEE Trans Geosci Remote Sens 40(11):2375–2383
9. Hooper A, Segall P, Zebker H (2007) Persistent scatterer interferometric synthetic aperture radar for crustal deformation analysis, with application to Volcán Alcedo, Galápagos. J Geophys Res 112(B07407).
10. Strozzi T, Wegmuller U, Keusen HR et al (2006) Analysis of the terrain displacement along a funicular by SAR interferometry. IEEE Geosci Remote Sens Lett 3(1):15–18
11. Perissin D, Teng W (2012) Repeat-pass SAR interferometry with partially coherent targets. IEEE Trans Geosci Remote Sens 50(1):271–280
12. Liu G, Buckley SM, Ding X et al (2009) Estimating spatiotemporal ground deformation with improved persistent-scatterer radar interferometry. IEEE Trans Geosci Remote Sens 47(9): 3209–3219
13. Zhang L, Lu Z, Ding X et al (2012) Mapping ground surface deformation using temporarily coherent point SAR interferometry: application to Los Angeles Basin. Remote Sens Environ 117:429–439
14. Colesanti C, Ferretti A, Novali F et al (2003) SAR monitoring of progressive and seasonal ground deformation using the permanent scatterers technique. IEEE Trans Geosci Remote Sens 41(7):1685–1701
15. Rodgers JL, Nicewander WA (1988) Thirteen ways to look at the correlation coefficient. Am Stat 42(1):59–66
16. Zhang L, Ding X, Lu Z (2011) Modeling PSInSAR time series without phase unwrapping. IEEE Trans Geosci Remote Sens 49(1):547–556
17. Ferretti A, Prati C, Rocca F (2000) Nonlinear subsidence rate estimation using permanent scatterers in differential SAR interferometry. IEEE Trans Geosci Remote Sens 38(5):2202–2212
18. Mora O, Mallorqui JJ, Broquetas A (2003) Linear and nonlinear terrain deformation maps from a reduced set of interferometric SAR images. IEEE Trans Geosci Remote Sens 41(10):2243–2253
19. Hooper A, Zebker HA (2007) Phase unwrapping in three dimensions with application to InSAR time series. J Opt Soc Am A 24(9):2737–2747
20. Gong S, Li C, Yang S (2009) The microscopic characteristics of Shanghai soft clay and its effect on soil body deformation and land subsidence. Environ Geol 56(6):1051–1056
21. Damoah-Afari P, Ding X (2005) Measuring ground subsidence in Shanghai using permanent scatterer technique. Paper presented at The 26th Asian Conference on Remote Sensing (ACRS 2005), Hanoi, Vietnam, 7–11 November, 2005
22. Perissin D, Teng W (2011) Time-series InSAR applications over urban areas in China. IEEE J Sel Top Appl Earth Obs Remote Sens 4(1):92–100
23. Wikipedia (2013) Shanghai Metro. http://en.wikipedia.org/wiki/Shanghai_Metro. Accessed 12 Nov, 2013
24. Wegmüller U, Walter D, Spreckels V et al (2010) Nonuniform ground motion monitoring with TerraSAR-X persistent scatterer interferometry. IEEE Trans Geosci Remote Sens 48(2):895–904
25. Damoah-Afari P, Xiaoli D, Zhiwei L, et al (2007) Six years of land subsidence in shanghai revealed by JERS-1 SAR data. IEEE Geoscience and Remote Sensing Symposium (IGARSS 2007), Barcelona, Spain, 23–28 July, 2007
26. Wikipedia (2013) Geography of Shanghai. http://en.wikipedia.org/wiki/Geography_of_Shanghai. Accessed 12 Nov, 2013

8

Leave the expressway or not? Impact of dynamic information

Hongcheng Gan · Xin Ye

Abstract This study investigates drivers' diversion decision behavior under expressway variable message signs that provide travel time of both an expressway route and a local street route. Both a conventional cross-sectional logit model and a mixed logit model are developed to model drivers' response to travel time information. It is based on the data collected from a stated preference survey in Shanghai, China. The mixed logit model captures the heterogeneity in the value of "travel time" and "number of traffic lights" and accounts for correlations among repeated choices of the same respondent. Results show that travel time saving and driving experience serve as positive factors, while the number of traffic lights on the arterial road, expressway use frequency, being a middle-aged driver, and being a driver of an employer-provided car serve as negative factors in diversion. The mixed logit model obviously outperforms the cross-sectional model in dealing with repeated choices and capturing heterogeneity regarding the goodness-of-fit criterion. The significance of standard deviations of random coefficients for travel time and number of traffic lights evidences the existence of heterogeneity in the driver population. The findings of this study have implications for future efforts in driver behavior modeling and advanced traveler information system assessment.

H. Gan (✉)
Transportation Center, Northwestern University, Evanston, IL, USA
e-mail: hongchenggan@hotmail.com

H. Gan
Department of Transportation Engineering, University of Shanghai for Science and Technology, Shanghai, China

X. Ye
Civil Engineering Department, California State Polytechnic University, Pomona, CA, USA

Keywords Travel decision · Mixed logit · Travel time · Repeated choices · Variable message sign · Stated preference

1 Introduction

The effectiveness of advanced traveler information systems (ATIS) depends highly on travelers' behavior in response to real-time information. It is well recognized that it is important to indentify the factors that influence travelers' decision behavior under ATIS [1–29]. Research results in this challenging field can facilitate better investment, design, and operation of ATIS technologies.

Internationally, variable message signs (VMSs), a common ATIS technology, has been widely used to manage the traffic on urban expressways with high demand. In developed countries, many metropolitan cities such Paris (France), Munich (Europe), Chicago (USA), and Tokyo (Japan), use VMS to enhance expressway management. In China, big cities such as Shanghai, Beijing, Guangzhou, Hangzhou, Ningbo, Chengdu, and Suzhou have installed a lot of VMS on urban expressways. In the real world, VMS information can be descriptive (e.g., statements of traffic conditions) or prescriptive (e.g., suggestions on what to do). It can be quantitative (e.g., travel time estimate, estimated delay, and length of queue) or qualitative (e.g., warnings of incidents, statements of level of service, and bad weather alerts).

However, the existing expressway VMSs usually can only provide information about expressway conditions and do not provide information about local streets (e.g., parallel arterial roads) due to technological reasons and/or institutional barriers (e.g., expressways and local streets are operated by different agencies). This may limit their effectiveness in diverting urban expressway traffic to local streets, since travelers are not given real-time information about local streets. In Shanghai, as the Traffic Police Department reported on newspapers, many outbound elevated roads (urban expressways) connecting the downtown and the suburb often have big delays, and their travel time is surprisingly much longer than the travel time of parallel arterial road under them during some traditional national holidays (e.g., the Qingming holiday during which people go to big cemeteries in neighbor cities to hold a memorial ceremony for their families they lost). This situation is partially due to the fact that drivers are not so confident that they will be better off after they divert to a local street since they are not given any real-time information about alternate routes. A feasible way to help drivers make more informed diversion decisions and alleviate expressway congestion is to update the existing expressway VMS service in Shanghai. In this context, Shanghai is planning to provide a new expressway VMS service which gives travel time information about both an urban expressway route and a competitive alternate arterial road route.

This study, therefore, will investigate the impact on drivers' diversion decision behavior of the new expressway VMS information that provide travel time of both an expressway route and a local street route in the context of Shanghai, China. Such expressway VMS was rarely addressed in the literature to the best knowledge of the authors.

In previous studies, many researchers used stated preference (SP) data from questionnaire surveys to model drivers' response behavior, e.g., [5–16]. Some other studies used SP data from travel simulator experiments (e.g., [17–25]). Others used revealed preference (RP) data to model drivers' response behavior (e.g., [26–29]).

When new ATIS features or options are to be addressed which do not exist in the market, only SP survey is available. In a typical SP survey, each respondent responds to several hypothetical scenarios, thus, the issue of correlations among repeated observations from the same respondent arises. This issue should be addressed carefully when developing driver response models [3, 18].

Despite the large number of publications on travel behavior under ATIS, relatively fewer studies accounted for correlations among repeated observations; see, for example, [3] for a recent review. With the increasing popularity of simulation-based estimation, panel data models that address repeated observations are gaining more attention. Methodologies that have been applied to address repeated observations include mixed logit models (e.g., [19, 23]), random effect models (e.g., [14, 29]), multinomial probit (e.g., [13, 18]), normal mixing distributions (e.g., [6]), generalized estimating equations (e.g., [5, 15]), and mixed linear models (e.g., [25]).

It is also desirable that a response behavior model is capable of capturing the heterogeneity in drivers' taste (preferences) [4, 19]. In the context of this study, the possible preference variations across individuals regarding travel time information and other alternative attributes will be appropriately addressed.

The mixed logit model provides the flexibility to cope with these above issues. In mixed logit models, an additional error term is added to the utility specification. The additional term captures heteroscedasticity among individuals and allows correlation over alternative and time. Recent advances in simulation-based estimation procedures make the mixed logit model more computationally feasible and attractive. This study, therefore, will use the mixed logit model to account for repeated choices and capture the heterogeneity of drivers' decision behavior.

Given the above context, two distinguishing features of this study are: (1) A mixed logit model is developed that addresses correlations among repeated choices from the same respondent and capture the heterogeneity in drivers' value of certain alternative attributes (i.e., travel time and number of traffic lights); it is also compared with the conventional cross-sectional logit model. (2) The type of expressway VMS information addressed by this study is travel time of both an expressway route and a local street route. Such expressway VMS information was rarely addressed in the literature.

This study will obtain a preliminary understanding of drivers' diversion decision behavior under Shanghai's new expressway VMS information and will have implications for further modeling efforts in drivers' decision behavior under ATIS.

The rest of this paper is organized as follows: First, the design of SP survey and collected data are described. Next, the mixed logit model for drivers' diversion decision behavior under VMS is developed and compared with the conventional cross-sectional logit model. Finally, concluding remarks are given.

2 Methodology

2.1 Survey method

Expressway VMS has been used for many years in Shanghai [30, 31]. However, they currently do not provide traffic information about local streets. Thus, only SP behavioral data can be collected by this study.

The SP experiment was conducted based on a hypothetical trip that is outlined by a dotted rectangle on the

Shanghai urban expressway network map shown in Fig. 1a. Symbol "$\underline{O}$" means trip origin and symbol "$\underline{D}$" means trip destination. Trip origin is Pudong International Airport, and trip destination is Wujiaochang central business district. The travel scenario contains an untolled expressway route and an alternate arterial road route (depicted in Fig. 1b). The VMS before the diversion point provides travel time of both the expressway and the arterial road.

Respondents were required to assume that they were making a trip during the off-peak period in a weekday afternoon. Respondents were told that once they diverted to the arterial road it would be impossible to get back on the expressway. The expressway is the usual route from Pudong Airport to Wujiaochang. The alternate arterial road route is an imaginary route. The arterial road route can be deemed as an alternate route that a real-world VMS-based ATIS recommends to drivers [5, 25]. Thus, our SP settings are reasonable, though the employed network at first sight seems simple. Normal travel time for the expressway is thirty minutes. A similar SP experiment design was adopted by Abdel-Aty et al. [6] which also specified a hypothetical journey consisting of a primary route and an imaginary alternate route.

The VMS messages designed in the SP survey have a wording style similar to the real-life Shanghai expressway VMS and consist of two parts: (1) travel time of the expressway and travel time of the arterial road and (2) cause of expressway delay (Fig. 1b).

The factors controlling the SP experiment are the following attributes: Travel time of the expressway route, Cause of expressway delay, and Number of traffic lights on the arterial road.

The attribute values are specified based on discussions with Shanghai expressway network traffic management center operators and VMS messages records. For a 30-minutes-around off-peak expressway journey, the range of [0, 10] minutes is considered reasonable for expressway delays by traffic management center operators. To this end, expressway travel time takes two values: "35 min" (i.e., a 5-min delay) and "40 min" (i.e., a 10-min delay). Cause of expressway delay contains two levels: "Congestion" and "Accident." The number of traffic lights on the arterial road takes two values: "10" and "20", with consideration of typical spacing of traffic lights in Shanghai.

The complete factorial design [32] was used to produce eight (2 × 2 × 2) SP choice scenarios which are in accordance with eight questions. In all SP scenarios, travel time of the local street route remains to be thirty minutes.

Given a specific VMS message, a respondent was asked to choose between "continue via expressway" and "divert to the arterial road."

2.2 Data collection and descriptive analysis

An SP questionnaire survey was conducted in April 2007 in the parking lot of Shanghai Pudong international airport. The collected data consisted of two parts: (a) driver characteristics, such as, gender, age, years of driving experience, frequency of using expressway, and driver type; (b) diversion decisions under VMS.

A total of 171 drivers participated in the survey. The experimenters read questions to respondents and recorded answers of the respondents. After removing the respondents, not fully completing the questionnaire, the data set available for model development contains 140 drivers and 1,120 (140 × 8) choice observations in total.

Table 1 shows driver characteristics of the sample.

In the sample, 74.3 % of respondents are male drivers. The majority of the sample is frequent expressway users (49.3 % + 23.6 %).

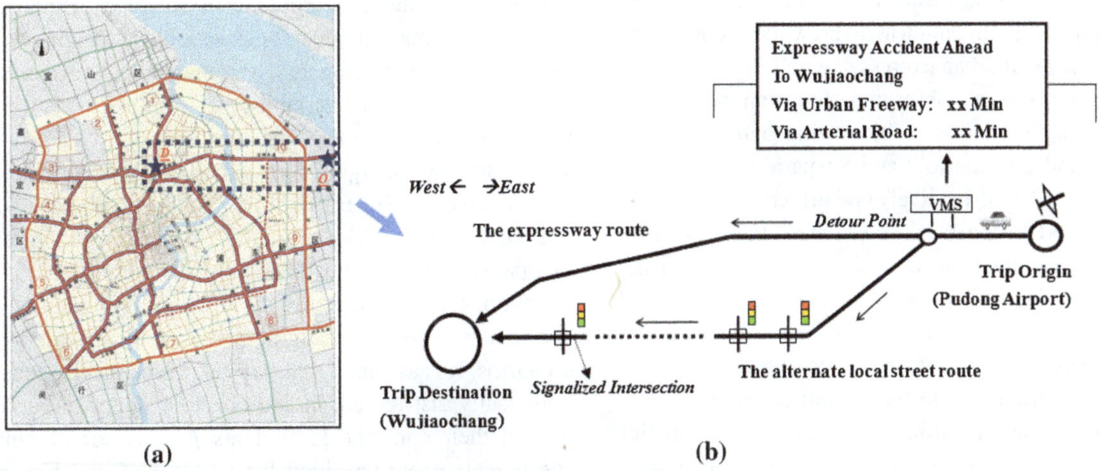

Fig. 1 SP survey. **a** Expressway network in Shanghai, **b** travel scenario

Table 1 Driver characteristics of the sample

Attribute	Range	Percentage (%)
Gender	Male	74.3
	Female	25.7
Age (years)	20–29	24.3
	30–39	41.4
	40–49	25.0
	50–64	9.3
Years of driving experience	<2	6.4
	2–5	41.4
	6–10	37.1
	>10	15.0
Driver type	Private car	42.1
	Employer-provided car	35.0
	Taxi	22.9
Expressway use frequency	Almost every day	49.3
	2–3 days per week	23.6
	Seldom	27.1

"Private car" is owned by a driver himself/herself

"Employer-provided car" is not owned by a driver but assigned by his/her employer for business purpose

In China, a person at a high hierarchy level in a company or governmental agency is allowed to use a car owned by his employer. The high proportion of employer-provided car drivers does reflect Shanghai situation.

The seemingly high proportion (22.9 %) of taxi drivers accords with TMC officials' suggestion that the proportion of taxi vehicles on the expressway originating from Pudong airport typically ranges from 10 % to 50 % varying with time of day. This estimation was also justified by real-world observations. Thus, taxi drivers are included in model estimation.

Overall, in case of expressway delays, the diversion (i.e., choosing the local street) percentage for all the eight SP choice scenarios is 47.3 %. That means almost half of the survey respondents stated their intention to divert while encountering delay on their original urban expressway route.

At the scenario level, observing diversion percentage variations among scenarios is interesting and insightful. For example, under Scenario 2 and Scenario 6, over 70 % (72.1 % and 70.7 %, respectively) of drivers express their intention to divert to the local street, presumably because the travel time saving from diversion is 10 min, and the number of traffic lights is only 10. Conversely, Scenario 3 and Scenario 8 only cause a bit more than 20 % (22.1 % and 23.6 %, respectively) of drivers to intend to divert, possibly because travel time saving from diversion is only 5 min but the number of traffic lights is 20. These statistics sheds some light on the relationship among travel time saving, number of traffic lights, and diversion percent.

2.3 Modeling methodology

2.3.1 *Cross-sectional model*

In our SP survey, drivers' response is binary choice in nature: drivers will either choose to divert to the arterial road or keep driving on the expressway. Thus, the binary logit model [33] is an appropriate modeling method for behavior analysis. It starts from an assumption that driver "i" makes decision based on one random utility function U_i^*, which can be parameterized as

$$U_i^* = \beta_0 + \boldsymbol{x}_i\boldsymbol{\beta} + \varepsilon_i. \tag{1}$$

In this equation, "i" is an index variable indicating each observation; $\boldsymbol{x}_i$ is a row vector of explanatory variables of interest (e.g., travel time saving, number of traffic lights, and demographic characteristics); β_0 is a constant and $\boldsymbol{\beta}$ is a column vector of coefficients associated with the explanatory variables; and ε_i is a random variable that takes account of unspecified explanatory variables for U_i^*, which is assumed to be independently standard logistically distributed. If we specify y_i as a dummy variable indicating whether driver "i" will divert to the arterial road ($y_i = 1$, divert; $y_i = 0$, not divert), then the probability of observing y_i for each observation "i" is

$$P_i = \left[\frac{\exp(\beta_0 + \boldsymbol{x}_i\boldsymbol{\beta})}{1 + \exp(\beta_0 + \boldsymbol{x}_i\boldsymbol{\beta})}\right]^{y_i} \left[\frac{1}{1 + \exp(\beta_0 + \boldsymbol{x}_i\boldsymbol{\beta})}\right]^{1-y_i}. \tag{2}$$

2.3.2 *Mixed logit model*

As per Train [34], mixed logit model with random parameters can accommodate correlation of utilities of the same driver. In a mixed logit model, the utility function is formulated as

$$U_{it}^* = \boldsymbol{x}_{it}\boldsymbol{\beta} + z_{it}\gamma_i + \varepsilon_{it}. \tag{3}$$

In the utility function, "i" is the driver index and "t" is the scenario index; $\boldsymbol{x}_{it}$ contains a vector of explanatory variables. $\boldsymbol{x}_{it}$ may include some variables changing across drivers but not changing across scenarios (e.g., driver's age and type). Those variables are called "individual variables" in this paper. The vector $\boldsymbol{x}_{it}$ may also include some variables changing across scenarios but not changing across drivers (e.g., travel time saving, number of traffic lights). Those variables are called "scenario variables" in this paper. The vector $\boldsymbol{x}_{it}$ also contains a constant "1" for the alternative specific constant in the utility function. "ε_{it}" is a random variable changing across both individuals and scenarios. It is assumed that "ε_{it}" is independently standard logistically distributed. In addition to a vector of variables $\boldsymbol{x}_{it}$ and their constant coefficients $\boldsymbol{\beta}$, a vector of random coefficients γ_i are specified for a vector of variables z_{it} in the utility function. The random coefficients γ_i vary across drivers but do not vary across scenarios for the same driver. Assume that γ_i are independently normally distributed and associated with a vector of expectations γ and a vector of standard deviations σ_γ. Then, one may first obtain the probabilistic function conditional on random parameters γ_i as

$$P(y_{it} = 1|\gamma_i) = \frac{\exp(\mathbf{x}_{it}\boldsymbol{\beta} + z_{it}\gamma_i)}{1 + \exp(\mathbf{x}_{it}\boldsymbol{\beta} + z_{it}\gamma_i)}, \qquad (4)$$

$$P(y_{it} = 0|\gamma_i) = \frac{1}{1 + \exp(\mathbf{x}_{it}\boldsymbol{\beta} + z_{it}\gamma_i)}. \qquad (5)$$

Here, y_{it} is a dummy variable indicating whether driver i will divert under scenario t ($y_{it} = 1$, divert; $y_{it} = 0$, not divert)

For an unconditional probabilistic function, the conditional probabilistic function needs to be integrated for all the scenarios over the probability density function of γ_i, $f(\gamma_i)$:

$$P_i = \int_{-\infty}^{+\infty} f(\gamma_i) \prod_{t=1}^{T} P(y_{it}|\gamma_i)\mathrm{d}\gamma_i. \qquad (6)$$

Here, T is the number of scenarios. The maximum simulated likelihood estimation method [35] can be employed to evaluate the integral.

The log-likelihood function for the entire sample can be formulated as

$$\mathrm{LL} = \sum_{i=1}^{N} \ln(P_i). \qquad (7)$$

Here, N is the number of observations. Then, the simulated log-likelihood function can be maximized for estimating all the model coefficients.

3 Model estimation results and discussion

3.1 Model estimation results

The model estimation procedure is executed via GAUSS 8.0 [36].The explanatory variables tested for the cross-sectional binary logit model include age, age square, gender, years of driving experience, driver type, expressway use frequency, travel time saving, cause of expressway delay, and number of traffic lights on the arterial road. For the mixed logit model, travel time saving, and number of traffic lights are variables taking random parameters in the utility specification. It is one of the interests of this study to explore whether there exists heterogeneity regarding these two variables.

Table 2 provides model estimation results for the cross-sectional binary logit model and the mixed logit model. All the variables remaining in the final cross-sectional binary logit model take statistically significant coefficients. The variables of statistical significance that enter the final cross-sectional binary logit include: (a) years of driving experience; (b) the dummy variable indicating driver seldom using expressway; (c) the dummy variable indicating driver using expressway every day; (d) the dummy variable indicating employer-provided car driver; (e) age and age square; (f) number of traffic lights on the arterial road (l); and (g) travel time saving (s).

Other attribute variables such as gender and cause of expressway delay do not obtain coefficient of statistical significance in the cross-sectional binary logit model.

3.2 Discussions about VMS impacts

Discussions of the coefficients of the final cross-sectional model are presented below.

3.2.1 Driving experience

Driving experience plays a positive role in diversion decision under VMS as shown by the positive coefficient of "years of driving experience." This is probably because drivers with rich-driving experience are more adaptable to expressway delays and more familiar with local streets and thereby more likely to divert in response to VMS. Drivers with less-driving experience may not feel comfortable with diversion-related vehicle operating such as finding an available inserting gap and making a lane change in dense traffic.

3.2.2 Expressway use frequency

The positive coefficient of "use expressway seldom" and the negative coefficient of "use expressway everyday" indicate that the increase of expressway use frequency decreases the probability of diverting to the alternate arterial road route under VMS. It is probably because drivers using expressways frequently have a big dependence on or a bias for expressways. Interestingly, similar findings were obtained in some earlier studies, e.g., [37, 38].

3.2.3 Driver type

Interestingly, employer-provided car drivers are less likely to divert in response to VMS, as indicated by the negative coefficient of the dummy "employer-provided car." This finding has implications for design and assessment of VMS systems since employer-provided cars represent a significant percentage of traffic in many Chinese cities (e.g., 5 %–20 % in Shanghai). Moreover, this finding coincides with the author's earlier study which found that employer-

Table 2 Estimation results of two alternative logit models

Variable	Cross-sectional model		p value	Mixed logit model		p value
	Coefficient	T test		Coefficient	T test	
Constant	3.2646	2.356	0.018	4.9583	1.447	0.148
Driving experience	0.0329	1.977	0.048	0.0358	0.870	0.384
Use expressway seldom	0.3659	1.944	0.052	0.2473	0.537	0.591
Use expressway everyday	−0.8287	−5.070	0.000	−1.4441	−3.507	0.000
Employer-provided car	−0.5928	−4.034	0.000	−0.9869	−2.713	0.007
Age	−0.2154	−2.954	0.003	−0.2928	−1.614	0.107
Age square/100	0.2934	3.041	0.002	0.3883	1.626	0.104
Number of traffic lights (l)	−0.0931	−6.916	0.000	−0.1436	−7.107	0.000
Travel time savings (s)	0.2738	10.083	0.000	0.4027	9.684	0.000
Standard deviation of coefficient for l	–	–		0.094	5.21	
Standard deviation of coefficient for s	–	–		0.1706	5.16	
Maximum log likelihood	−662.13			−595.39		
Log likelihood only with constant	−774.72			−774.72		
Adjusted goodness-of-fit index	0.1350			0.2186		

provided cars are less likely to divert from expressway to arterial roads in response to VMS displaying a color-coded level of service map [16].

3.2.4 Age

The specification of age and quadratic term of age is to quantify the potential non-linear effect of age on diversion behavior. This kind of specification is often used in social sciences (e.g., [39]). "Age" and "age square" receive negative and positive coefficients, respectively, indicating that young and old drivers are more likely to divert under VMS, while middle-aged drivers are less likely to divert. Based on estimation results of the cross-sectional model, mid-age drivers, at the age of 37 (i.e., $[0.2154/(0.29349 \times 2)] \times 1 \approx 37$), are the least likely to divert.

3.2.5 Number of traffic lights

Negative effects of number of traffic lights on drivers' diversion are reflected in the negative coefficient of "number of traffic lights on the alternate route." This result is reasonable since more traffic lights on the alternate route means more frequent stops and a lower comfort level of driving and will naturally decrease the probability of diverting to the a driver to on the alternate route under VMS. This finding coincides with some previous studies (e.g., [5]).

3.2.6 Travel time saving

Travel time saving measures how much travel time drivers can save through diverting to the local street. The positive effects of travel time saving on diversion behavior are evidenced by the positive coefficient of "travel time saving." This indicates that explicitly displaying the travel time of the expressway, and the arterial road alternate route by VMS is meaningful and will positively influence drivers' diversion decision behavior.

3.3 Discussions about heterogeneity

The second block of Table 2 lists the model estimation results for the mixed logit model in which random coefficients are specified to accommodate heterogeneity in driver behavior and correlation of repeated choices by the same driver. Two random coefficients are specified for two scenario variables: time saving and number of traffic lights.

The standard deviation of the random coefficient for travel time savings is estimated at 0.1706 and appears highly significant (t value 5.16). This evidences that the random coefficient for travel time savings is not a constant but a variable varying among the driver population. This indicates that there exist random preferences for "travel time saving" across the driver population. This actually reveals the heterogeneity in value of time among the driver population. For the specific VMS information addressed by this study, this finding shows that although the travel time saving (as indicated by estimated travel time for the expressway route and the arterial road route) basically has a positive role in encouraging drivers to divert from the expressway to the local street, the value of travel time information is perceived differently by different people. For example, drivers under time pressure to get to a meeting on time will more value the travel time saving than those not having such time restrictions. For these drivers, the coefficient of the variable travel time saving should be larger than other drivers, i.e., being more sensitive to travel time savings indicated by VMS. In reality it is also possible that some people just have a preference for expressways or it is just a routine (inertia) for them to choose expressways; thus for these people the coefficient of the variable travel time saving should be relatively small, i.e., being less sensitive to travel time savings.

The standard deviation of the random coefficient for number of traffic lights is estimated at 0.0940 and appears highly significant (*t* value 5.21). The estimation result evidences that the coefficient for number of traffic lights is not a constant but a variable varying among the driver population. This indicates that there exists a significant degree of heterogeneity in the perception of the negative impacts of number of traffic lights. This reflects the real Shanghai situation. In Shanghai, most urban expressways are elevated roads and their competitive alternate routes are the parallel arterial roads under them. Under normal traffic conditions, an elevated road usually has much shorter travel time than an alternate arterial road route. The superiority of an elevated road is due to the fact that the elevated road typically has a good geometrical alignment and has a higher free-flow speed without intersections while the arterial road typically has a not so good geometrical alignment, has many signal-controlled intersections, and often has abrupt (unreasonable) changes in road markings which may affect the comfort of driving and cause delays. Given these facts, many less-experienced drivers are likely to not feel comfortable with the arterial road that has many traffic lights and prefer to use the elevated road even when VMS indicates the arterial road is faster. Yet, drivers with rich driving experience may be more confident of their ability of manipulating vehicles on the arterial road and are more adaptable to expressway delays and more willing to divert to the arterial road under VMS.

The above estimation results show that the developed mixed logit model can be successfully applied to model our SP panel data in which correlation of random utilities for the same driver needs to be accommodated. As a result, all the *t* test values of coefficients for the individual variables are smaller in the mixed logit model than those in the cross-sectional logit model. Comparison of adjusted ρ^2 values between the mixed logit model and the cross-sectional model suggests that the mixed logit model performs obviously better than the cross-sectional model (0.2186 vs. 0.1350).

4 Concluding remarks

A conventional cross-sectional logit model and a mixed logit model are developed to model drivers' decision behavior under Shanghai's new expressway VMS information which provides travel time of an expressway and an alternate arterial road route. This is based on the data collected from the SP survey that explores drivers' diversion response to the new expressway VMS information. The mixed logit model has a utility specification that accounts for preference variations across individuals regarding travel time and number of traffic lights and correlations among repeated choices. Several substantive conclusions have been obtained in this study as summarized below.

(1) The new expressway VMS information service has significant impacts on driver diversion decisions. Travel time saving and driving experience serve as positive factors, while the number of traffic lights on the arterial road, expressway use frequency, being a middle-aged driver, and being a driver of an employer-provided car serve as negative factors in diversion.

(2) There exists an obvious heterogeneity in value of travel time among the driver population, as evidenced by the significance of the standard deviation of the random coefficient for travel time saving in the mixed logit model.

(3) There exists an obvious heterogeneity in the perceived importance of "number of traffic lights" among the driver population as evidenced by the significance of the standard deviation of the random coefficient for number of traffic lights in the mixed logit model.

(4) The mixed logit model is successfully applied to model our SP panel data in which correlation of random utilities for the same driver needs to be accommodated, which is indicated by the fact that the mixed logit model obviously outperforms the conventional cross-sectional logit model regarding the goodness-of-fit criterion.

This study highlights the importance of capturing the heterogeneity of driver preferences and recognizing potential correlations between the individual's choices using appropriate modeling techniques such as the mixed logit model used in this study.

The estimated route choice probability model may be incorporated within a dynamic traffic assignment and simulation framework to assess network-level impacts of the enhanced expressway VMS information and estimate VMS benefits.

Acknowledgments This work was supported by a project (No. 51008195) funded by National Natural Science Foundation of China, a Shanghai First-Class Academic Discipline Project (No. S1201YLXK) funded by Shanghai Government, a project (No. 14XSZ02) funded by University of Shanghai for Science and Technology, and a project funded by Key Laboratory of Road and Traffic Engineering of the Ministry of Education, Tongji University.

References

1. Mahmassani HS (2011) Impact of information on traveler decisions. In: Transportation Research Board 90th annual meeting, Washington, DC
2. Chorus CG (2007) Traveler response to information. Doctoral thesis, TU Delft, TRAIL Research School
3. Abdel-Aty M, Abdalla MF (2006) Examination of multiple mode/route-choice paradigms under ATIS. IEEE Trans Intell Transp Syst 7(3):332–348
4. Lappin J, Bottom J (2001) Understanding and predicting traveler response to information: a literature review. USDOT, Washington, DC
5. Abdel-Aty M, Abdalla MF (2004) Modeling drivers diversion from normal routes under ATIS using generalized estimating equations and binomial probit link function. Transportation 31(3):327–348
6. Abdel-Aty MA, Kitamura R, Jovanis PP (1997) Using stated preference data for studying the effect of advanced traffic information on drivers' route choice. Transp Res C 5(1):39–50
7. Wardman M, Bonsall PW, Shires JD (1997) Driver response to variable message signs: a stated preference investigation. Transp Res C 5(6):389–405
8. Chatterjee K, Hounsell NB, Firmin PE, Bonsall PW (2002) Driver response to variable message sign information in London. Transp Res C 10(2):149–169
9. Peeta S, Ramos JL, Pasupathy R (2000) Content of variable message signs and on-line driver behavior. Transp Res Rec J Transp Res Board 1725(1):102–108
10. Gao S, Frejinger E, Ben-Akiva M (2011) Cognitive cost in route choice with real-time information: an exploratory analysis. Procedia Soc Behav Sci 17:136–149
11. Lai K, Wong W (2000) SP approach toward driver comprehension of message formats on VMS. J Transp Eng 126(3):221–227
12. Khattak A, Schofer J, Koppelman F (1993) Commuters' en-route diversion and return decisions: analysis and implications for advanced traveler information systems. Transp Res A 27(2): 101–111
13. Jou RC, Lam SH, Liu YH, Chen KH (2005) Route switching behavior on freeways with the provision of different types of real-time traffic information. Transp Res A 39(5):445–461
14. Gan HC, Ye X (2012) Urban freeway users' diversion response to variable message sign displaying the travel time of both freeway and local street. IET Intell Transp Syst 6(1):78–86
15. Gan HC, Bai Y, Wei J (2013) Why do people change routes? Impact of information services. Ind Manag Data Syst 113(3):403–422
16. Gan HC, Ye X, Fan BQ (2008) Drivers' en-route diversion response to graphical variable message sign in Shanghai, China. In: Proceedings of the 10th international conference of applications of advanced technologies in transportation, Greece
17. Mahmassani HS, Stephan DG (1988) Experimental investigation of route and departure time choice dynamics of urban commuters. Transp Res Rec J Transp Res Board 1203:69–84
18. Mahmassani HS, Liu YH (1999) Dynamics of commuting decision behavior under advanced traveler information systems. Transp Res C 7(2–3):97–107
19. Srinivasan KK, Mahmassani HS (2003) Analyzing heterogeneity and unobserved structural effects in routeswitching behavior under ATIS: a dynamic kernel logit formulation. Transp Res B 37(9):793–814
20. Yang H, Kitamura R, Jovanis PP, Vaughn KM, Abdel-Aty MA (1993) Exploration of route choice behavior with advanced traveler information using neural network concepts. Transportation 20(2):199–223
21. Bonsall P, Firmin P, Anderson M, Palmer I, Balmforth P (1997) Validating the results of a route choice simulator. Transp Res C 5(6):371–387
22. Koutsopoulos HN, Lotan T, Yang Q (1994) A driving simulator and its application for modeling route choice in the presence of information. Transp Res C 2(2):91–107
23. Ben-Elia E, Shiftan Y (2010) Which road do I take? A learning-based model of route-choice behavior with real-time information. Transp Res A 44(4):249–264
24. Pace RD, Marinelli M, Bifulco GN, Dellorco M (2011) Modeling risk perception in ATIS context through fuzzy logic. Procedia Soc Behav Sci 20:916–926
25. Chen WH, Jovanis PP (2003) Driver en route guidance compliance and driver learning with advanced traveler information systems: analysis with travel simulation experiment. Transp Res Rec J Transp Res Board 1843(1):81–88
26. Hato E, Taniguchi M, Sugie Y, Kuwahara M, Morita H (1999) Incorporating an information acquisition process into a route choice model with multiple information sources. Transp Res C 7(2–3):109–129
27. Emmerink RHM, Nijkamp P, Rietveld P, Ommeren J (1996) Variable message signs and radio traffic information: an integrated empirical analysis of drivers' route choice behavior. Transp Res A 30(2):135–153
28. Peng ZR, Guequierre N, Blakeman JC (2004) Motorist response to arterial variable message signs. Transp Res Rec J Transp Res Board 1899(1):55–63
29. Tsirimpa A, Polydoropouloua A, Antoniou C (2007) Development of a mixed multi-nomial logit model to capture the impact of information systems on travelers' switching behavior. J Intell Transp Syst Technol Plan Oper 11(2):79–89
30. Gan HC (2010) Graphical route information panel for the urban freeway network in Shanghai, China. IET Intell Transp Syst 4(3):212–220
31. Gan HC, Sun LJ, Chen JY, Yuan WP (2006) Advanced traveler information system for metropolitan expressways in Shanghai, China. Transp Res Rec J Transp Res Board 1944:35–40
32. Louviere JJ, Hensher DA, Swait JD (2000) Stated choice methods: analysis and applications. Cambridge University Press, Cambridge
33. Ben-Akiva M, Lerman SR (1985) Discrete choice analysis. MIT Press, Cambridge
34. Train KE (2009) Discrete choice methods with simulation. Cambridge University Press, Cambridge
35. Bhat CR (2001) Quasi-random maximum simulated likelihood estimation of the mixed multinomial logit model. Transp Res B 35(7):677–693
36. GAUSS 8.0 (2006) Aptech systems. Maple Valley, Washington
37. Hato E, Taniguchi M, Sugie Y (1995) Influence of traffic information on drivers' route choice. In: Proceedings of the 7th world conference on transportation research. Sidney, Australia, pp 27–40
38. Kitamura R, Jovanis PP, Abdel-Aty M, Vaughn KM, Reddy P (1999) Impacts of pretrip and en-route information on commuters' travel decisions: summary of laboratory and survey-based experiments from California. In: Emmerink R, Nijkamp P (eds) Behavioural and network impacts of driver information systems. Ashgate, Aldershot, pp 241–267
39. Wooldridg JM (2002) Econometric analysis of cross section and panel data. MIT Press, Cambridge

9

Railroad capacity tools and methodologies in the U.S. and Europe

Hamed Pouryousef · Pasi Lautala · Thomas White

Abstract A growing demand for passenger and freight transportation, combined with limited capital to expand the United States (U.S.) rail infrastructure, is creating pressure for a more efficient use of the current line capacity. This is further exacerbated by the fact that most passenger rail services operate on corridors that are shared with freight traffic. A capacity analysis is one alternative to address the situation and there are various approaches, tools, and methodologies available for application. As the U.S. continues to develop higher speed passenger services with similar characteristics to those in European shared-use lines, understanding the common methods and tools used on both continents grows in relevance. There has not as yet been a detailed investigation as to how each continent approaches capacity analysis, and whether any benefits could be gained from cross-pollination. This paper utilizes more than 50 past capacity studies from the U.S. and Europe to describe the different railroad capacity definitions and approaches, and then categorizes them, based on each approach. The capacity methods are commonly divided into analytical and simulation methods, but this paper also introduces a third, "combined simulation–analytical" category. The paper concludes that European rail studies are more unified in terms of capacity, concepts, and techniques, while the U.S. studies represent a greater variation in methods, tools, and objectives. The majority of studies on both continents use either simulation or a combined simulation–analytical approach. However, due to the significant differences between operating philosophy and network characteristics of these two rail systems, European studies tend to use timetable-based simulation tools as opposed to the non-timetable-based tools commonly used in the U.S. rail networks. It was also found that validation of studies against actual operations was not typically completed or was limited to comparisons with a base model.

Keywords Railroad capacity · Simulation · Railroad operation · The U.S. and European railway characteristics

H. Pouryousef (✉) · P. Lautala
Civil and Environmental Engineering Department, Michigan Technological University, 1400 Townsend Drive, Houghton, MI 49931, USA
e-mail: hpouryou@mtu.edu

P. Lautala
e-mail: ptlautal@mtu.edu

T. White
Transit Safety Management, 3604 220th Pl SW, Mountlake Terrace, WA 98043, USA
e-mail: taw@vtd.net

1 Introduction

Typically, the capacity of a rail corridor is defined as the number of trains that can safely pass a given segment within a period of time. The capacity is affected by variations in system configurations, such as track infrastructure, signaling system, operation philosophy, and rolling stock.

The configuration differences between European and the U.S. rail systems may lead to different methodologies, techniques, and tools to measure and evaluate the capacity levels. There are high utilization corridors in Europe where intercity passenger, commuter, freight, and even high-speed passenger services operate on shared tracks, and all train movements follow their predefined schedule in highly structured daily timetables that may be planned for a full year in advance. On the contrary, the prevalent operations pattern on current shared corridors in the U.S. follows unstructured (improvised) philosophy, where schedules and routings (especially for freight trains) are often adjusted on a daily or weekly basis. Recently, the U.S. has placed an increasing emphasis on either the development of new higher speed passenger services, or to incrementally increase the speeds of current passenger services on selected shared corridors [1]. At the same time, the slower speed freight rail transportation volumes are also expected to increase [2]. These increases in volumes and operational heterogeneity can be expected to add pressure for higher capacity utilization of the U.S. shared-use corridors. Capacity measurement and analysis approaches (and their methods and tools) will play a crucial part in preparing the U.S. network for these changes. To maximize the efficiency of future improvements, such as new passenger and high-speed rail services, the accuracy and applicability of capacity tools and methods in the U.S. environment need to be carefully evaluated. Whether the analytical and operational approaches utilized in Europe would provide any benefits for the U.S. shared-use corridors should also be reviewed.

This paper starts by identifying the various definitions of capacity and by discussing the similarities and differences between the U.S. and European rail systems that may affect both the methods and outcomes of capacity analysis. It will also identify different approaches to conduct the analysis and concludes with an examination of several past capacity studies from both continents.

2 What is capacity?

2.1 Capacity concept and definitions

The definition used for rail capacity in the literature varies based on the techniques and objectives of the specific study. For instance, Barkan and Lai [3] defined capacity as "a measure of the ability to move a specific amount of traffic over a defined rail corridor in the U.S. rail environment with a given set of resources under a specific service plan, known as level of service (LOS)". They listed several infrastructure and operational characteristics which affect capacity levels, including length of subdivision, siding length and spacing, intermediate signal spacing, percentage of number of tracks (single, double, and multi-tracks), and heterogeneity in train types (train length, power-to-weight ratios). In another paper, Tolliver [4] introduced freight rail capacity as the number of trains per day for typical track configurations depending on several factors, such as track segment length, train speed, signal aspects and signal block length, directional traffic balance, and peaking characteristics. The American Railroad Engineering and Maintenance-of-Way Association (AREMA) offers a simplified approach for line capacity that estimates practical capacity by multiplying theoretical capacity (C_t) and dispatching efficiency (E) of the line ($C = C_t \times E$). AREMA's method for calculating theoretical capacity and dispatching efficiency requires consideration of various factors, such as number of tracks, the operations rules (single or bi-direction operation), stopping distance between trains (or headway), alignment specifications (grade, curves, sidings, etc.), trains specifications (type of train, length, weight, etc.), maintenance activities requirements, and the signaling and train control systems [5]. A capacity modeling guidebook for the U.S. shared-use corridors, released by the Transportation Research Board (TRB), defines capacity as "the capability of a given set of facilities, along with their related management and support systems, to deliver acceptable levels of service for each category of use." Similar to the other capacity definitions, TRB notes that different parameters and variables should be considered in the capacity analysis, such as train dispatching patterns, train type and consist, signaling system, infrastructure, track maintenance system, etc. [6].

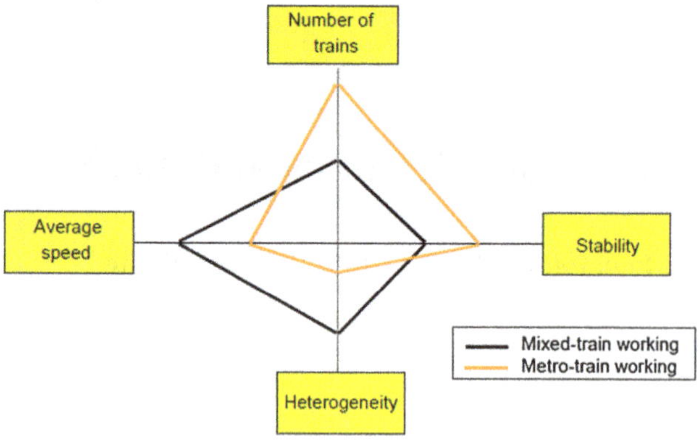

Fig. 1 Capacity balance according to UIC code 406 definition [7]

In Europe, the most common method for capacity analysis is provided by the International Union of Railways (UIC) code 406. According to UIC 406, there is no single way to define capacity, and the concerns and expectations vary between different points of view by railroad customers, infrastructure and timetable planners, and railroad operators. UIC also emphasizes that the capacity is affected by interdependencies and the interrelationships between the four major elements of railway capacity including average speed, stability,[1] number of trains, and heterogeneity,[2] as shown in Fig. 1 [7]. According to the figure, a rail

[1] The state of keeping the same train schedule by providing time margins/buffers between trains arrival/departures; despite of minor delay which may occur during operation.

[2] Diversity level of train types which are in operation along a shared-use corridor.

line with various types of trains on the same track (mixed traffic operations or shared-use corridor) has a higher heterogeneity level compared to the urban metro (subway) system with dedicated right-of-way and homogeneous operations. While the average speed of a mixed traffic corridor might be higher than a dedicated metro line, the various train types reduce the stability of train schedules, as well as the total number of trains that can operate on the corridor, due to increased headway requirements.

According to UIC, the absolute maximum capacity, or "Theoretical Capacity", is almost impossible to achieve in practice, and it is subject to:

- Absolute train-path harmony (the same parameters for majority of trains)
- Minimum headway (shortest possible spacing between all trains)
- Providing best quality of service [7].

In addition to the UIC literature, research conducted as part of European Commission's "Improve Rail" project produced a definition of ultimate capacity that was similar to the UIC's theoretical capacity definition, but placed higher emphasis on the train schedules and running time [8].

2.2 Capacity metrics

The literature categorizes the main metrics of capacity level measurements into three groups: throughput (such as number of trains, tons, and train-miles), LOS (terminal/station dwell, punctuality/reliability factor, and delay), and asset utilization (velocity, infrastructure occupation time, or percentage) [9]. In 1975, The Federal Railroad Administration (FRA) introduced a parametric approach developed by "Peat, Marwick, Mitchell and Company" to measure capacity in the U.S. rail network based on delay units (hours per 100 train-miles) [4]. The European rail operators typically use throughput metrics (number of trains per day or hours) to measure the capacity levels, although punctuality and asset utilization metrics are also applied as secondary units [8, 10].

3 Differences between the U.S. and European rail systems

The U.S. and European rail networks have several similarities, such as mixed operations on shared-use corridors, and using modern signaling and traffic control systems (e.g., developing ETCS in Europe and PTC in the U.S.). On the other hand, significant differences also exist and they may change the preferred methodologies, tools, and the outcomes of capacity analysis. Figure 2 and the following discussion uses the literature review to highlight several key differences between infrastructure, signaling, operations, and rolling stock in Europe and the U.S.

3.1 Infrastructure characteristics

- *Public versus private ownership of infrastructure* The ownership of rail infrastructure is one of the important differences between Europe and the U.S. rail networks.

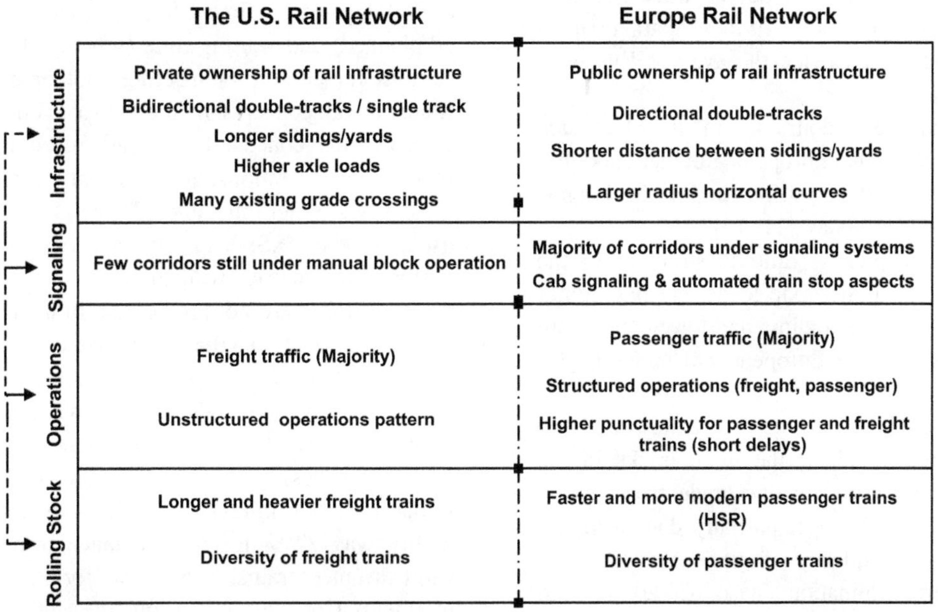
The U.S. Rail Network
Europe Rail Network
Infrastructure
Private ownership of rail infrastructure
Bidirectional double-tracks / single track
Longer sidings/yards
Higher axle loads
Many existing grade crossings
Public ownership of rail infrastructure
Directional double-tracks
Shorter distance between sidings/yards
Larger radius horizontal curves
Signaling
Few corridors still under manual block operation
Majority of corridors under signaling systems
Cab signaling & automated train stop aspects
Operations
Freight traffic (Majority)
Unstructured operations pattern
Passenger traffic (Majority)
Structured operations (freight, passenger)
Higher punctuality for passenger and freight trains (short delays)
Rolling Stock
Longer and heavier freight trains
Diversity of freight trains
Faster and more modern passenger trains (HSR)
Diversity of passenger trains

Fig. 2 The main differences in the U.S. and Europe rail systems

More than 90 % of the infrastructure is owned and managed by private freight railroads in the U.S., while in Europe almost all infrastructure is owned and managed by governments or public agencies. In addition, operations and infrastructure are vertically separated in Europe, while in the U.S. the majority of operations (mainly freight) are controlled by the same corporations who own the infrastructure. The ownership and vertical separation have wide impact in the railroad system. Perhaps the greatest effect is on the prioritization of operations and accessibility for operating companies, but other aspects, such as operations philosophy, maintenance strategy and practices, signaling and train control systems, rolling stock configuration, and capital investment strategies are also affected [4, 11].

- *Single versus double track* More than 46 % of rail corridors in Europe are at least double track [12, 13], while approximately 80 % of the U.S. rail corridors are single track [2, 4].
- *Directional versus bidirectional* Most of the U.S. double tracks operate in bidirectional fashion and use crossovers along the corridor, while directional operation with intermediate sidings and stations is the common approach in Europe [4].
- *Distance between sidings* The distances between stations and sidings in the European rail network are generally shorter than in the U.S. The average distance between sidings/stations throughout the European network (total route mileage vs. number of freight and passenger stations) is approximately four miles between sidings/stations in both UK and Germany [13, 14]. In the U.S., the distance between sidings varies greatly between corridors. On double track sections, passing sidings are typically further apart than in Europe, often more than twice the average European distance [11, 15].
- *Siding length* Siding/yard tracks in the U.S. are typically longer than the European rail network, but in many cases are still not sufficient for the longest freight trains operating today [11, 16].
- *Track conditions* Typically, railroad structure in the U.S. is designed for higher axle loads, but has tighter horizontal curves (smaller radius) and lower maximum speed operations than the European rail network [11, 16].
- *Grade crossings* There are approximately 227,000 active grade crossings along the main tracks in the U.S. [17, 18], while there are few grade crossings on the main corridors in Europe, partially due to higher train speeds. High frequency of grade crossings and difficulty of their elimination cause operational and safety challenges for increased train speeds in the U.S. [19].

3.2 Signaling characteristics

- *Manual blocking versus signaling systems* Manual blocking is absent on main passenger corridors in the U.S. today, but relatively common on lower density branch ones, including some of the lines proposed for passenger corridors. In Europe, most shared-use corridors are equipped with one of the common signaling systems [20].
- *Cab signaling* A more significant difference is the extensive use of cab signaling and enforced signal systems, such as ETMS and ATS in Europe. Implementation of automatic systems is limited in the U.S., despite the current effort to introduce the positive train control (PTC) on a large portion of corridors [11].

3.3 Operation characteristics

- *Improvised versus structured operation* While some specific freight trains (mainly intermodal) have tight schedules, the U.S. operations philosophy is based on the improvised pattern with no long-term timetable or dispatching plan. On the passenger side, the daily operation patterns of many Amtrak and commuter trains are also developed without details, anticipating improvised resolution of conflicts among the passenger trains, or between passenger and freight trains. In Europe, almost all freight and passenger trains have a regular schedule developed well in advance, known as structured operations [21].
- *Freight versus passenger traffic* The majority of the U.S. rail traffic is freight, while the majority of European rail traffic is passenger rail [4, 22].
- *Delay versus waiting time* Delay (deviation of train arrival/departure time from what was predicted/planned) and waiting time (scheduled time spent at stations for passing or meeting another train) are two fundamental concepts in the railroad operations. The waiting time concept is typically used in Europe to manage rail operations, due to the structured operations pattern with strict timetables. Delay is more commonly used in the U.S. capacity analysis as the main performance metric, while it is limited in Europe to the events that are not predictable in advance [21].
- *Punctuality* The punctuality criteria of trains are quite different in the U.S. and Europe. Amtrak's trains are considered on-time if they arrive within 15 min of a scheduled timetable for short-distance journeys (less than 500 miles) or within 30 min for long-distance trains (over 500 miles). In 2011, Amtrak's train punctuality was 77 % for long-distance trains, 84 % for short-distance trains, and 92 % for Acela trains on Northeast Corridor. According to Amtrak, more than 70 % of passenger train delays were caused either by

the freight trains performance or infrastructure failure [23]. The passenger trains in Europe have shorter average delay per train. For instance, Network Rail in the UK reported that approximately 90 % of all short-distance passenger trains had less than 5 min deviation from planned timetable, while for long-distance trains, the same was true for deviation less than 10 min [24]. In Switzerland, more than 95 % of all passenger trains are punctual with an arrival delay of 5 min or less [25]. The punctuality of European freight trains in 2003 was reported to be approximately 70 % [26].

3.4 Rolling stock characteristics

- *Train configuration (length and speed)* Typically freight trains in the U.S. are longer and heavier than freight trains in Europe. Based on the Association of American Railroads (AAR), the typical number of cars in a U.S. freight train varies between 63–164 cars in the West and 57–110 cars in the East, while the typical number in Europe is 25–40. From speed perspective, the average speed of intercity passenger trains in Europe is significantly faster than in the U.S. [2, 11, 16]. Freight trains also typically operate at higher speeds and with less variability in Europe.
- *Diversity of freight versus passenger trains* The U.S. rail transportation is more concentrated on the freight trains than Europe, and there is a great diversity between the types, lengths, etc., of freight trains. On the passenger side, Europe has more diverse configurations (such as speed, propulsion, train type, power assignment, HSR services, diesel, and electric multiple unit (EMU) trains) in comparison to the U.S. [2, 20].

While the principles of rail capacity remain the same in all rail networks, the above characteristics have an effect on capacity and its utilization. What remains unclear is how these differences have been considered in various capacity analysis tools and methodologies used and how much they limit the applicability of the U.S. tools in the European environment and vice versa. This paper introduces some of the common tools and methodologies, including examples of their use in past studies, but excludes any direct comparisons between the capabilities of individual tools. A more detailed (case study based) comparative analysis of selected U.S. and European simulation tools and methodologies is provided by the authors in separate papers [27, 28].

4 Capacity measurement, analytical, simulation, and combined approaches

Generally speaking, there are two main approaches to improve the capacity levels: either by applying new capital investment toward upgraded or expanded infrastructure or by improving operational characteristics and parameters of the rail services [29]. In either approach, it is necessary to assess and analyze the benefits, limitations, and challenges of the approach, often done through capacity analysis. The literature classifies capacity analysis approaches and methodologies in several different ways. Although the approaches differ, the input typically includes infrastructure and rolling stock data, operating rules, and signaling features. Abril et al. [30] classified the capacity methodologies as analytical methods, optimization methods, and simulation methods. Pachl [31] divided the capacity methodologies into two major classes: analytic and simulation. Similar categorization was used in research conducted by Murali on delay estimation technique [32]. Khadem Sameni, and Preston et al. [9] categorized capacity methods to timetable-based and non-timetable-based approaches. The capacity guidebook developed by TRB also divides capacity evaluation methods into two approaches: simple analysis, and complex simulation modeling [6]. Finally, in research conducted at the University of Illinois, Sogin, Barkan et al. [3, 33] divided capacity methods to theoretical (analytical), parametric, and simulation methods. Overall, the analytical and simulation methods are the most common methods found in the literature. For our review, we divided methods into three groups: analytical, simulation, and combined. Although the term "combined methodology" was not used commonly in the reviewed literature, it was added as a new class to address the fact that many reviewed studies took advantage of both analytical and simulation methods.

4.1 Analytical approach

The *analytical approach* typically uses several steps of data processing through mathematical equations or algebraic expressions and is often used to determine theoretical capacity of the segment/corridor. The outcomes vary based on the level of complexity of the scenario and may be as simple as the number of trains per day, or a combination of several performance indicators, such as timetable, track occupancy chart, fuel consumption, speed diagrams, etc. Analytical methods can be conducted without software developed for railroad applications, such as Microsoft Excel, but there are also analytical capacity tools specifically developed for rail applications. One example is SLS PLUS in Germany, which is used in the German rail network (DB Netz AG) for capacity estimation through analytical determination of the performance, asynchronous simulation, and manual timetable construction [34]. Figure 3 presents the different levels of analytical approach and how complexity can be added to the process to provide more detailed results. In some cases, analytical models are

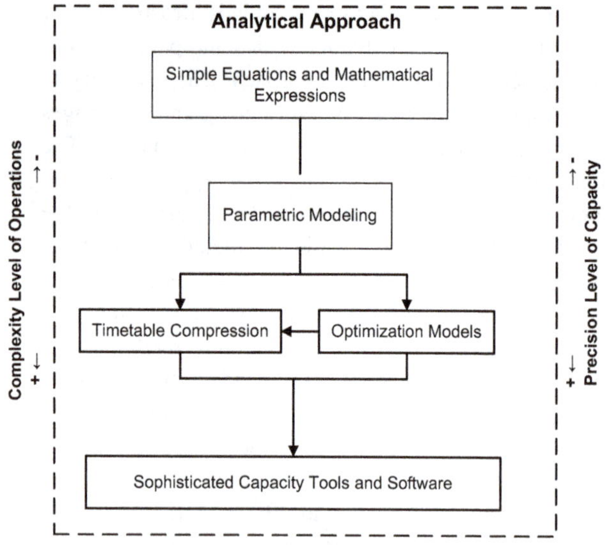

Fig. 3 Levels of analytical approaches for capacity analysis

called optimization methods or parametric models, taking advantage of different modeling features, such as probabilistic distribution or timetable optimization. The latter method, timetable optimization, is typically achieved using specialized software or simulation tools [30, 31].

Timetable compression method is one of the main analytical approaches in Europe to improve the capacity levels, especially on the corridors with pre-determined timetables (structured operation pattern). A majority of techniques and tools for improving the capacity utilization in Europe, including the UIC method (leaflet 406), are partly developed based on timetable compression [7, 10, 35–37]. The UIC's method modifies the pre-determined timetable and reschedules the trains as close as possible to each other [30]. Figure 4 provides an example of the methodology where a given timetable along a corridor with quadruple tracks (Scenario a) is first modified by compressing the timetable (Scenario b) and then further improved by optimizing the order of trains (Scenario c). As demonstrated in the figure, the third scenario could provide a higher level of theoretical capacity in comparison to the Scenarios a and b [10]. It should be emphasized that due to the unstructured nature of the U.S. rail operation philosophy, timetable compression technique has not been practically applied yet in the U.S. rail environment.

4.2 Simulation approach

Simulation is an imitation of a system's operation which should be as close as possible to its real-world equivalent [30]. In this approach, the process of simulation is repeated several times until an acceptable result is achieved by the software. The data needed for the simulation are similar to the analytical methods, but typically at a higher level of detail. The simulation practices in rail industry started in the early 1980s through the development of models and techniques, such as dynamic programming and branch-and-bound, proposed by Petersen, as well as heuristic methods developed by Welch and Gussow [30]. Today, the simulation process utilizes computer tools to handle sophisticated computations and stochastic models in a faster and more efficient way. The simulation approaches use either *general simulation* tools, such as AweSim, Minitab, and Arena [32, 38]; or *commercial railroad simulation* software specifically designed for rail transportation, such as RTC, MultiRail, RAILSIM, OpenTrack, RailSys, and CMS [9, 30]. The use of general simulation tools requires the user to develop all models, equations, and constraints step by step (often manually). This requires more expertise, creativity, and effort, but it can also offer more flexible and customization when it comes to results and outputs. The commercial railroad simulation tools offer an easier path toward development of different scenarios, in addition to providing a variety of outputs in a user-friendly way, but the core decision models and processes are not easily

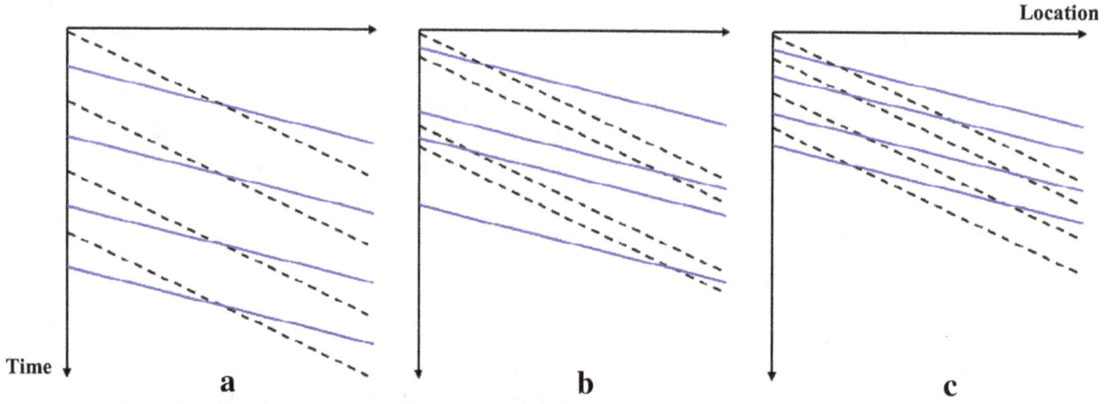

Fig. 4 Actual timetable for a quadruple-track corridor (**a**) compressed timetable with train order maintained (**b**) compressed timetable with optimized train order (**c**) (*Note* chart layout follows typical European presentation and *solid and dot lines* represent different types of trains) [10]

customizable or reviewable, which may reduce the flexibility of applying these tools.

The commercial railroad simulation software typically revolves around two key simulation components: (1) train movement and (2) train dispatching. The first component uses railroad system component data provided as an input, such as track and infrastructure characteristics (curvature and grades), station and yard layout, signaling system, and rolling stock characteristics, to calculate the train speed along the track. Train dynamics is typically determined based on train resistance formulas, such as Davis equation and train power/traction. The dispatching simulation component typically emulates (or attempts to emulate) the action of the dispatcher in traffic management, but in some cases, it can be also used as part of a traffic management software to help traffic dispatchers to manage and organize the daily train schedules (Fig. 5) [21].

According to Pachl, the simulation method can also be divided into asynchronous and synchronous methods. Asynchronous simulation software is able to consider stochastically generated train paths within a timetable, following the scheduling rules and the train priorities. In synchronous simulation, the process of rail operations is followed in real-time sequences, and the results are expected to be closely aligned with real operations. In contrast to the asynchronous method, synchronous methods cannot directly simulate the scheduling, or develop a timetable, without use of additional computer tools and programs to create a timetable [31]. The outputs of simulation software typically include several parameters such as delay, dwell time, waiting time, elapsed time (all travel time), transit time (time between scheduled stops), trains speed, and fuel consumption of trains [21, 30].

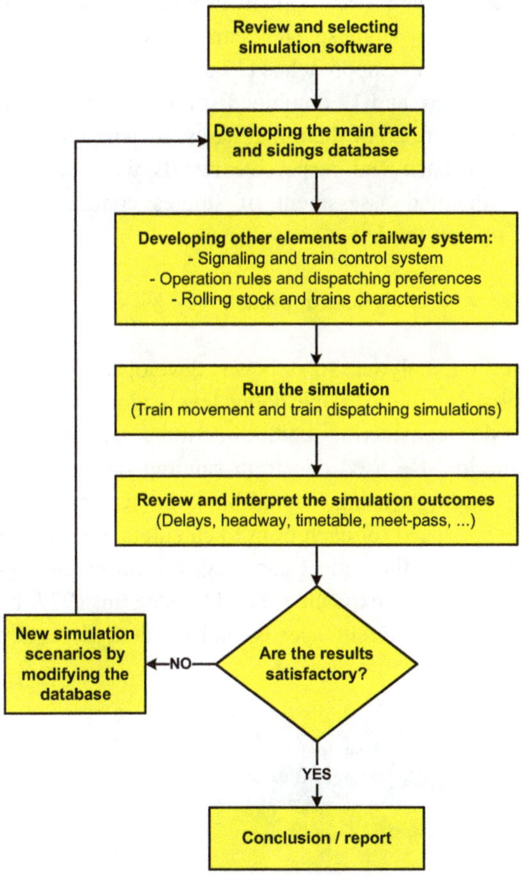

Fig. 5 Steps for railway capacity analysis in commercial simulation approach

4.2.1 Simulation methods: timetable based versus non-timetable based

The commercial railroad simulation software can be classified in two groups: non-timetable based and timetable based. The non-timetable-based simulations are typically utilized by railroads that use the improvised (unstructured) operation pattern without an initial timetable, such as the majority of the U.S. rail networks. In this type of simulation, after loading the input data in the software, the train dispatching simulation process uses the departure times from the initial station that are provided as part of the input data. The software may encounter a problem to assign all trains and request assistance from the user to resolve the issue by manually adjusting the train data, or by modifying the schedule constraints [9, 21].

The simulation procedure in timetable-based software (typically used in Europe) is based on the initial timetable of trains and the objective is to improve the timetable as much as possible. The UIC's capacity approach is often one of the main theories behind the timetable-based simulation approach. The simulation process in this methodology begins with creating a timetable for each train. In the case of schedule conflict between the trains, the user must adjust the timetable until a feasible schedule is achieved. However, the user actions are more structured compared to the improvised method, and is implemented as part of the simulation process [21]. There are several common software tools in this category, such as MultiRail (U.S.), RAILSIM (U.S.), OpenTrack (Switzerland), SIMONE (The Netherlands), RailSys (Germany), DEMIURGE (France), RAILCAP (Belgium), and CMS (UK) [9, 30]. A comprehensive capability review of various simulation tools is outside the scope of this paper, but three simulation packages (RTC, RailSys, and OpenTrack) are briefly introduced to demonstrate some key differences between non-timetable-based and timetable-based software.

The rail traffic controller (RTC), developed by Berkeley Simulation Software, is the most common software in the non-timetable-based category, used extensively by the U.S. rail industry [9]. RTC was launched in the U.S. (and North American) rail market in 1995 and has since been continuously developed and upgraded. Since majority of the U.S. train services (particularly freight trains) have frequent adjustments in their daily schedules, RTC has several features and tools for simulating the rail operations in non-scheduled environment, including train movement animation, automated train conflict resolution, and randomization of train schedule. The dispatching simulation component of RTC is based on a decision support core, called "meet-pass N-train logic". For any dispatching simulation practice, "meet-pass N-train logic" will decide when the given trains should exactly arrive and depart from different sidings, based on the defined train priorities and preferred times of departure. The simulation outcomes may include variation between the simulated departure times and preferred times [39]. Besides its decision core fitting the U.S. operational philosophy, RTC has other system characteristics, such as attention to grade crossing, that make it well suited to the U.S. market.

RailSys, developed by Rail Management Consultants GmbH (RMCon) in Germany, is an operation management software package that includes features, such as timetable construction/slot management, track possession planning, and simulation. It has been in the market since 2000 and it is one of the commonly used timetable-based simulation software in Europe. The capacity feature of RailSys uses the UIC code 406 which is based on the timetable compression technique [40, 41]. OpenTrack is another common simulation package in Europe. It was initially developed by Swiss Federal Institute of Technology-Zurich (ETH-Zurich) and has since 2006 been supplied by OpenTrack Railway Technology Ltd. OpenTrack is also a timetable-based simulation tool with several features, such as automatic conflict resolution based on train priority, routing options and delay probabilistic functions, as well as several outputs and reporting options, such as train diagram, timetable and delay statistics, station statistics, and speed/time diagram [42, 43].

4.3 Combined analytical–simulation approach

In addition to the analytical and simulation approaches, a *combined analytical–simulation method* can also be used to investigate the rail capacity. Parametric and heuristic modelings (in analytical approach) are more flexible when creating new aspects and rules for the analysis. On the other hand, updating the railroad component input data and criteria tends to be easier in the simulation approach, and the process of running the new scenarios is generally faster, although simulation may place some limitations when adjusting the characteristics of signaling or operation rules. A combined simulation–analytical methodology takes advantage of both methodologies' techniques and benefits, and the process can be repeated until an acceptable set of outputs and alternatives is found (Fig. 6). There are several ways to combine analytical and simulation tools. For instance, finding a basic and reasonable schedule of trains through simulation, followed by analytical schedule can be considered as one example of combined analytical–simulation approach. Another example would be application of a simplistic analytical model to provide the basic inputs, such as determining the type of signaling system, or developing train schedule, followed by more extensive and detailed analysis in commercial rail simulation tools.

5 Review of capacity studies in the U.S. and Europe

The approaches, methodologies, and tools highlighted in previous section have been applied in numerous U.S. and European capacity studies. The team reviewed 51 total studies using all three approaches (17 analytical studies, 22 simulation studies, and 12 combined simulation–analytical approaches). Then, 25 of them that had sufficient details of the study approach and respective results were used to conduct a detailed assessment of studies conducted in Europe versus in the U.S.

5.1 Studies with analytical approach

One of the first analytical models was developed by Frank [44] in 1966 by studying the delay levels along a single track corridor considering both directional and bidirectional scenarios. He used one train running between two consecutive sidings (using manual blocking system) and a single average speed for each train to calculate the number of possible trains (theoretical capacity) on the given segment. Petersen [45] expanded Frank's idea in 1974 by considering two different speeds, independent departure

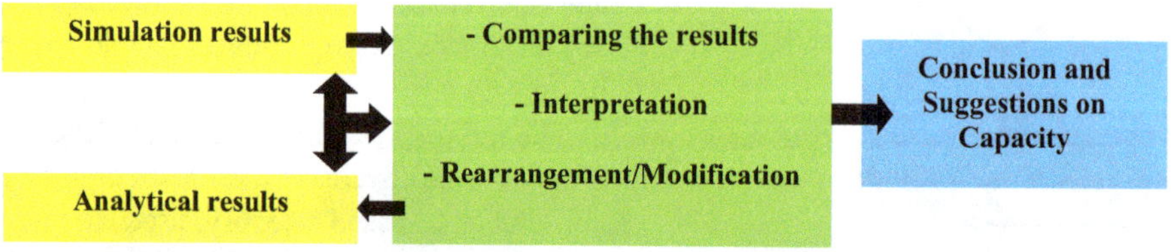

Fig. 6 Basic diagram of combined analytical–simulation approach for capacity analysis

times, equal spacing between sidings, and constant delays between two trains. Higgins et al. [46] developed a model in 1998 for urban rail networks to evaluate the delays of trains by considering different factors such as trains' schedule, track links, sidings, crossings, and the directional/bidirectional operation patterns throughout the network.

De Kort et al. [47] analyzed the capacity of new corridors in 2003 by applying an optimization method and considering uncertainty of demand levels on the planned route. Ghoseiri et al. [48] introduced a multi-objective train scheduling model of passenger trains along single and multiple tracks of rail network, based on minimizing the fuel consumption cost as well as minimizing the total passenger time of trains. Burdett and Kozan [49] developed analytical techniques and models in 2006 to estimate the theoretical capacity of a corridor based on several criteria, such as mixed traffic, directional operation pattern, crossings and intermediate signals along the track, length of the trains, and dwell time of trains at sidings or stations. Wendler [50] used queuing theory and the semi-Markov chains in 2007 to provide a technique of predicting the waiting times of trains based on the arrival times, minimum headway of trains, and the theory of blockings. Lai and Barkan [3] introduced an enhanced technique of capacity evaluation tools in 2009 based on the parametric modeling of capacity evaluation, which was initially developed by CN Railroad. The railroad capacity evaluation tool (RCET), developed by Lai and Barkan, can evaluate the expansion scenarios of network by estimating the line capacity and investment costs, based on the future demand and available budget.

Lindner [51], recently, reviewed the applicability of timetable compression technique, UIC code 406, to evaluate the corridor and station capacity. He used several case studies and examples to conclude that UIC code 406 is a good methodology for evaluating the main corridor capacity, but it may encounter difficulties with node (station) capacity evaluation. Corman et al. [52] conducted another study in 2011 to analyze an innovative approach of optimization of multi-class rescheduling problem. The problem focused on train scheduling with multiple priority classes in different steps, using the branch-and-bound algorithm.

In addition to specific studies on railroad capacity, a book edited by Hansen and Pachl [53], containing several articles and sections conducted by different railroad studies mostly by European universities and academic centers, was released in 2008 as one of the latest resources of timetable optimization and train rescheduling problem. The book covers articles on various topics, such as cyclic timetabling, robust timetabling, use of simulation for timetable construction, statistical analysis of train delays, rescheduling, and performance evaluation.

5.2 Studies with simulation and combined approach

Studies using analytical approach preceded simulation and combined approaches. One of the first general simulation studies was conducted by Petersen et al. in 1982 by dividing a given corridor into different track segments where each segment represented the distance between two siding/switches [54]. Kaas [55] developed another general simulation model in 1991, called "Strategic Capacity Analysis for Network" (SCAN), by defining different factors of simulation which could determine the rail network capacity. In another study, Dessouky et al. (1995) [56] used a general simulation model for analyzing the track capacity and train delay throughout a rail network. Their model included both single and double track corridors, as well as other network parameters, such as trains length, speed limits, and train headways. Sogin et al. [57] recently used RTC to simulate several case studies at University of Illinois, Urbana-Champaign. One of their studies evaluated the impact of passenger trains along U.S. shared-use double track corridors, considering different speed scenarios. They concluded that increasing speed gap between the trains can result in higher delays.

The Missouri DOT used the combined analytical–simulation approach in 2007 to analyze the rail capacity on the Union Pacific (UP) corridor between St. Louis to Kansas City to improve the passenger train service reliability and to reduce the freight train delay. Six different alternatives were generated based on a theory of constraints (TOC) analysis[3] and then compared with each other using the Arena simulation method. A set of recommendations and capital investment for each proposed alternative were proposed with respect to delay reduction [38].

In another project, Washington DOT (WSDOT) conducted a master plan in 2006 to provide a detailed operation and capital plan for the intercity passenger rail program along Amtrak Cascades route. The capacity of the corridor was also evaluated using the combined simulation–analytical approach. First, analytical methods were used to determine the proposed infrastructure. Then the proposed traffic and infrastructure were simulated with RTC software to test the proposed infrastructure and operational results. After running simulation on RTC software, a heuristic (analytical) method, called root cause analysis (RCA), was applied to evaluate the simulation output. The objective of RCA method was to identify the real reason of a delay along the rail corridor by comparing

[3] TOC is a management technique that focuses on each system constraints based on five-step approach to identify the constraints and restructure the rest of the system around it. These steps are: 1) identify the constraints, 2) decision on how to exploit the constraints, 3) subordinate everything around the above decision, 4) elevate the system's constraints, and 5) feedback, back to step 1.

the output reports of each delayed train with other train services and to re-adjust the simulation outputs to be more accurate, in addition to locating infrastructure bottlenecks which caused the capacity issues and delays [58].

The Swedish National Rail Administration (Banverket) carried out a research project in 2005 to evaluate the application of the UIC capacity methodology (timetable compression) for the Swedish rail network. RailSys software was used for the simulations and the research team analytically evaluated the capacity consumption, its relationship with time supplements (or buffer times), and the service punctuality. The research concluded that the buffer times are absolutely necessary for the service recovery, in case of operation interruption. When there is no buffer time, the service punctuality can be significantly degraded due to increased capacity consumption. Banverket also confirmed the validity of the framework and the results of the UIC's approach and asked their experts and consultants to implement this analytical approach in their network [36].

In research conducted through combined analytical–simulation approach, Medeossi et al. [59] applied stochastic approach on blocking times of trains to improve the timetable planning using OpenTrack simulation software. They redefined timetable conflicts by considering a probability for each train conflict as a function of process time variability. The method repeatedly simulated individual train runs on a given infrastructure model to show the occupation staircase of trains in different color spectrums, while each color represents the probability of trains' conflict which should be resolved.

Recently, a new "Web-based Screening Tool for Shared-Use Rail Corridors" was developed in the U.S. by Brod and Metcalf [60] to perform a preliminary feasibility screening of proposed shared-use passenger and freight rail corridor projects. The outcomes can be used to either reject projects or move them to more detailed analytical/simulation investigations. The concept behind the tool is based on a simplified simulation technique which does not provide optimization features or complex simulation algorithms. The tool requires development of basic levels of infrastructure, rolling stock, and operation rules (trains schedule) of the given corridor; and a conflict identifier assists the user in identifying locations for a siding or yard extension needed to resolve the conflict between existing and future train services.

5.3 Detailed assessment of selected studies

Only a subsection of reviewed studies offered sufficiently detailed explanation of the study approach and respective results. These studies were broken into several categories and subcategories for a comparison between the studies conducted in Europe versus in the U.S. Table 1 and the following discussion summarize the approach, tools used,

Table 1 Category/subcategory breakdown of 25 selected studies in the U.S. and Europe [2, 3, 9, 10, 21, 25, 29, 32, 35, 36, 38, 57–70]

Category/subcategory		The U.S. (14 studies)	Europe (11 studies)
Capacity approach	Analytical	4 studies [2, 3, 29, 32]	–
	Simulation	5 studies [57, 61, 62, 64, 65]	5 studies [10, 25, 36, 67, 69]
	Combined analytical–simulation	5 studies [9, 21, 38, 58, 60]	6 studies [35, 59, 63, 66, 68, 70]
Tools/software	Only mathematical/parametric modeling	3 studies [2, 3, 29]	–
	General simulation software	3 studies [32, 38, 60]	–
	Timetable-based simulation software	–	11 studies [10, 25, 35, 36, 59, 63, 66–70]
	Non-timetable-based simulation software	8 studies [9, 21, 57, 58, 61, 62, 64, 65]	–
Purpose of research	New methodology development/ methodology approval	5 studies [3, 9, 21, 29, 60]	7 studies [10, 25, 35, 36, 59, 66, 68]
	Master plan/capacity analysis	3 studies [2, 38, 58]	–
	Academic research/project	6 studies [32, 57, 61, 62, 64, 65]	4 studies [63, 67, 69, 70]
Type of outcomes/solutions	Delay analysis/improvement	3 studies [32, 57, 61]	1 study [69]
	Infrastructure development,	1 study [2]	–
	Rescheduling/operation changes	2 studies [21, 62]	4 studies [25, 35, 63, 67]
	Combination of above solutions	4 studies [38, 58, 64, 65]	2 studies [68, 70]
	New tools/methodology approval	4 studies [3, 9, 29, 60]	4 studies [10, 36, 59, 66]
Validation of simulation results	No comparison	6 studies [2, 9, 29, 57, 61, 62]	3 studies [25, 35, 36]
	Base model	7 studies [3, 21, 38, 58, 60, 64, 65]	7 studies [10, 59, 63, 66, 67, 69, 70]
	Base and alternative results	1 study [32]	1 study [68]

study purpose, types of outcomes, and validation methods of the 25 studies selected for more detailed comparison.

Approach Most studies used either simulation or combined analytical–simulation approaches. However, research conducted by AAR [2], University of Illinois at Urbana-Champaign (UIUC) [3, 29], and University of Southern California (USC) [32] applied analytical-only methodologies.

Tools and software All European studies used timetable-based simulation software (e.g., RailSys, OpenTrack, ROMA), while the U.S. studies relied on other tools like optimization/parametric modeling (UIUC and USC) [2, 3, 29], general simulation software (e.g., Arena) [38], web-based screening tools [60], and non-timetable-based rail simulation software (RTC).

Purpose of Research Three main purposes were identified for studies: (1) introducing new methodology for capacity evaluation, (2) evaluating the capacity status of a given corridor as part of a corridor master plan development, and (3) academic research on various capacity issues. The majority of European studies (Denmark, Austria, Germany, the Netherlands, and Sweden) were conducted by industry or academic research teams to justify and evaluate the UIC's approach (UIC code 406) for capacity evaluation [10, 35, 36, 67, 70], while the objectives of the U.S. studies included all three subcategories.

Type of outcomes or solutions The outcomes and solutions obtained from the U.S. studies included variety of different types such as delay analysis (UIUC by using RTC and USC by using Awesim/Minitab), rescheduling and recommendations related to current operations (UIUC and White) [21, 62], infrastructure development, and combination of all outcomes mentioned above (typically as part of a master plan). In addition, new tools and parametric models were also developed as the final outcome of three U.S. studies (mainly by UIUC). The outcomes of European studies were not so diverse, as they either approved the application of UIC's capacity methodology to be used on their network [10, 36], or suggested network rescheduling and operational changes (the timetable compression concept) [25, 35, 63, 67, 70]. One of the common conclusions of various studies was the identification of operational heterogeneity as a major reason of delay, especially in the U.S. rail network with unstructured operation pattern.

Validation of simulation results None of the studies using analytical method compared the results to a real-life scenario, but some of the simulation-based studies validated the results with one of the following three types of comparisons:

- *No comparison* No specific information or comparison was provided between simulated results and actual practices. As presented in Table 1, approximately one-third of the studies (9 out of 25) did not validate the simulation results, either because the study was not based on actual operational data, or comparison was not conducted as part of the research.
- *Base model* Only the results of a base model were compared with the real data. More than half of the studies (14 out of 25) compared the simulation results only with the base model.
- *Base and alternative results* In addition to base model comparison, the alternative outcomes were compared with the real data. Only two studies belonged to this category.

6 Summary and conclusions

This paper has provided an overview of capacity definitions, alternative analysis approaches, and tools available to evaluate capacity. It has also highlighted the key similarities and differences between the U.S. and European rail systems and how they affect related capacity analysis. Finally, the paper has reviewed over 50 past capacity studies and selected 25 of them for more detailed investigation,

The review revealed no single definition of railroad capacity. Rather, the definition varies based on the techniques and objectives of the specific study. The capacity analysis approaches and methodologies can also be classified in several ways, but are most commonly divided into analytical and simulation methods. This paper also introduced a third "combined" approach that uses both analytical and simulation approaches.

While the objective of capacity analysis is common, there are several differences between the U.S. and European rail systems that affect the approaches, tools, and outcomes of analysis. Europe tends to use a structured operations philosophy and thus uses often timetable-based simulation approaches for analysis, while the improvised U.S. operations warrant non-timetable-based analysis. Other factors, such as differences in ownership, type and extent of double track network, distance between and length of sidings, punctuality of service, dominating type of traffic (passenger vs. freight), and train configuration also affect the analysis methods and tools.

The review of over 50 past studies revealed that a majority of analyses (approximately 65 % of studies) utilized either simulation or combined simulation–analytical methods, while the remainder relied on analytical methods. Although the general simulation tools and modeling approaches have been used, most studies use commercial simulation software either in the U.S. (non-timetable based) or in Europe (timetable based). Based on the more detailed review of 25 of the studies, European capacity analysis tends to be linked to the UIC 406 method, while the U.S. does not seem to have as extensive principles as the European case studies, but the methodologies vary more from one study to another. The outcomes of European studies were also less diverse than in the U.S., and commonly suggested rescheduling and operation changes as the solutions for capacity improvement. Also the studies showed limited effort in comparing the simulation results to the actual conditions (the validation step), especially after recommended improvements were implemented. Only two studies did the full validation, 14 out of 25 only compared the results with the base model, and the remaining one-third of the studies had no validation process. Overall, it was found that there was no major divergence between approaches or criteria used for capacity evaluation in the U.S. and Europe. However, there are differences in the tools used in these two regions, as the tool designs follow the main operational philosophy of each region (timetable vs. non-timetable) and include features that concentrate on other rail network characteristics for the particular region.

References

1. FRA (2009) Vision for high-speed rail in America. Federal Railroad Administration (FRA), Washington, DC
2. Cambridge Systematics I (2007) National rail freight infrastructure capacity and investment study. Association of American Railroads (AAR), Cambridge
3. Lai Y-C, Barkan C (2009) Enhanced parametric railway capacity evaluation tool. J Transp Res Board 2117:33–40. doi:10.3141/2117-05 (National Academies, Washington, DC)
4. Tolliver D (2010) Railroad planning & design, CE 456/656 Lecture Notes. North Dakota State University
5. AREMA (2006) Manual for railway engineering. Volume 4-systems management. American Railway Engineering and Maintenance-of-Way Association, Lanham
6. Fox J, Hirsch P, Kanike O, Simpson DP, Cebula AJ, Bing AJ, Horowitz ESB (2014) Capacity modeling guidebook for shared-use passenger and freight rail operations, NCHRP Report 773. NCHRP, Washington, DC
7. UIC (2004) UIC CODE 406 R-capacity, 1st edn. International Union of Railways (UIC), Paris
8. IMPROVERAIL (2003) IMPROVEd tools for RAILway capacity and access management-deliverable 6 methods for capacity and resource management. Competitive and Sustainable Growth (GROWTH) Programme—European Commission, Brussels-Belgium
9. Sameni MK, Dingler M, Preston JM, Barkan CPL (2011) Profit-generating capacity for a freight railroad. In: TRB 90th Annual Meeting, TRB, Washington, DC
10. Alex Landex EA (2006) Evaluation of railway capacity. In: Annual Transport Conference, Aalborg University, Denmark
11. White T (2012) Comparison of U.S. and European railroads, Lecture notes for CE-4490. Michigan Tech. University-CEE Department, Houghton
12. Railway-Technical (2008) Railway statistics for Britain. http://www.railway-technical.com/statistics.shtml. Accessed May 2012
13. ERRAC (2004) A comparison of member state public research programmes with the ERRAC SRRA 2020. European Rail Research Advisory Council (ERRAC), Brussels
14. DB-AG (2011) Station categories 2012. DB-AG, Berlin-Germany (in German)
15. Amtrak (2011) Annual report fiscal year 2011. Amtrak, Washington, DC
16. Teixeira P (2009) Railway engineering principles, Technical University of Lisbon, CTIS Master's course in MIT-Portugal Program, Lisbon
17. FRA (2008) Highway-rail grade crossing safety fact sheet. FRA Public Affairs, Washington, DC
18. DOT (2011) National transportation statistics. U.S. Department of Transportation, Washington, DC
19. NetworkRail (2010) Curtail crazy driving at level crossings or risk more lives, says rail chief. http://www.networkrailmediacentre.co.uk/Press-Releases/CURTAIL-CRAZY-DRIVING-AT-LEVEL-CROSSINGS-OR-RISK-MORE-LIVES-SAYS-RAIL-CHIEF-13c1/SearchCategoryID-2.aspx.Accessed on 2012
20. Verkehrswissenschaftliches (2008) Influence of ETCS on the line capacity. International Union of Railways (UIC), Paris
21. White T (2005) Alternatives for railroad traffic simulation analysis. J Transp Res Board 1916:34–41 (TRB-Washington, DC)
22. Systra (2007) Technical classification and main technologies for the HSR systems. Complementary studies of Tehran-Qom-Isfahan HSR, Paris
23. Amtrak (2012) Monthly performance report for December 2011. AMTRAK, Washington, DC
24. Wearden G (2007) UK train times back on track. The Guardian, Friday 23 November 2007
25. Marco Luethi AN, Weidmann U (2007) Increasing railway capacity and reliability through integrated real-time rescheduling. Swiss Federal Railways, SBB AG, Zurich
26. CER (2004) Rail freight quality: meeting the challenge, A Report on the first year of the CER-UIC-CIT Freight Quality Charter. Community of European Railway and Infrastructure Companies (CER), Bruxelle, Belgium
27. Pouryousef H, Lautala P (2013) Evaluating the results and features of two capacity simulation tools on the SHARED-use corridors. In: ASME/ASCE/IEEE 2013 Joint Rail Conference (JRC2013), Knoxville, TN, USA
28. Pouryousef H, Lautala P (2014) Evaluating two capacity simulation tools on shared-use U.S. rail corridor. In: 2014 Annual Meeting of the Transportation Research Board, TRB, Washington, DC
29. Lai Y-C, Barkan C (2011) A comprehensive decision support framework for strategic railway capacity planning. J Transp Eng 137:738–749. doi:10.1061/(ASCE)TE.1943-5436.0000248 (ASCE 137 (October 2011))
30. Abril M, Barber F, Ingolotti L, Salido MA, Tormos P, Lova A (2007) An assessment of railway capacity. Technical University of Valencia, Valencia, Spain
31. Pachl J (2002) Railway operation and control. VTD rail publishing, Mountlake Terrace
32. Pavankumar Murali MMD, Ordonez F, Palmer K (2009) A delay estimation technique for single and double-track railroads. Department of Industrial & Systems Engineering, University of Southern California, Los Angeles, CA
33. Sogin SL, Barkan CPL (2012) Railroad capacity analysis. In: Railroad engineering education symposium (REES) by AREMA, Kansas City

34. Schultze GI (2012) SLS PLUS (Software Manual). Schultze + Gast Ingenieure, Berlin
35. Robert Prinz JH (2005) Implementation of UIC 406 capacity calculation method at Austrian Railway (OBB). Austrian Railway (OBB), Vienna
36. Banverket (2005) Application of the UIC capacity leaflet at Banverket. Banverket, Swedish National Rail Administration
37. Melody Khadem Sameni MD, Preston JM (2010) Revising the UIC 406 method: revenue generating capacity- JRC2010-36281. In: Joint Rail Conference (JRC) 2010, Urbana, IL, USA, 2010
38. Noble JS, Charles N (2007) Missouri freight and passenger rail capacity analysis. Missouri Department of Transportation (MoDOT), Jefferson City
39. RTC (2011) RTC online help (Manual). Berkeley Simulation Software, CA, USA
40. RMCon (2012) RailSys 8 classic, Railsys 8 Enterprise, network-wide timetable and infrastructure management. Rail management consultants GmbH (RMCon), Hannover
41. RMCon (2010) RailSys user manual. Rail management consultants GmbH (RMCon), Hannover
42. Huerlimann D, Nash AB (2012) Manual of opentrack-simulation of railway networks, Version 1.6. OpenTrack Railway Technology Ltd. and ETH Zurich, Zurich, Switzerland
43. Huerlimann D (2012) OpenTrack—simulation of railway networks. OpenTrack Railway Technology Ltd., Zurich
44. Frank O (1966) Two-way traffic in a single line of railway. Oper Res 14:801–811
45. Petersen ER (1974) Over the road transit time for a single track railway. Transp Sci 8:65–74
46. Higgins A, Kozan E (1998) Modeling train delays in urban networks. Transp Sci 32(4):251–356
47. de Kort AF, Heidergott B, Ayhan H (2003) A probabilistic approach for determining railway infrastructure capacity. Eur J Oper Res 148:644–661
48. Ghoseiri K, Szidarovszky F, Asgharpour MJ (2004) A multi-objective train scheduling model and solution. Transp Res Part B 38:927–952
49. Burdett RL, Kozan E (2006) Techniques for absolute capacity determination in railways. Transp Res Part B 40:616–632
50. Wendler E (2007) The scheduled waiting time on railway lines. Transp Res Part B 41:148–158
51. Lindner T (2011) Applicability of the analytical UIC Code 406 compression method for evaluating line and station capacity. J Rail Transp Plan Manag 1:49–57
52. Corman F, D'Ariano A, Hansen IA, Pacciarelli D (2011) Optimal multi-class rescheduling of railway traffic. J Rail Transp Plan Manag 1:14–24
53. Hansen IA, Pachl J (2008) Railway timetable & traffic, analysis, modeling, simulation, 1st edn. Eurailpress, Hamburg
54. Petersen ER, Taylor AJ (1982) A structured model for rail line simulation and optimization. Transp Sci 16:192–206
55. Kaas AH (1991) Strategic capacity analysis of networks: developing and practical use of capacity model for railway networks. ScanRail consult, Technical University of Denmark
56. Dessouky MM, Leachman RC (1995) A simulation modeling methodology for analyzing large complex rail networks. Simulation 65(2):131–142
57. Sogin SL, Barken CPL, Lai Y-C, Saat MR (2012) Impact of passenger trains in double track networks. In: 2012 Joint rail conference, Philadelphia, Pennsylvania
58. Transit Safety Management Inc.; HDR Engineering (2006) Amtrak cascades operating and infrastructure plan—technical report. Vol 1. Washington State Department of Transportation-Freight Systems Division
59. Medeossi G, Longo G, Fabris Sd (2011) A method for using stochastic blocking times to improve timetable planning. J Rail Transp Plan Manag 1:1–13
60. Brod D, Metcalf AE (2014) Web-based screening tool for shared-use rail corridors (NCFRP REPORT 27). National cooperative freight research program-Transportation research board (TRB), Washington, DC
61. Sogin SL, Barkan CPL, Lai Y-C, Saat MR (2012) Measuring the impact of additional rail traffic using highway & railroad metrics. In: 2012 Joint rail conference (JRC), Philadelphia, Pennsylvania
62. Dingler M, Lai Y-C, Barkan CPL (2009) Impact of operational practices on rail line capacity: a simulation analysis. In: 2009 Annual AREMA conference, Chicago, IL
63. Siggel S (2006) Analysis of capacity and optimization of infrastructure of the line minot (North Dakota)—Havre (Montana). Braunschweig Technical University, Braunschweig-Germany
64. Atanassov I, Dick CT, Barkan CPL (2014) Siding spacing and the incremental capacity of the transition from single to double track. In: 2014 Joint rail conference (JRC), Colorado Springs, CO, USA
65. Shih M-C, Dick CT, Sogin SL, Barkan CPL (2014) Comparison of capacity expansion strategies for single-track railway lines with sparse sidings. In: TRB 2014 Annual meeting, Washington, DC, USA
66. Schlechte T, Borndörfer R, Erol B, Graffagnino T, Swarat E (2011) Micro–macro transformation of railway networks. J Rail Transp Plan Manag 1:38–48
67. Gille A, Siefer T (2013) Sophisticated capacity determination using simulation. In: Transportation research board 2013 Annual meeting, Washington, DC, USA
68. Medeossi G, Longo G (2013) An approach for calibrating and validating the simulation of complex rail networks. In: TRB 2013 Annual meeting, Washington, DC, USA
69. Sipila H (2014) Evaluation of single track timetables using simulation. In: 2014 Joint rail conference (JRC), Colorado Springs, CO, USA
70. Goverde RMP, Corman F, D'Ariano (2014) A investigation on the capacity consumption at Dutch railways for various signalling technologies and traffic conditions. In: TRB 2014 Annual Meeting, Washington, DC, USA

10

Can a polycentric structure affect travel behaviour? A comparison of Melbourne, Australia and Riyadh, Saudi Arabia

M. Alqhatani · S. Setunge · S. Mirodpour

Abstract This study models the impact of the shift from a monocentric private-car-oriented city to polycentric public-transport-oriented city. Metropolitan areas have suffered traffic problems—in particular increase in travel time and travel distance. Urban expansion, population growth and road network development have led to urban sprawl in monocentric cities. In many monocentric cities, travel time and distance has steadily increased and is only expected to increase in the future. Excessive travel leads to several problems such as air pollution, noise, congestion, reduction in productive time, greenhouse emissions, and increased stress and accident rates. This study examines the interaction of land use and travel. A model was developed and calibrated to Melbourne and Riyadh conditions and used for scenario analysis. This model included two parts: a spatial model and a transport model. The scenario analysis included variations of residential and activity distribution, as well as conditions of public transport service.

Keywords Monocentric · Polycentric · Private mode · Public transport · Four-step transport modelling

1 Introduction

Since the end of World War II, economic growth and advancement in transport technologies have resulted in rapid urbanisation. This rapid urbanisation has promoted the shift from compact monocentric city to urban sprawl which has caused not only traffic congestion problems, but also longer trip distances, increased trip times and traffic accidents [1].

Many researchers have suggested that monocentric urban structures fail to optimise existing transport network utilisation. CBD workers arrive at similar times each day generating an inward commuting flow from the outer suburbs during morning peak hours and an outward commuting flow during evening peak hours [2]. Decentralisation of employment can be made possible by re-organisation of the suburban structure, by shifting from single core city centre to multiple suburban activity centres (employment, shopping, recreation, etc.) located in the periphery of the city.

These suburban activity centres become strong alternatives to the CBD, potentially combining the advantage of sprawl locations (low density, lower land price and less traffic congestion) with the advantages of subcentre locations (economy, urbanisation and personal interaction) which are connected by a good transport system [3, 4].

In Australia, there is much interest in encouraging a change in urban structure because it delivers significant transport improvements, particularly public transport. All major cities in Australia have developed spatial plans that encourage transit-oriented development and in-fill development (the so-called 'urban consolidation') [2].

These transport and land use policies look to bring residences closer to public transport and to key activity centres, in an attempt to improve public transport share and to respond to concerns about traffic congestion [5]. However, there are other approaches towards the mix of activity transport and land use, and interest in employment decentralisation has been encouraged in Melbourne and Riyadh by the future master plan 2030.

M. Alqhatani (✉) · S. Mirodpour
School of Civil, Environmental and Chemical Engineering, RMIT University, GPO Box 2476, Melbourne, VIC 3001, Australia
e-mail: mohammed.alqhatani@student.rmit.edu.au

S. Setunge
Ministry of Higher Education, Riyadh, Saudi Arabia

In Melbourne, activity centres were identified and classified into the following categories: 25 principal activity centres, 79 major activity centres and 10 specialised activity centres [6]. The latest focus advocates a polycentric city with new six subcentres in addition to the CBD areas, called Central Activity Districts, such as Box Hill, Broadmeadows, Footscray, Frankston and Ringwood [7]. In contrast, Riyadh's land use and transport polices have adopted a polycentric model, identifying six subcentres with traditional centres in the future master plan 2030 [8].

While the benefits of polycentric urban structure over monocentric urban structure have been well examined and qualitatively reported on in the literature, examination of the shift from monocentric to polycentric urban structure and quantification of its effects on travel remain a key gap in the existing literature on this topic. This study endeavours to discover how this shift reduces travel.

In conjunction with changes to urban structure, public transport has been proposed as a key solution that can decrease congestion and trip length, and urban form (population and employment distribution) is recognised as a useful way to reduce trip length. Some cities, such as Riyadh, have so far limited public transport service. However, Riyadh is also proposing to restructure to a polycentric city. It is, therefore, useful to examine the significance of a public transport system for a polycentric structure urban policy.

Riyadh and Melbourne have been selected as case studies. Riyadh and Melbourne are similar in urban form and population number, yet different in transport systems. Private mode is predominantly used in Riyadh and Melbourne. However, Melbourne has a good public transport system while Riyadh is constrained by the limited scale of its public transport network.

The objective of this paper is to quantify the impact of polycentric urban structure policy in Melbourne and Riyadh, as well as the significance of public transport in supporting the polycentric urban structure.

2 Literature review

It has been understood that monocentric employment structures lead to long travel times for commuters, as well as strong flows in one direction. Concentration of flows into a small area creates traffic congestion for all modes during peak times. On the other hand, polycentric structures are linked to decreased travel times and distance, through a better mix and balance of employment and residential areas [2].

2.1 Polycentric structure examples

In cities of Australia and New Zealand, residential dispersion has occurred but this has not led to a similar form of employment dispersion. This is partly due to the development of white-collar office jobs, as seen in the Melbourne CBD, where between 1996 and 2006 100,000 new white-collar jobs were created (Mees et al. [10]).

The overcentralisation of office employment in Australian CBD areas has been linked to significant levels of long travel distance and commuting time, traffic congestion, poor balance between employment–housing structures, significant levels of subsidy for public transport and excessive costs for office leases. Urban traffic congestion has grown significantly, and is recognised as a key policy area and optimal growth and decentralisation policies have been proposed as a response to these issues [2].

Review of research into employment decentralisation policies and effect on transport has been developed [2]. Transport outcomes are different in different cities, with decentralisation to rail-based localities in Japan and Singapore being extremely successful in decreasing traffic congestion and increasing the performance of transit systems. Additionally, decentralisation in London over the 1960s–1970s has also decreased traveller flows into the CBD.

In New South Wales (NSW), the government has promoted a secondary centre in Parramatta. Parramatta started with 10,000 jobs in 1971 and reached 40,000 jobs by 2005, and Parramatta contributed to reduction in travel time. In a recent master plan, the NSW government (City of Cities) proposed to strengthen two new regional subcentres which are Liverpool and Penrith [2].

However, in San Francisco, employment decentralisation has had little impact on reduction of mean distances or travel times for commuters, with a lack of development of self-containment. Within Stockholm, the development of urban fringe residential areas for workers has led to increased mean commuting distances, and ultimately led to higher staff turnover [2].

While there appears to be positive outcomes to decentralisation, there have been negative outcomes as well. Little is known about how long it takes workers and households to transition to new spatial arrangements after a workplace relocation, say, by moving house, changing to a job located elsewhere or changing schools for children. There is a risk that decentralisation can lead to higher proportion of suburb-to-suburb commutes which may entail increases in car use as more workers travel farther [9].

2.2 Polycentric structure of Melbourne and Riyadh

The polycentric urban structure policy of Melbourne and Riyadh need to be better understood based on the specific context of these two cities. Most of the studies conducted for Melbourne and Riyadh have been qualitative rather than quantitative studies. In Melbourne, the Department of

Planning and Community Development [11] mentioned that shifting from one main centre to more alternative centres has a similar CBD area function, which will reduce congestion and allow residents to spend less time travelling to and from work. Meanwhile, in Riyadh, the Arriyadh Development Authority (ADA) [8] mentioned that 'sub-centres will create the opportunity for a balance between work and residence in the new growth sectors of city suburbs. The sub-centres will have significant effects to the pattern of the daily trips between the suburbs and the city centre by creating new workplaces. This also will reduce the total of number daily trip, because may residents will be living and working in the same section of the city and therefore the movement will be directed away from the city centre area, instead to the current pattern of going to the city centre. There has been no clear view about how this shift's dynamic process works.

This study aims to contribute to the better understanding of the impact of polycentric structure. A model analysis was conducted to quantify the impact of the shift from monocentric structure to polycentric structure in Melbourne and Riyadh. The analysis focusses on the impact of variants in activity and residential redistribution as shown in Fig. 1.

3 Methodology

3.1 Data collection

One of the major tasks in this study was to collect data used to develop a land use/transport model. Data collection was divided into two groups: land use and demographic information, and transport data. In Riyadh, the demographic and population data were collected by the Municipality of Riyadh (MOR) in 2008; however, the land use data were conducted by the ADA in 2002.

In Melbourne, the demographic and population data were collected by the Department of Transport of Victoria in 2008; however, the land use data were collected by the Local Municipalities Councils (2009).

The explanatory variables applied to the model are based on data availability. The variables were formed in relation to the socioeconomic characteristics of households, demographic data, land use and urban form. The data for Riyadh were sourced from the 2008 survey dataset, created by the MOR; and for Melbourne the data were sourced from the Department of Transport (DOT), Melbourne, Victoria. The data were within the city's Traffic Analysis Zones (TAZs), with Riyadh featuring 2,166 TAZs and Melbourne having 2,253 TAZs.

Riyadh's trip data were obtained from the 2008 MOR report, while for Melbourne they were sourced from the 2008 DOT data. The data focused on trip purposes, including home-based work (HBW); home-based education including primary, secondary and high school; home-based recreation; home-based shopping; home-based other; and non-home based.

The origin and destination (OD) trip data for Riyadh were also sourced from MOR (2008) and for Melbourne they were similarly sourced from DOT (2008). Trip data are available for the AM peak period (2 h) in Riyadh, this period was measured across 2 h in Melbourne, and the OD trip matrix was organised by car, public transport and walking modes. Trip distance is calculated as the shortest distance between two centroid points which were sourced

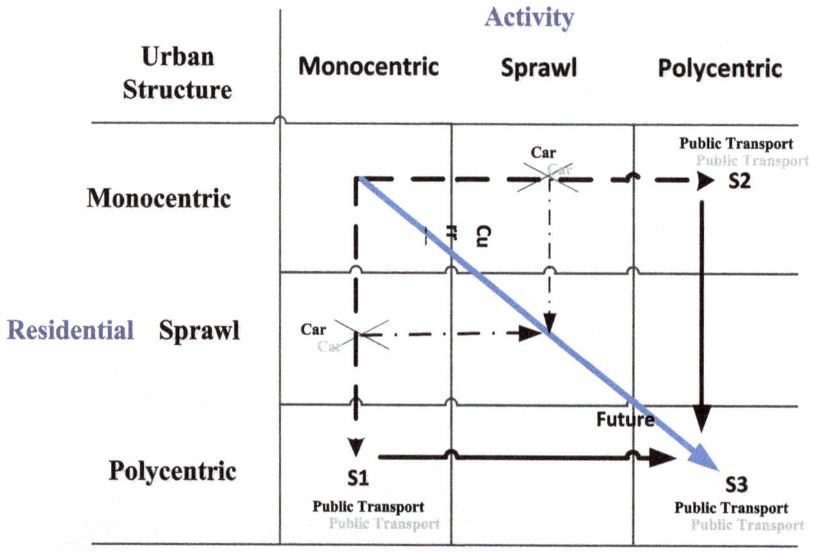

Fig. 1 Urban structure and population distribution scenarios process

by MOR and DOT in 2008 in Riyadh and Melbourne, respectively.

At present, Riyadh has no tram and train service but it has been designed to be used in the upcoming years. Public transport data have been collected from MOR (2008). Also, the data of walk trip distance were extracted from Koushki's study (1989) and used to predict the walk trip mode share in Riyadh.

Socioeconomic data for Melbourne were applied at the TAZ level and used as an independent variable of the models in this study, and the socioeconomic variables included the number of employed residents, the number of households, age group (0–17, 18–64, and 65+), the number of jobs and the number of students.

Finally, land use data were also applied as an independent variable, with independent variables organised into two categories. Firstly, socioeconomic variables and land use variables were broken into building gross floor area (GFA), and secondly the number of buildings and activities in each zone was measured. The GFA for Riyadh was sourced from the ADA (2002), and for Melbourne it was sourced from the Local Municipalities Councils (2009). The number of buildings and activities in each zone measured the amount of retail, shopping areas and universities, with the last measure providing strong analysis for trip attraction in Riyadh. Land use data for Melbourne were not readily available for this study and were not used.

3.2 Land use/transport interaction (LUTI) modelling

This section describes the land use model and transport model used in the analysis.

3.2.1 Land use model

The land use model is a spatial model which is based on several input variables such as population, employment, housing and businesses. The interaction of spatial settlement is created by the attractiveness of the spatial cells. The interaction between the spatial model and the transport network can be calculated by the accessibility indices, which describe accessibility of different regions in the city, in turn representing access to employment, shopping and recreational facilities. Geographic Information System was used for this purpose.

3.2.2 Transport model

A four-step transport model which models trip generation, distribution, choice of mode and traffic assignment was used to model transport network performance (Fig. 2).

The first module is trip generation, which makes use of land use and socioeconomic data, such as demographic and population data, to determine the number of trips produced by and attracted to traffic zones. The second module is trip distribution which determines the OD of trips that have been estimated in the first module. The third module, model split, organises the trip into different modes of transport (i.e. private mode, train, tram, bus, cycling and walking). The fourth module is traffic assignment which allocates trips to different modes in the transportation network [12]. This study did not involve traffic assignment due to limitations in resources and data. Base year travel time and cost estimates were assumed in this study.

3.2.3 Model calibration and validation

The model calibration process refers to an estimate of the model parameters to fit the model results to a set of observed data, while the validation process refers to an evaluation of the results of the model outputs using the calibrated model parameters compared to the observed outcomes. In this study, part of the land use transport datasets were used for calibration and the remainder for validation. The data were divided into two groups: 80 % for calibration and 20 % for validation by applying cross-classification method because past data for both the cities were unavailable.

In Melbourne, the following Table 1 compares Trips, VKT and mode share estimated by the model and estimated based on the available data. The model was moderately accurate. The model was considered adequate for the purpose of this study given the resources available. In the case of Riyadh meanly all trips are made by cars with limited PT services at that amount. In the future PT will be improved and some trips will use PT.

Figures 3, 4 and 5 show the trip length distribution based on the model and available data.

3.2.4 Model application

The model was then applied to scenario analysis. This analysis examined both future HBW and NHBW trips. The explanatory variables for HBW and NHBW for Melbourne and Riyadh are as follows (Table 2).

Four scenarios were set for the scale and distribution activity and residential areas, as follows (Fig. 6):

Scenario 0 exhibits monocentric structure (existing structure) without change in structure and network.

Scenarios 1, 2 and 3 exhibit polycentric structures with variations in the redistribution of employment and residences, as follows:

- Scenario 1 has 5 new subcentres with redistribution of population around the new subcentres.
- Scenario 2 has 5 new subcentres with redistribution of employment around the new subcentres.

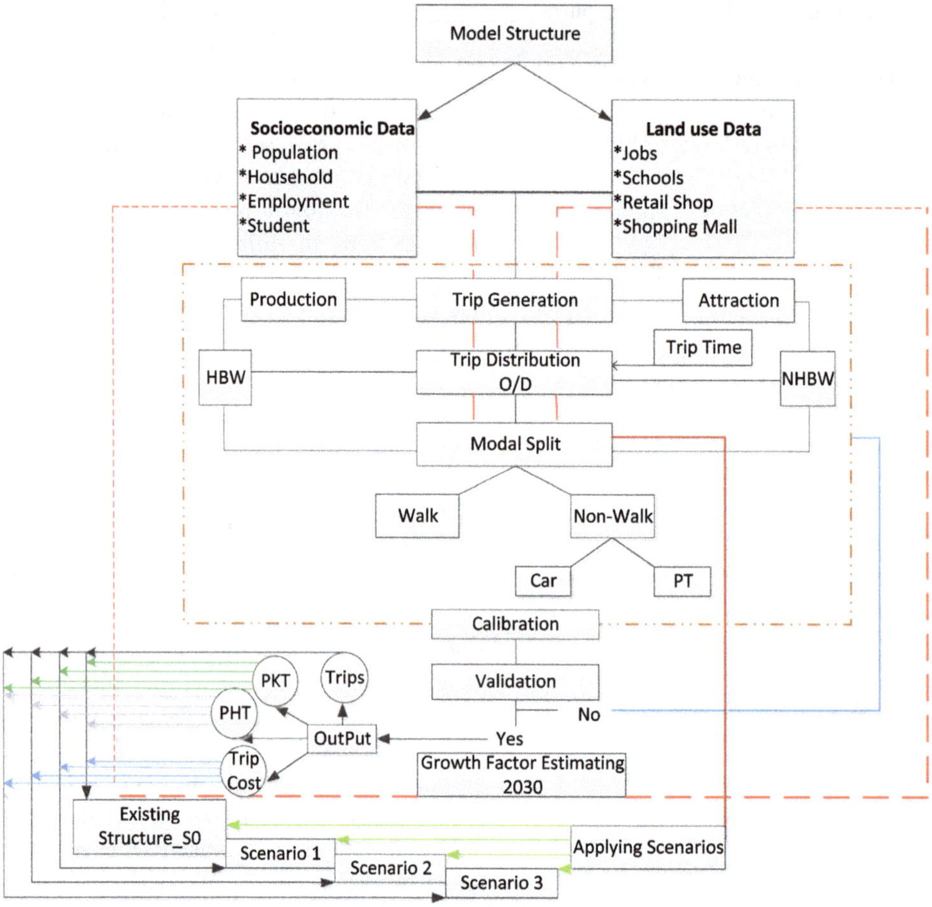

Fig. 2 Land use and transport model interaction process

Table 1 The comparison of VKT and mode share between observed data and model

Indicator	Melbourne model	
	Estimated from available data	Estimated by model
VKT	25,530,322	30,075,288
Mode share		
Car	1,996,470	1,769,266
Transit	270,884	493,184
Walk	171,469	171,911

- Scenario 3 has 5 new subcentres with redistribution of both population and employment around the new subcentres (see Fig. 2).

Additionally, different transport network structures and their impact on the traffic congestion, as well as spatial fragmentation, and trip times and distances were analysed.

Scenario 0 is paternal after historical growth from 2008 year conditions, while Scenarios 1, 2 and 3 were formulated based on varying redistribution patterns. The base year was set as 2008 and the future analysis year was set as

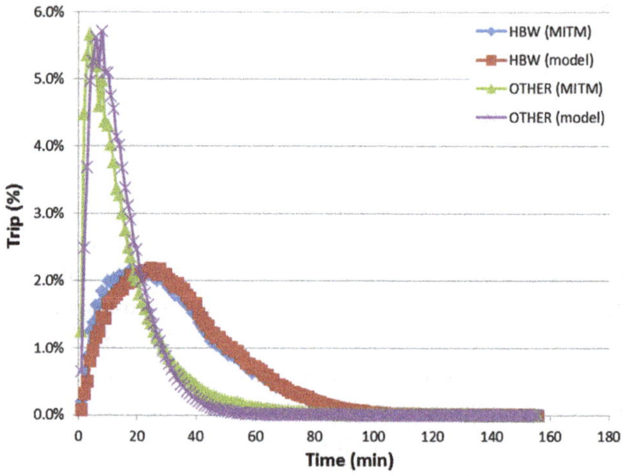

Fig. 3 Distribution of trip length in Melbourne between the model and observed data

2030. For redistributing activity or residence, it was assumed that 7.5 % at the incremental in activity/residence will add to the CBD, while the remaining 92.5 % will be shared amongst the new subcentres. The area outside the CBD or subcentres will remain as 2008 (see Tables 3 and 4).

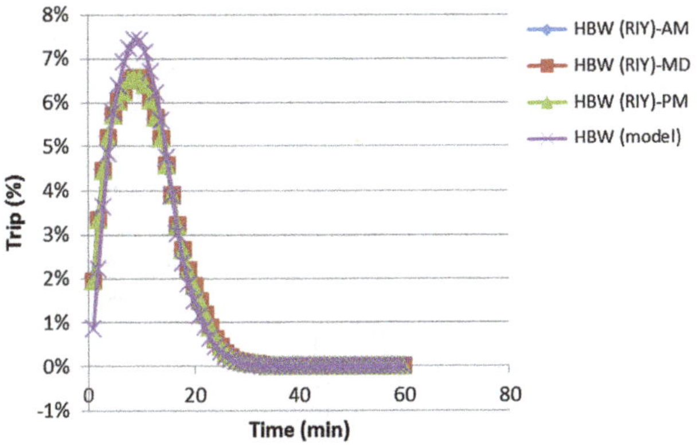

Fig. 4 Distribution of HBW trip length in Riyadh between the model and observed data

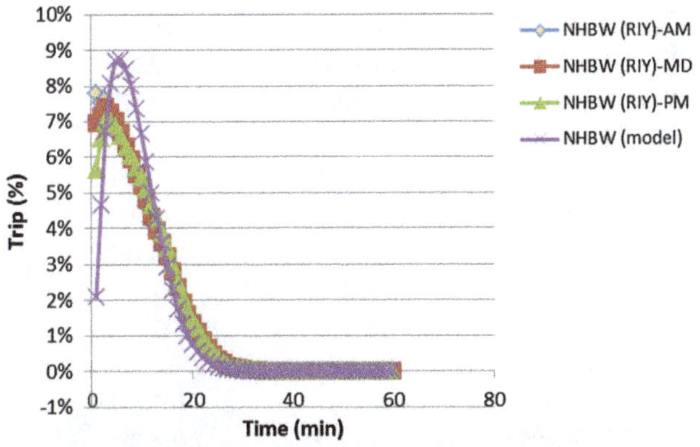

Fig. 5 Distribution of NHBW trip length in Riyadh between the model and observed data

Table 2 The explanatory variables which are used for Melbourne and Riyadh

Trip purpose	Explanatory variables for Melbourne	Explanatory variables for Riyadh
HBW	Number of workers	Number of workers
	Number of households	Number of households
	Number of jobs	Number of jobs
NHBW	Number of workers	Number of workers
	Number of households	Number of households
	Number of student	Number of student
	Number of jobs	Number of jobs
	Number of students in school	Number of students in school
	Number of retail shops	Number of retail shops
		Number of malls
		Number of Students in Universities

The growth in total population or activity was based on historical trends as sourced from ADA in Riyadh and the Australia Bureau of Statistics in Melbourne.

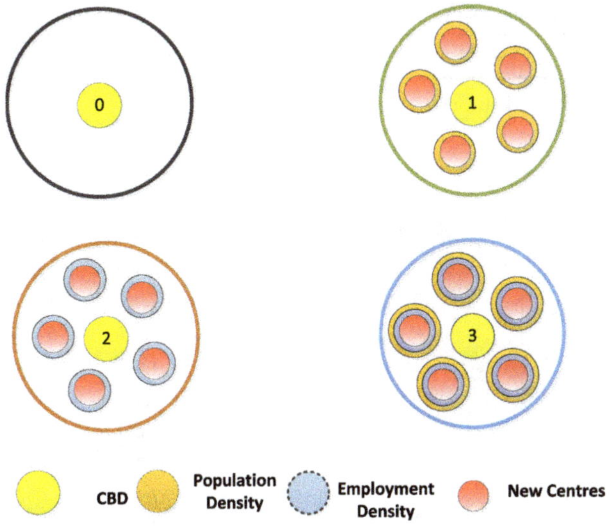

Fig. 6 Four scenarios in different structures

4 Model result

The comparative modelling suggests significant changes in travel behaviour under idealised decentralisation scenarios in both Riyadh and Melbourne.

4.1 Trip rates

4.1.1 HBW purpose

In Melbourne, car trip rates declined in scenarios 1 and 3, and not in scenario 2. PT trip rates decreased in scenarios 2 and 3; however, it increased in scenario 1. Walk trip rates increased in scenarios 1 and 3, yet there was a slight decrease in scenario 2 (Table 5).

In Riyadh, car trip rates declined in all scenarios. Scenario 3 decreased to a greater degree compared to scenarios 1 and 2. PT trip rates increased in scenarios 1, 2 and 3. Walk trip rates increased in all scenarios. Scenario 3 had the highest walk trip rates compared to all scenarios (Table 6).

4.1.2 NHBW purpose

In Melbourne, car trip rates declined in all scenarios. PT trip rates declined in scenario 2, and increased slightly in scenarios 1 and 3. Walk trip rates increased in all scenarios (Table 5).

In Riyadh, car trip rates declined in all scenarios. PT trip rates declined in scenarios 2 and 3, and increased slightly in scenario 1. Walk trip rates increased in all scenarios; scenario 3 had the highest walk trip rate (Table 6).

Table 3 The distribution of residential and activities into six centres by different scenarios in Riyadh

Factors	Variables	Centre	2008	2030 (Scenario 0)	Scenario 1 2030	Scenario 2 2030	Scenario 3 2030
Residential	Households	CBD	2,096	2,351	*250,837*	2,351	*250,837*
		1	1,744	78,513	*506,069*	78,513	*506,069*
		2	3,089	225,178	*507,414*	225,178	506,069
		3	2,268	43,460	*506,593*	43,460	*506,593*
		4	917	22,825	*505,242*	22,825	*505,242*
		5	2,507	22,451	*506,832*	22,451	*506,832*
		Other	763,055	3,151,264	*763,055*	3,151,264	*763,055*
		Total	775,676	3,546,042	*3,546,042*	3,546,042	*3,546,042*
	Workers	CBD	4,884	5,478	*584,701*	5,478	*584,701*
		1	4,070	183,181	*1,179,653*	183,181	*1,179,653*
		2	7,207	525,360	*1,182,790*	525,360	*1,182,790*
		3	5,290	101,379	*1,180,873*	101,379	*1,180,873*
		4	2,143	53,342	*1,177,726*	53,342	*1,177,726*
		5	5,846	52,347	*1,181,429*	52,347	*1,181,429*
		Other	1,779,119	7,345,202	*1,779,119*	7,345,202	*1,779,119*
		Total	1,808,559	8,266,288	*8,266,288*	8,266,288	*8,266,288*
	Students in residences	CBD	2,960	3,320	*384,874*	3,320	*384,874*
		1	3,459	155,681	*778,753*	155,681	*778,753*
		2	2,457	179,105	*777,751*	179,105	*777,751*
		3	1,663	31,870	*776,957*	31,870	*776,957*
		4	903	22,477	*776,197*	22,477	*776,197*
		5	5,945	53,233	*781,239*	53,233	*781,239*
		Other	1,782,550	6,803,567	*6,803,567*	1,782,550	*1,782,550*
		Total	1,808,387	7,401,914	*7,401,914*	7,401,914	*7,401,914*
Activities	Jobs	CBD	4,830	5,417	5,417	*507,841*	*507,841*
		1	682	30,695	30,695	*1,018,785*	*1,018,785*
		2	1,058	77,124	77,124	*1,019,161*	*1,019,161*
		3	10,854	208,008	208,008	*1,028,957*	*1,028,957*
		4	1,366	34,002	34,002	*1,019,469*	*1,019,469*
		5	7,047	243,101	243,101	*1,025,150*	*1,025,150*
		Other	1,782,550	6,803,567	6,803,567	*1,782,550*	*1,782,550*
		Total	1,808,387	7,401,914	7,401,914	*7,401,914*	*7,401,914*
	Students in schools	CBD	1,087	1,219	1,219	*341,717*	*341,717*
		1	4,646	209,105	209,105	*695,367*	*695,367*
		2	429	31,272	31,272	*691,150*	*691,150*
		3	1,286	24,645	24,645	*692,007*	*692,007*
		4	1,708	42,515	42,515	*692,429*	*692,429*
		5	3,381	30,274	30,274	*694,102*	*694,102*
		Other	1,239,874	4,707,615	4,707,615	*1,239,874*	*1,239,874*
		Total	1,252,411	5,046,646	5,046,646	*5,046,646*	*5,046,646*
	No. of retail shops	CBD	378	424	424	*19,125*	*19,125*
		1	61	2,745	2,745	*38,058*	*38,058*
		2	21	1,531	1,531	*38,018*	*38,018*
		3	65	1,246	1,246	*38,062*	*38,062*
		4	17	423	423	*38,014*	*38,014*
		5	217	7,943	7,943	*38,214*	*38,214*
		Other	66,561	261,739	261,739	*66,561*	*66,561*
		Total	67,320	276,051	276,051	*276,051*	*276,051*

Italics indicate significant change to Scenario 0

Table 4 The distribution of residential and activities into six centres by different scenarios in Melbourne

Factors	Variable	Centre	2008	2030 (Scenario 0)	Scenario 1 2030	Scenario 2 2030	Scenario 3 2030
Residential	Households	CBD	9,936	12,859	*54,090*	12,859	*54,090*
		1	5,862	7,586	*114,775*	7,586	*114,775*
		2	3,154	5,821	*112,067*	5,821	*112,067*
		3	10,627	13,754	*119,541*	13,754	*119,541*
		4	15,455	20,002	*124,369*	20,002	*124,369*
		5	24,484	57,460	*133,397*	57,460	*133,397*
		Other	1,357,873	1,898,631	*1,357,874*	1,898,631	*1,357,874*
		Total	1,427,391	2,016,113	*2,016,113*	2,016,113	*2,016,113*
	Workers	CBD	10,818	14,001	*67,228*	14,001	*67,228*
		1	6,517	8,434	*145,661*	8,434	*145,661*
		2	2,367	4,368	*141,511*	4,368	*141,511*
		3	12,845	16,625	*151,990*	16,625	*151,990*
		4	19,595	25,360	*158,740*	25,360	*158,740*
		5	27,477	64,486	*166,622*	64,486	*166,622*
		Other	1,767,762	2,466,240	*1,767,762*	2,466,240	*1,767,762*
		Total	1,847,381	2,599,514	*2,599,514*	2,599,514	*2,599,514*
	Student in residences	CBD	947	1,226	*26,449*	1,226	*26,449*
		1	2,033	2,631	*65,548*	2,631	*65,548*
		2	2,651	4,893	*65,572*	4,893	*65,572*
		3	4,851	6,278	*67,768*	6,278	*67,768*
		4	7,778	10,067	*70,694*	10,067	*70,694*
		5	12,045	28,268	*74,964*	28,268	*74,964*
		Other	764,020.4	1,081,081.2	*763,449.2*	1,081,081.2	*763,449.2*
		Total	794,325.4	1,134,444.2	*1,134,444.2*	1,134,444.2	*1,134,444.2*
Activities	Jobs	CBD	219,548	280,283	280,283	*277,229*	*277,229*
		1	13,637	18,354	18,354	*160,739*	*160,739*
		2	8,249	13,008	13,008	*150,662*	*150,662*
		3	20,333	26,150	26,150	*162,714*	*162,714*
		4	21,409	27,534	27,534	*163,781*	*163,781*
		5	23,832	44,978	44,978	*166,234*	*166,234*
		Other	1,548,321	2,214,875	2,214,875	*1,543,823*	*1,543,823*
		Total	1,855,329	2,625,182	2,625,182	*2,625,182*	2,625,182
	Student in school	CBD	63,056	81,610	81,610	*96,218*	*96,218*
		1	20,536	26,579	26,579	*108,510*	*108,510*
		2	20,396	37,645	37,645	*102,362*	*102,362*
		3	15,984	20,687	20,687	*97,909*	*97,909*
		4	10,534	13,633	13,633	*92,449*	*92,449*
		5	30,472	71,514	71,514	*112,424*	*112,424*
		Other	943,266	1,295,707	1,295,707	*937,503*	*937,503*
		Total	1,104,244	1,547,375	1,547,375	*1,547,375*	*1,547,375*
	No. of retail shops	CBD	15,607	20,199	20,199	*23,407*	*23,407*
		1	1,822	2,358	2,358	*21,611*	*21,611*
		2	1,468	2,709	2,709	*20,724*	*20,724*
		3	2,340	3,028	3,028	*21,592*	*21,592*
		4	5,327	6,894	6,894	*24,578*	*24,578*
		5	6,215	14,587	14,587	*25,470*	*25,470*
		Other	229,663	316,760	316,760	*229,153*	*229,153*
		Total	262,442	366,535	366,535	*366,535*	*366,535*

Italics indicate significant change to Scenario 0

Table 5 Comparison of trip by modes between baseline 2030 and polycentric scenarios in Melbourne

	S1 Change % vs. S0	S2 Change % vs S0	S3 Change % vs S0
HBW			
Car trip	−1.23	2.34	−0.71
PT trip	1.91	−7.71	−6.19
Walk trip	17.93	−5.09	59.24
NHBW			
Car trip	−2.85	−0.13	−4.42
PT trip	6.37	−0.21	5.80
Walk trip	6.86	1.26	15.60

Table 6 Comparison of trip by modes between baseline 2030 and polycentric scenarios in Riyadh

	S1 Change % vs. S0	S2 Change % vs S0	S3 Change % vs S0
HBW			
Car trip	−7.83	−3.35	−20.32
PT trip	12.35	5.63	9.98
Walk trip	29.55	19.17	89.05
NHBW			
Car trip	−3.04	−5.01	−16.25
PT trip	4.02	−0.13	−1.07
Walk trip	10.28	59.52	63.03

4.2 Mode share

4.2.1 HBW purpose

In Melbourne, car mode share declined in scenarios 1 and 3, compared to the scenario 0. However, scenario 2 had a higher rate of increase compared to all scenarios. PT share decreased in scenarios 2 and 3; however, PT share of scenario 1 increased compared to scenario 0. Walk trip increased in scenarios 1 and 3; however, scenario 2 was similar to scenario 0 (Table 7).

In Riyadh, all scenarios resulted in a slight decrease in car mode share. PT mode increased in all scenarios. Walk mode share increased in all scenarios, and scenario 3 had the highest increase (Table 8).

4.2.2 NHBW purpose

In Melbourne, car trip share declined in all scenarios compared to baseline year 2030. Scenario 3 had a larger decrease compared to scenarios 1 and 2. PT mode share increased in all scenarios. Also, the walk trip share increased in all scenarios compared to scenario 0. Scenario 3 PT mode share was higher than that of scenarios 1 and 2.

Riyadh's result in mode share was similar to Melbourne's in terms of the decline in car mode share and increase in walk share. However, the PT share increased in scenario 1 and declined to a degree in scenario 3. In scenario 2 there was no change compared to scenario 0.

4.3 Trip distance

Table 1 shows the comparative analysis of trip distance between scenarios by HBW and NHBW purposes in both Melbourne and Riyadh.

4.3.1 HBW purpose

In the case of Melbourne, there was a reduction in car and PT trip distance for scenarios 1, 2 and 3. There was an increase in walk trips distance (Table 9).

For Riyadh, there was a decrease in trip distance by car, but there was an increase in trip distance by PT and walk (Table 10).

4.3.2 NHBW purpose

Similar to HBW, there was a reduction in car and PT trip distance for scenarios 1, 2 and 3 there was an increase in walk trips distance in the case of Melbourne (Table 9).

For Riyadh, there was decrease in trip distance by car, but there was an increase in trip distance by PT and walk for scenarios 1 and 2. In the case of scenario 3, there was a decrease in trips distance for all modes (car, PT and walk) (Table 10).

4.4 Travel time

This section will display the trip time based on the different scenarios in both Riyadh and Melbourne for HBW and NHBW trip purpose.

4.4.1 HBW purpose

Table 11 shows that travel time by car has declined in all scenarios except scenario 2 for HBW purpose, compared to scenario 0. PT travel time decreased in all scenarios. Walk travel time increased in all scenarios.

In the case of Riyadh, car travel time decreased in all scenarios, particularly scenario 3. PT travel time increased in scenario 1 and 2, but decreased in scenario 3. Walk travel time increased in all three scenarios.

Table 7 Comparison of mode shares for baseline 2030 and polycentric scenarios in Melbourne

	2008 (%)	Scenario 0 (%)	Scenario 1 (%)	Scenario 2 (%)	Scenario 3 (%)
HBW					
Car share	84.0	74.2	73.5	76.0	73.8
PT share	14.8	24.5	25.0	22.8	23.1
Walk share	1.3	1.2	1.5	1.2	3.0
NHBW					
Car share	81.0	71.9	69.9	71.8	68.8
PT share	9.6	17.5	18.7	17.5	18.6
Walk share	9.4	10.6	11.4	10.8	12.6

Table 8 Comparison of mode shares for baseline 2030 and polycentric scenarios in Riyadh

	2008 (%)	Scenario 0 (%)	Scenario 1 (%)	Scenario 2 (%)	Scenario 3 (%)
HBW					
Car share	99.3	67.4	62.6	65.3	56.1
PT share	–	31.6	36.1	33.5	35.2
Walk share	0.7	1.0	1.4	1.2	8.8
NHBW					
Car share	97.1	64.0	62.1	60.8	55.1
PT share	–	30.6	31.9	30.6	30.3
Walk share	2.9	5.4	6.0	8.6	14.5

Table 9 Comparison of transport performance between baseline 2030 and polycentric scenarios in Melbourne

	S1 Change % vs. S0	S2 Change % vs S0	S3 Change % vs S0
HBW			
Car PKT	−5.56	−1.41	−12.69
PT PKT	−5.93	−13.44	−23.54
Walk PKT	24.24	4.37	56.65
NHBW			
Car PKT	−8.61	−1.35	−14.40
PT PKT	−0.62	−1.01	−4.55
Walk PKT	22.48	4.91	29.87

Table 10 Comparison of transport performance between baseline 2030 and polycentric scenarios in Riyadh

	S1 Change % vs. S0	S2 Change % vs S0	S3 Change % vs S0
HBW			
Car PKT	−6.19	−1.60	−27.75
PT PKT	10.72	12.76	4.07
Walk PKT	26.07	5.07	83.46
NHBW			
Car PKT	−2.22	−3.27	−39.44
PT PKT	4.26	7.96	−11.52
Walk PKT	1.37	10.43	−27.02

4.4.2 NHBW purpose

In Melbourne, car travel time declined in all scenarios compared to scenario 0. PT travel time also decreased in all scenarios. Walk travel time increased in all three scenarios.

In Riyadh, Table 12 shows that car travel time decreased in all scenarios, particularly scenario 3. PT travel time changed slightly in scenarios 1 and 2 but decreased notably in scenario 3. Walk travel time increased in scenarios 1 and 2, but decreased in scenario 3.

This means that all services such as school, medical and retail lay close together, which assisted in reducing PT trip time. Walk trip time improved, particularly in scenario 3. This confirmed our expectation explained above, which is that self-containment leads to increasing walk trip. In Riyadh, car trip time declined in all scenarios; however, scenario 3 had a higher rate of decline in car trip time compared to scenarios 1 and 2. Also, this means that the provision of services may have helped in the reduction of car trip time. PT trip time fell in scenarios 2 and 3. The latter had a higher rate of reduction due to the

Table 11 Comparison of transport performance by hours between baseline 2030 and polycentric scenarios in Melbourne

	S1 Change % vs. S0	S2 Change % vs S0	S3 Change % vs S0
HBW			
Car PHT	−3.24	0.91	−5.37
PT PHT	−2.22	−3.88	−6.95
Walk PHT	0.11	0.02	0.50
NHBW			
Car PHT	−3.58	−0.16	−5.86
PT PHT	−0.21	−0.44	−1.05
Walk PHT	0.94	0.16	1.42

Table 12 Comparison of transport performance by hours between baseline 2030 and polycentric scenarios in Riyadh

	S1 Change % vs. S0	S2 Change % vs S0	S3 Change % vs S0
HBW			
Car PHT	−5.72	−1.27	−20.15
PT PHT	2.36	1.14	−1.87
Walk PHT	0.35	0.05	5.72
NHBW			
Car PHT	−1.79	−2.60	−26.23
PT PHT	0.59	−0.83	−7.30
Walk PHT	0.12	0.99	−2.39

concentration of all services near the residential structure. Walk trip time would increase slightly in scenarios 1 and 2, at 0.12 and 0.99, respectively. However, scenario 3 fell slightly at −2.39. This means that some services in and around suburban activity centres mean that many more people will be able to walk less than 2 km.

5 Conclusion

In this paper, we have reported on comparative analysis through modelling that investigated the shift from monocentric to polycentric structure, and from private mode- to PT mode-oriented city for Riyadh and Melbourne. The results indicate that planned and concentrated employment and population in key activity centres may deliver significant benefits to reducing car trip distance. The findings of the combined and coordinated redistribution of activity and residences would achieve the best possible transport outcome, with regards to reducing car trips, car mode share, car travel distance and car travel. It also reduced travel consumption in general, including PT travel. It also promoted walk trips.

The finding of the study pointed that combination and coordination of activity and residences redistribution into polycentric structure for Melbourne and Riyadh will bring about significant benefits and will play a key role in achieving a more sustainable transport outcome.

It is recommended that urban restructure polices should focus on both activity and residence re-alignment.

This study did not include network assignment. Further research is needed to examine in detail the impact of urban restructure with consideration to the capacity of the actual network.

References

1. Weisbrod G, Vary D, Treyz G (2003) Measuring economic costs of urban traffic congestion to business. Transp Res Record 1839(1):98–106
2. Burke M, Dodson J, and Gleeson B (2010) Employment decentralisation in South East Queensland: scoping the transport impacts. Research paper, 2010. 29
3. Anas A, Arnott R, Small KA (1998) Urban spatial structure. J Econ Lit 36(3):1426–1464
4. McMillen DP, McDonald JF (1998) Population density in suburban Chicago: a bid-rent approach. Urban Stud 35(7):1119–1130
5. Cervero R (2004) Transit-oriented development in the United States: Experiences, challenges, and prospects. Transp Res Board 102:3–12
6. DPCD (2007) Melbourne 2030 Audit analysis of progress and findings from the 2006 Census Melbourne. Government of Victoria. Department of Planning and Community Development, 2007
7. DPCD (2011) Central activities areas. Government of Victoria. Department of Planning and Community Development, 2011
8. ADA (2004) Masterplan phase 2. High commission for the development of Riyadh, Arriyadh Development Authority 2004
9. Aguiléra A, Wenglenski S, Proulhac L (2009) Employment suburbanisation, reverse commuting and travel behaviour by residents of the central city in the Paris metropolitan area. Transp Res Part A 43(7):685–691
10. Mees P, O'Connell G, Stone J (2008) Travel to work in Australian capital cities, 1976–2006. Urban Policy Res 26:363–378
11. DPCD (2008) Melbourne 2030 a Planning Update: Melbourne @5 Million Melbourne, Victoria Government. Department of Transport, Victoria, Melbourne, 2008
12. de Dios Ortúzar J (2001) Modelling transport. Wiley, Chichester

11

Monitoring subsidence rates along road network by persistent scatterer SAR interferometry with high-resolution TerraSAR-X imagery

Bing Yu · Guoxiang Liu · Rui Zhang · Hongguo Jia · Tao Li · Xiaowen Wang · Keren Dai · Deying Ma

Abstract Ground subsidence is one of the key factors damaging transportation facilities, e.g., road networks consisting of highways and railways. In this paper, we propose to apply the persistent scatterer synthetic aperture radar interferometry (PS-InSAR) approach that uses high-resolution TerraSAR-X (TSX) imagery to extract the regional scale subsidence rates (i.e., average annual subsidence in mm/year) along road networks. The primary procedures involve interferometric pair selection, interferogram generation, persistent scatterer (PS) detection, PS networking, phase parameterization, and subsidence rate estimation. The Xiqing District in southwest Tianjin (China) is selected as the study area. This district contains one railway line and several highway lines. A total of 15 TSX images covering this area between April 2009 and June 2010 are utilized to obtain the subsidence rates by using the PS-InSAR (PSI) approach. The subsidence rates derived from PSI range from −68.7 to −1.3 mm/year. These findings show a significantly uneven subsidence pattern along the road network. Comparison between the PSI-derived subsidence rates and the leveling data obtained along the highways shows that the mean and standard deviation (SD) of the discrepancies between the two types of subsidence rates are 0.1 and ±3.2 mm/year, respectively. The results indicate that the high-resolution TSX PSI is capable of providing comprehensive and detailed subsidence information regarding road networks with millimeter-level accuracy. Further inspections under geological conditions and land-use categories in the study area indicate that the observed subsidence is highly related to aquifer compression due to groundwater pumping. Therefore, measures should be taken to mitigate groundwater extraction for the study area.

Keywords Subsidence · Road network · Persistent scatterer interferometry · TerraSAR-X · Highway · Railway

1 Introduction

As the primary mode of transport, the road network, which consists of highways and railways, handles the majority of traffic volume [1, 2]. Maintaining the sustainability and stability of these transportation infrastructures is crucial to maintaining traffic safety [3]. Previous investigations indicate that ground subsidence caused by tectonic movements or anthropic activities, such as groundwater overuse, is a major concern in land-use planning, infrastructure sustainability evaluation, and engineering construction [4]. Subsidence (especially subsidence troughs) along highways and railways impairs the sustainability and stability of these transportation infrastructures, thus resulting in great risk in traffic safety [3]. Therefore, the effective and accurate monitoring of subsidence along road networks is necessary in preventing these negative effects. This is particularly important in urban areas where highway and railway lines are highly integrated and dense.

In the last decades, the subsidence of highways and railways has generally been measured by conventional point-based surveying techniques, e.g., leveling, wire-flex extensometer, and global positioning system [4]. Nowadays, the

B. Yu · G. Liu (✉) · R. Zhang · H. Jia · T. Li · X. Wang · K. Dai · D. Ma
Department of Remote Sensing and Geospatial Information Engineering, Southwest Jiaotong University, 111 North 1st Section, 2nd Ring Road, Chengdu 610031, Sichuan, China
e-mail: rsgxliu@swjtu.edu.cn

B. Yu
e-mail: rsbingyu@gmail.com

differential synthetic aperture radar interferometry (DInSAR) [5, 6] has exhibited great potential as a newly developed geodetic technique to map ground subsidence. The spatial resolution and coverage of DInSAR are advantageous compared with conventional methods. However, the DInSAR technique is inevitably affected by spatiotemporal decorrelation [7], topographic errors [5], and atmospheric delay [8], which significantly reduce the accuracy of DInSAR measurements. To overcome these drawbacks, Ferretti et al. [9] proposed the permanent scatterer synthetic aperture radar interferometry (PS-InSAR) method. Numerous PS-InSAR (PSI) approaches have been developed since then and have been tested by using moderate spatiotemporal resolution synthetic aperture radar (SAR) data, such as ERS-1/2 and ENVISAT ASAR C band SAR images [10–12]. In spite of the differences of data processing strategies among the different PSI approaches, PSI techniques generally track the deformations of stable point-like targets, i.e., the persistent scatterers (PSs) [9]. The PSI method employs strategies of specialized modeling, and estimates the deformation rate and topographic error at a given target on the basis of multi-temporal InSAR phase observations of the target. Therefore, it is possible to separate the deformation information from other phase components (e.g., topographic errors and atmospheric delay). The PSI method has been proved very useful for monitoring subsidence in urban areas [11–14]. More advantages of PSI approach can be found in [9–14].

In recent years, the InSAR techniques (both DInSAR and PSI) have been used to monitor deformations related to highways and railways. In 2006, the U.S. Department of Transportation reported the results of InSAR applications for deformation monitoring of slopes nearby the highway transportation projects connecting different states [15]. Shan et al. [16] used the InSAR technique to extract deformations of an expressway and road area in an isolated permafrost area of China in 2012. Froese et al. [17] used the InSAR technique to manage risks associated with ground movements along the railway corridors. Ge et al. [18] mapped the ground subsidence along the Beijing–Tianjin high-speed railway using InSAR technique, and studied the impact of the subsidence on the railway. These previous studies of using InSAR technique to monitor deformations related to highways and railways utilized the moderate-resolution SAR images as mentioned before, and they analyzed the capability of InSAR applications in these monitoring activities.

The recent X-band (that has a wavelength of 3.1 cm) radar sensor onboard the German satellite TerraSAR-X (TSX) is capable of producing SAR images at high spatial and temporal resolutions of approximately 1–3 m and at 11 days (i.e., satellite repeat cycle), respectively [13, 14]. Thus, data availability for subsidence detection by PSI is extended [13, 14, 20]. First of all, the high spatial resolution of TSX can remarkably increase PS density, and thus leading to more detailed monitoring of ground movements. Furthermore, the short repeat cycle and short radar wavelength of TSX make it highly sensitive to ground movements, which means higher capability of tracking the slowly-accumulated ground subsidence. In this paper, we propose to apply PSI approach with the use of high-resolution TSX images to monitor the comprehensive subsidence rates of the road networks in urban areas. The PSI approach includes interferometric pair selection, interferogram generation, PS detection, PS networking and neighborhood differencing, phase parameterization, and subsidence estimation strategies. The Xiqing district in southwest Tianjin (China), which contains multiple highways and one railway line, is selected as the study area. A total of 15 TSX images obtained from this area between April 29, 2009 and June 2, 2010 are utilized to extract subsidence rates along the road network by using the TSX PSI approach. To assess the accuracy of the subsidence derived by using the PSI approach, the subsidence of the road network as measured by leveling is used as reference datum for comparative analysis.

2 Study area and data source

2.1 Study area

We select the Xiqing district in Tianjin (China) as the study area to investigate the applicability and potential of the TSX PSI approach in subsidence inversion along road networks. Figure 1 shows that Tianjin is located in the northeastern part of China, bordering Beijing at northwest, the Hebei province at northeast, and the Bohai Bay at east [19, 20]. Being one of the largest and most important industrial cities in north China, Tianjin suffers from water shortage due to its natural geographic condition and semi-arid climate [19]. Thus, a large amount of groundwater (especially the deep phreatic water) has been exploited to meet industrial and agricultural needs, resulting in severe land subsidence in several areas of Tianjin [19].

As shown in Fig. 1, the study area (Xiqing) is located in the southwestern part of Tianjin, bordering the Tianjin urban area at northeast. Figure 2 shows the averaged SAR amplitude image of the study area. In recent years, several new industrial parks have been constructed and launched for daily production in Xiqing. To meet the transportation needs from the increased industrial production, several new highways were constructed in the recent years. As illustrated in Fig. 2, multiple highways and one railway line are located in the study area, forming a road network. As a result of the progress in industrial and agricultural production, much of the groundwater in this area is being exploited to meet the increasing water needs for production activities. This excessive exploitation of groundwater may result in uneven subsidence.

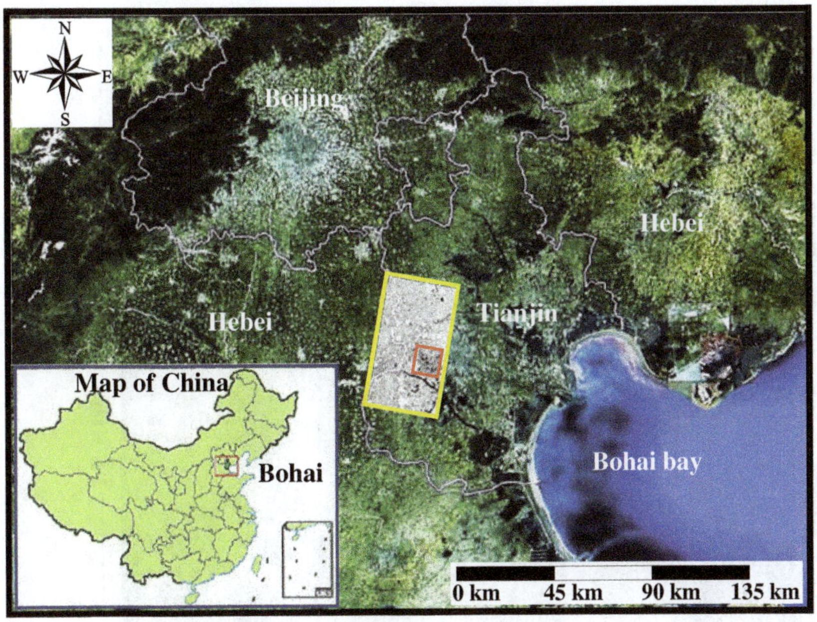

Fig. 1 Location map (modified from [20]) of the study area

Uneven subsidence can cause deformation on the highway surface and the railway tracks [1, 2]. The dynamic and static loads from heavy vehicles (carrier trucks and trains) should also be considered. These additional burdens lead to roadbed compression (i.e., subsidence), resulting in the distortion of the road surface and railway tracks.

2.2 Data source

To obtain the subsidence rates of the road network in the study area, the 15 TSX SAR images acquired between March 27, 2009 and June 2, 2010 are used in PSI processing. All images are provided in single look complex format, with a pixel spacing of 1.36 m in slant range (i.e., 2.07 m in ground range) and 1.90 m in azimuth. Figure 2 shows the averaged TSX SAR amplitude image of the study area, which is approximately 11×12 km^2. The roads, highways, railway line, buildings, fishponds, and crop parcels can easily be identified from the high-resolution TSX SAR amplitude image. As annotated in Fig. 2, 20 leveling points (LPs) were deployed along four highways, namely, Xiqing road (*AA′*), Zhongbei road (*BB′*), Jinjing road (*CC′*), and South Haitai road (*DD′*). Three leveling campaign epochs were conducted on all of the LPs. The acquired leveling data were used to validate the subsidence rates derived using TSX PSI.

The image acquired on September 30, 2009 was selected as the reference image and all other images were coregistered and resampled into the same grid space as the reference image. In this study, we first generate 105 interferometric pairs through a full combination of the 15 TSX images. The spatial (perpendicular) and temporal baselines (SB and TB) of the interferometric pairs range from 3.8 to 320 m and from 11 to 429 days, respectively. To reduce the spatial decorrelation and residual topographic errors (both of them are positively related to spatial baselines of the interferometric pairs) in the interferometric phases, only interferometric pairs with SBs shorter than 150 m are considered for processing. A total of 53 interferograms are generated. Table 1 lists the general information (acquisition dates of master and slave images, SBs, and TBs) of the interferograms. We generate 53 differential interferograms by using the two-pass differential InSAR method with the use of the digital elevation model (DEM) derived by the shuttle radar topography mission (SRTM). One potential disadvantage of differential InSAR with high-resolution TSX SAR images is lacking of external DEMs with comparable resolution, which is possible to induce extra topographic errors. Taking into consideration of the difference between resolution of the external DEM and the TSX SAR imagery, the external DEM was first oversampled to the same grid space as the reference TSX SAR image. Moreover, we rejected the interferometric pairs with longer perpendicular baselines to further confine the related topographic errors.

3 PSI procedures

3.1 Detection of PSs

We detect PSs by following method proposed by Ferretti et al. [9]. A pixel is a PS if it satisfies the following empirical criteria:

Fig. 2 Averaged amplitude image of the study area with annotation of highways, railway line, and the leveling points (LPs). The four highways and one railway line as annotated by *AA′*, *BB′*, *CC′*, *DD′*, and *EE′* are selected for further analysis

$$\begin{cases} D_{\text{amp}} = \frac{\sigma_{\text{amp}}}{\bar{a}} \le 0.25, \\ \bar{a} \ge \bar{A} + \sigma_A, \end{cases} \tag{1}$$

where D_{amp} is the amplitude dispersion index (ADI) [8]; $\bar{a}$ and σ_{amp} are the mean and standard deviation (SD), respectively, of the amplitude time series at a pixel; and $\bar{A}$ and σ_A are the overall mean and SD of the amplitude values of all the pixels in the mean amplitude image. The first criterion in Eq. (1) indicates that the pixels with smaller ADI are more temporally stable in radar backscattering than those with higher ADI, whereas the second criterion postulates that pixels with higher amplitude values tend to be more temporally coherent [12].

3.2 PS networking, phase parameterization, and subsidence estimation

After PS detection, all PSs are connected to form a Delaunay triangulation network, which is taken as the subsidence observation network [11, 20]. Phase modeling is based on the concept of neighborhood differencing [11, 12, 20] applied to each of the links in the Delaunay triangulation network.

Given N differential interferograms, the phase values at two neighboring PSs (e.g., p and q) extracted from the ith differential interferogram can be expressed by [12]:

$$\Phi_i^p = -\frac{4\pi B_i^{\mathrm{T}}}{\lambda} v_p \cos\theta_p - \frac{4\pi B_{i,p}^{\perp}}{\lambda R_p \sin\theta_p}\varepsilon_p + \overset{\frown}{\phi}{}_i^p - 2k_p\pi, \tag{2}$$

$$\Phi_i^q = -\frac{4\pi B_i^{\mathrm{T}}}{\lambda} v_q \cos\theta_q - \frac{4\pi B_{i,p}^{\perp}}{\lambda R_q \sin\theta_q}\varepsilon_q + \overset{\frown}{\phi}{}_i^q - 2k_q\pi, \tag{3}$$

where Φ_i^* is the wrapped phase at the PSs; v_* and ε_* are the subsidence rates and the elevation residuals (due to uncertainties in the SRTM DEM used) at the PSs, respectively; B_i^{T} is the TB of the ith interferogram; $B_{i,*}^{\perp}$ is the SB of the PSs in the ith interferogram; λ is the radar wavelength (3.1 cm for the TSX system); R_* and θ_* are the sensor-to-target range and the radar incident angle at the PSs, respectively; $\overset{\frown}{\phi}{}_i^*$ is the residual phase consisting of the nonlinear subsidence, atmospheric artifacts, orbit errors,

Table 1 Fifty-three short baseline interferograms

Master/slave dates in YMD	SB (m)	TB (days)	Master/slave dates in YMD	SB (m)	TB (days)	Master/slave dates in YMD	SB (m)	TB (days)
090429/090510	17	11	090510/100702	−109	418	090828/090930	−80	33
090429/090521	51	22	090521/090704	−81	44	090828/100107	77	132
090429/090704	−30	66	090521/090806	73	77	090828/100220	−53	176
090429/090806	125	99	090521/091102	54	165	090828/100325	111	209
090429/090828	−115	121	090521/091205	62	198	090828/100702	23	308
090429/091102	106	187	090521/100107	−89	231	090930/100220	26	143
090429/091205	113	220	090521/100325	−55	308	090930/100702	103	275
090429/100107	−37	253	090521/100702	−143	407	091102/091205	7	33
090429/100325	−3	330	090704/090828	−84	55	091102/100107	−144	66
090429/100702	−91	429	090704/091102	136	121	091102/100325	−110	143
090510/090521	33	11	090704/091205	143	154	091205/100107	−150	33
090510/090704	−48	55	090704/100107	−7	187	091205/100325	−117	110
090510/090806	107	88	090704/100220	−138	231	100107/100220	−130	44
090510/090828	−132	110	090704/100325	26	264	100107/100325	33	77
090510/091102	88	176	090704/100702	−61	363	100107/100702	−53	176
090510/091205	95	209	090806/091102	−18	88	100220/100702	77	132
090510/100107	−55	242	090806/091205	−11	121	100325/100702	−87	99
090510/100325	−21	319	090806/100325	−129	231			

and decorrelation noises; and $2k_{*}\pi$ denotes the integer ambiguity of the wrapped observation phases.

After neighborhood differencing, the phase increment ($\widehat{\Phi}_i$) between the two adjacent PSs of each link can be derived by using Eqs. (2) and (3) and represented as a function of the subsidence rate increment (υ), the elevation residual increment (ξ), and the residual phase increment ($\widehat{\phi}_i^{p,q}$) [12]. N differential equations can be obtained by using N differential interferograms. In each link, υ and ξ can be estimated by maximizing the following objective function [12]:

$$\begin{cases} \gamma = \max\left[\left|\frac{1}{N}\sum_{i=1}^{N}(\cos\omega_i + j\,\sin\omega_i)\right|\right] \\ \omega_i = \widehat{\phi}_i - \dfrac{4\pi B_i^{\mathrm{T}}}{\lambda}\upsilon\,\cos\bar{\theta} - \dfrac{4\pi\bar{B}_i^{\perp}}{\lambda\bar{R}\sin\bar{\theta}}\xi, \end{cases} \quad (4)$$

where $\bar{R}$, $\bar{\theta}$, and $\bar{B}_i^{\perp}$ are the mean sensor-to-target range, the mean radar incident angle, and the mean perpendicular baseline between two PSs, respectively; γ is the model coherence (MC) of the link; and $j = \sqrt{-1}$. υ and ξ can be derived by searching within a given solution space (e.g., −5 to 5 mm/year for υ and −20 to 20 m for ξ when high-resolution TSX images are used) to maximize the MC. In this study, the searching procedure was carried out with step values of 0.01 mm/year and 0.02 m for υ and ξ, respectively.

Once the subsidence rate and elevation residual increments of all the links are estimated by using Eq. (4), the Delaunay triangulation network can be treated by the weighted least squares (LS) adjustment to estimate the subsidence rates and the elevation residuals of all the PSs. The square of the maximized MC value of each link is considered the weight [12]. An LP with subsidence rate obtained through leveling measurements can be considered as a reference point for the LS adjustment. In this paper, we focus on analyzing the PSI-retrieved subsidence along road networks. A detailed discussion on PSI approach is beyond the scope of this work and can be found in [11–13].

4 Experimental results and analysis

4.1 Subsidence rate map and interpretation

The subsidence rates at all PSs were extracted from 53 differential interferograms by using the PSI approach presented in Sect. 3. One corner reflector with leveling measurements known was taken as the reference point (RP) while carrying out the LS adjustment. Figure 3 shows the generated subsidence rate map of the entire study area and the position of the RP (the black square). The map comprises two layers: the color-coded layer; and the base map layer (the averaged amplitude image). The color-coded layer, which includes a scale bar, presents the magnitude and distribution of the estimated subsidence rates of all the PSs. Very dense PSs were detected in the study area due to the high-resolution of TSX imagery. The total number of PSs is 686,598, while the averaged density of PSs is 5,201 per square kilometer. The color map shows the uneven subsiding pattern, with subsidence rate magnitude ranging from −69 to 0 mm/year. The mean subsidence rate in this area is −33.7 mm/year. The results indicate that the study area is a predictably active subsidence zone. Simultaneous inspections of Figs. 2 and 3 reveal that apparent subsidence was observed along the highways and the railway line.

Closer inspections of Fig. 3 reveal that the subsidence pattern in the study area is highly associated with local land-use categories. Four typical areas are selected as examples for further analysis. The four selected areas are identified by the dashed rectangles, namely, P_1, P_2, P_3, and P_4. Figure 4 shows the relevant optical images. To provide a good presentation of subsidence related to P_1, P_2, P_3, and P_4, we list their maximum, minimum, and mean subsidence rates in Table 2. The highest subsidence rate is observed in P_1, with the maximum and mean magnitudes of −68.8 and −50.6 mm/year, respectively. P_3 and P_4 have relatively higher subsidence rates with the mean values of −33.9 and −44.8 mm/year, respectively. P_2 has relatively lower subsidence rates as compared with P_1, P_3, and P_4. The maximum and mean subsidence rates of P_2 are −40.7 and −26.2 mm/year, respectively.

According to our in situ investigations, P_1 contains a thermal power plant (marked by the dashed oval), multiple industrial factories (marked by the dashed rectangle) and some residential quarters (marked by the dashed polygon). P_3 and P_4 mainly comprise industrial parks, whereas P_2 consists of residential quarters. The observed subsidence in these areas is caused by groundwater extraction for productive and domestic water needs. The aquifer in the study area was formed in the Neogene Period. This layer mainly belongs to the Neogene period and the Quaternary sedimentary formation period [4, 19]. Geologically, the aquifer is a thick clayey stratum consisting of grits and loose, or semi-loose argillaceous sediments [19]. Excessive exploitation of groundwater can result in severe water table depression. This depression further leads to interstitial water runoff and the consequence is aquifer compression which accounts for the subsidence. The heterogeneous subsidence observed in the study area (as shown in Fig. 3) is attributed to the various groundwater pumping intensities in different regions.

4.2 Closer inspections of the subsidence related to the road network

We select four highways and one railway line to further analyze the subsidence phenomena related to the road network (see Fig. 2). The selected highways and railway line, namely, Xiqing road, Zhongbei road, Jinjing road,

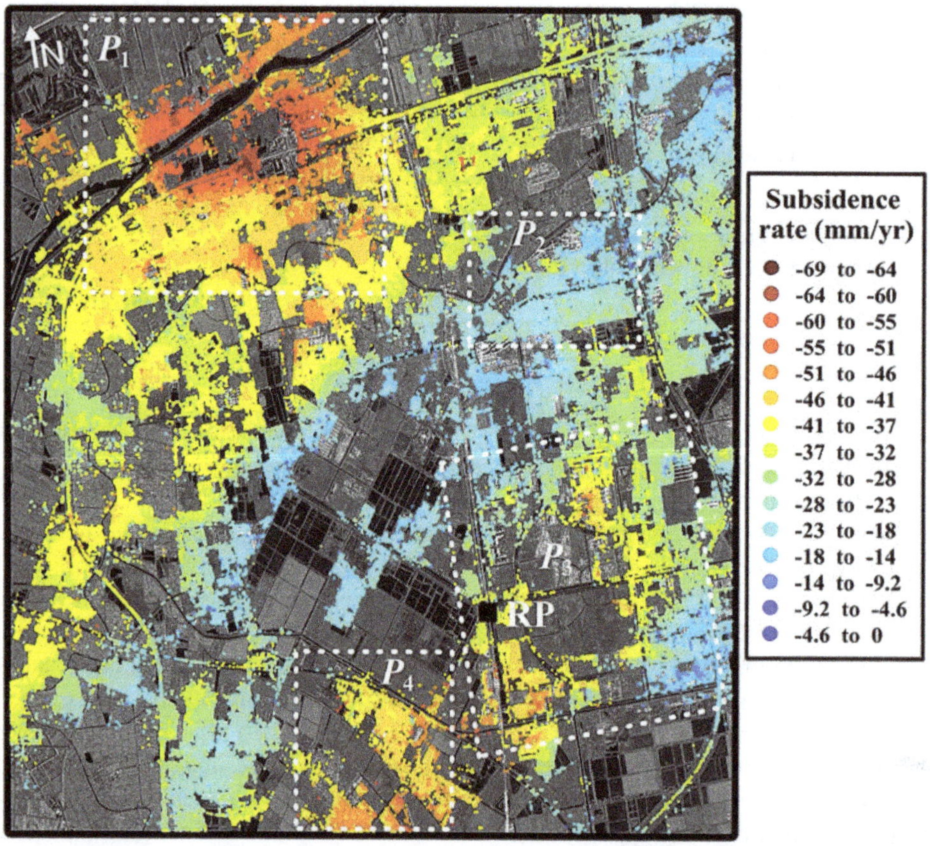

Fig. 3 Subsidence rate map of the study area

Fig. 4 Optical images related to P_1, P_2, P_3, and P_4

Table 2 Maximum, minimum, and mean subsidence rates of P_1, P_2, P_3, and P_4

Name	Maximum (mm/year)	Minimum (mm/year)	Mean (mm/year)
P_1	−68.8	−37.0	−50.6
P_2	−40.7	−14.9	−26.2
P_3	−52.5	−12.5	−33.9
P_4	−62.1	−28.0	−44.8

South Haitai road, and Jinpu railway, are annotated by AA', BB', CC', DD', and EE', respectively. We first estimated the full resolution (i.e., the original resolution as per the TSX image) subsidence rates along the highways and the railway line by using the Kriging interpolation method. We then extracted the subsidence rate profiles from the centerline of each highway and from the railway line. Figure 5a–e shows the subsidence rate profiles related to each highway and the railway line. Figure 5 exhibits the uneven subsidence distribution along the transportation facilities being studied. Table 3 lists the maximum, minimum, and mean subsidence rate values of each highway and the railway line. The Jinpu railway and the Xiqing highway have subsidence rates that are significantly higher than those of other highways.

Further inspections of Figs. 2 and 4 indicate that dense industrial parks and residential quarters are located along both sides of the selected highways and railway line. The local land-use category of each highway and railway line is annotated in Fig. 5. Figure 5 shows that the subsidence troughs revealed by the profiles are mostly located near industrial areas, whereas the sections with relatively lower subsidence rates are near residential areas, agricultural areas, and commercial and storage areas. A closer inspection of Figs. 2, 4, and 5a, e shows that the maximum subsidence rates along the Xiqing road and the Jinpu railway are observed near the thermal power plant. As discussed in Sect. 4.1, groundwater pumping due to

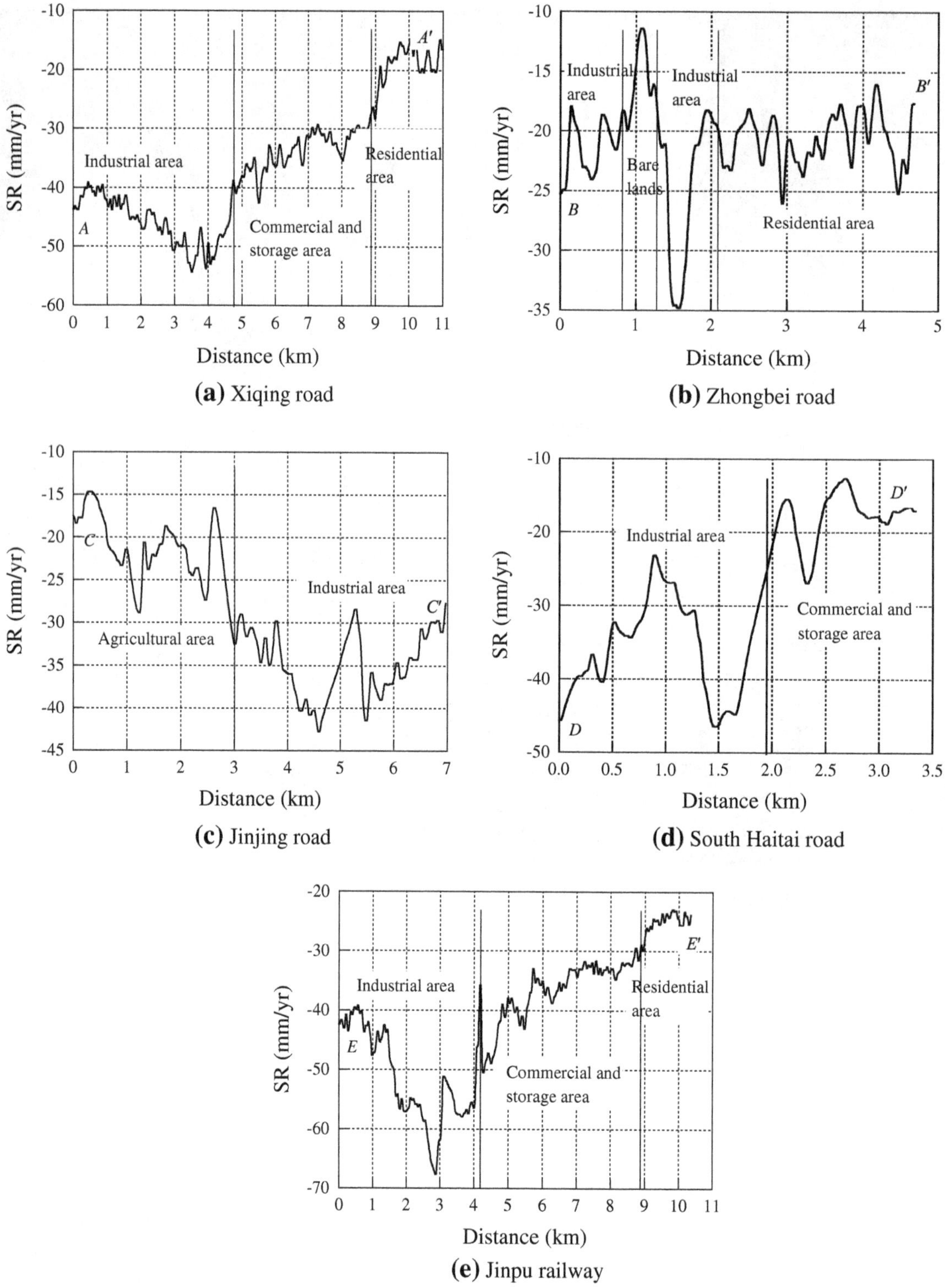

Fig. 5 Subsidence rate profiles extracted from the selected highways and the railway line

productive and domestic water needs in the residential and industrial regions is accountable for the subsidence in and around these areas. Water pumping has indeed affected the neighboring highways and the railway line. Therefore, measures should be taken to optimize groundwater use planning and to restrict groundwater pumping. However, groundwater pumping is not the only factor affecting the transportation facilities being studied. Another factor responsible for the uneven subsidence of the highways and the railway line is the dynamic load from heavy vehicles,

Table 3 Maximum, minimum, and mean subsidence rates of the selected highways and the railway line

Name	Maximum (mm/year)	Minimum (mm/year)	Mean (mm/year)
Xiqing road (*AA*′)	−54.5	−14.6	−36.3
Zhongbei road (*BB*′)	−34.8	−11.4	−21.1
Jinjing road (*CC*′)	−43.1	−14.6	−28.8
South Haitai road (*DD*′)	−34.8	−11.9	−21.3
Jinpu railway (*EE*′)	−67.8	−22.8	−41.0

Table 4 Maximum slope changes of subsidence rates (mm/year) along different sections of the selected highways and the railway line

Name	Industrial area	Commercial area	Agricultural area	Residential area
Xiqing road (*AA*′)	14.3	8.2	–	9.7
Zhongbei road (*BB*′)	23.3	–	–	9.2
Jinjing road (*CC*′)	14.4	–	16.7	–
South Haitai road (*DD*′)	23.8	14.5	–	–
Jinpu railway (*EE*′)	22.3	12.3	–	2.8

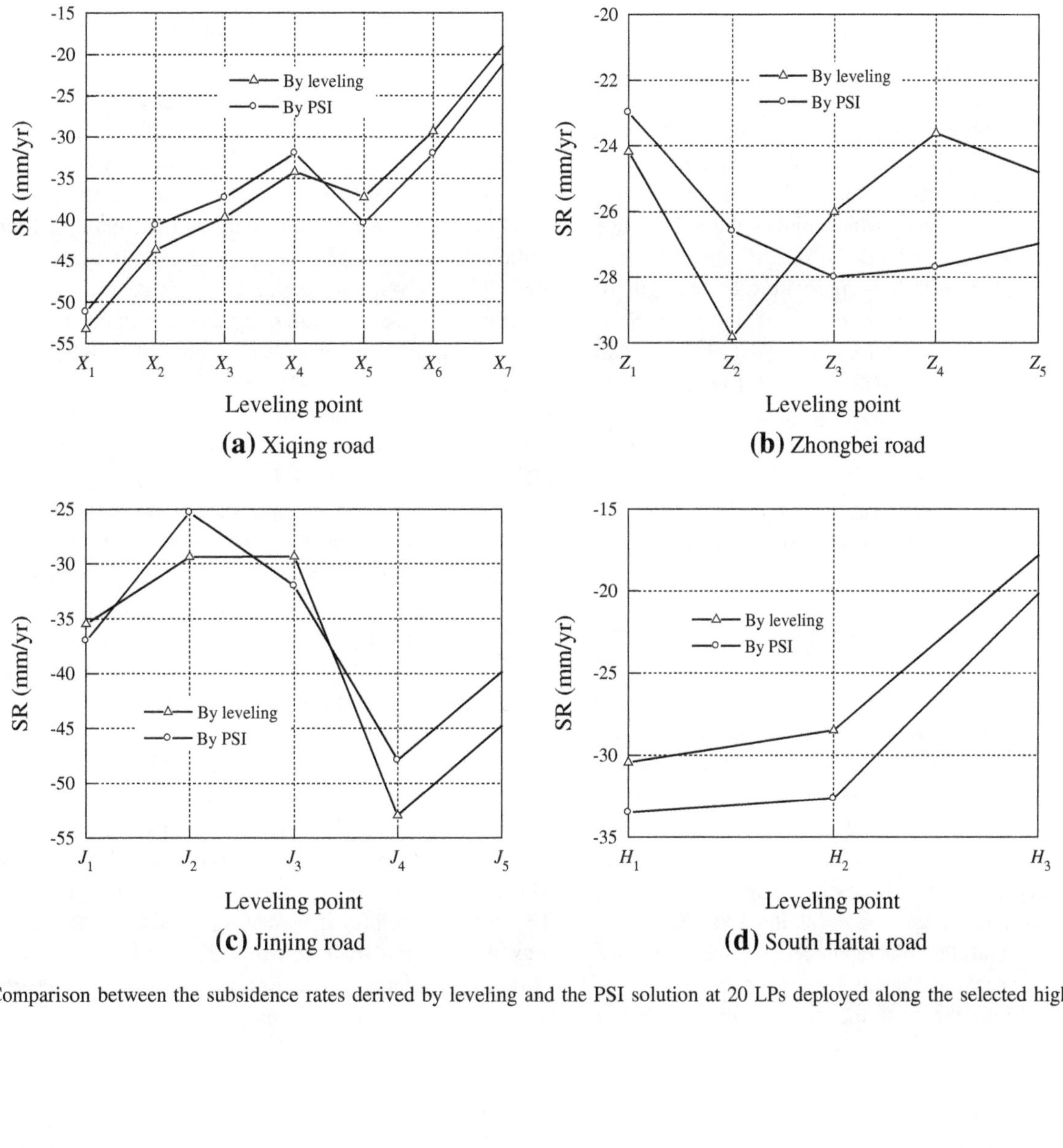

Fig. 6 Comparison between the subsidence rates derived by leveling and the PSI solution at 20 LPs deployed along the selected highways

such as trucks and trains. The uneven external forces from dynamic loads can result in heterogeneous surface deformations on the highways and railway line.

For more comprehensive understanding of the uneven subsidence, the maximum slope changes of subsidence rates (mm/year) along different sections (being divided by the black vertical lines in Fig. 5) of the selected highways and the railway line are estimated from the subsidence rate profiles and are shown in Table 4. The maximum slope changes in industrial areas show the highest magnitudes ranging from 14.3 to 23.8 mm/year, and those in residential areas have the lowest magnitudes, while the subsidence rates in agricultural and commercial areas reveal medium maximum slope changes. Such circumstances demonstrate that the subsidence in industrial areas is highly dominated by the heterogeneous pattern, while the subsidence in areas with other land-use categories (e.g., agricultural land, residential land, and commercial land) is relatively less heterogeneous.

It is worth noting that dense PSs were identified along the studied highway and railway lines in our study area (see Fig. 3). One reasonable explanation is that the street lamps, stones, and fences distributed along the highway and railway lines can be easily and individually identified by the high-resolution TSX imagery (i.e., TSX SAR images have small pixel size). This characteristic is advantageous of providing more detailed subsidence information in PSI analysis, which makes the subsidence troughs along the road network be clearly observed (see Fig. 3). Moreover, the TSX imagery has wavelength of 3.1 cm, which is shorter than other regular used SAR systems such as ALOS PALSAR L band (23.6 cm) and ENVISAT ASAR and ERS1/2 C band (5.6 cm) SAR imagery. Shorter wavelength corresponds to relatively higher observation sensibility to ground movements, resulting in more precise monitoring of the land subsidence. The accuracy of the subsidence rates derived by TSX PSI will be validated by comparing with leveling measurements (see Sect. 4.3).

4.3 Subsidence validation with leveling data

For validation purposes, we used the in situ leveling data to assess the accuracy and reliability of the subsidence rate measurements derived by using the TSX PSI. We compared the subsidence rates derived by TSX PSI with those derived through leveling, i.e., subsidence rates obtained at the LPs (as marked in Fig. 2) by three precise leveling campaign epochs. Figure 6 shows the comparison between the two types of subsidence rates at the LPs. Statistical analysis shows that the discrepancies between the two types of subsidence rates at all LPs range from −5.1 to 4.2 mm/year, while those at the LPs on each highway range from −3.0 to 3.2 mm/year, −3.2 to 4.1 mm/year, −5.1 to 2.7 mm/year, and −2.4 to 4.2 mm/year, respectively. The mean and SD of the discrepancies at all LPs are 0.1 and ±3.2 mm/year, respectively. The results show that the high-resolution TSX PSI is useful in monitoring subsidence related to highways and railways at millimeter-level accuracy.

5 Conclusions

We propose to apply PSI approach using high-resolution TSX imagery to extract the regional scale subsidence related to the road network. The Xiqing district in southwest Tianjin (China) is selected as the study area, and the subsidence rates along the highways and the railway line in this area are obtained by using the PSI approach utilizing 15 TSX SAR images acquired between April 29, 2009 and June 2, 2010. The quality of the subsidence rates derived using the PSI approach is assessed with use of ground truth data obtained by precise leveling.

The subsidence rate map shows that the study area is highly affected by uneven subsidence. The maximum, minimum, and mean subsidence rates are −68.7, −1.3, and −33.7 mm/year, respectively. The subsidence rate map, the optical images, and our in situ investigations in the study area demonstrate that the subsidence pattern is highly related to land-use categories (i.e., industrial lands, residential lands, agricultural lands, commercial lands, etc.). The uneven subsidence in this area is attributed to aquifer compression caused by the unbalanced pumping of groundwater in different types of lands. Further inspections of the subsidence rate profiles of the transportation facilities and the land-use categories around them indicate that groundwater pumping has indeed affected the road network being studied. Therefore, measures should be taken to optimize groundwater use planning and to mitigate groundwater extraction. The comparison between the subsidence rates derived using PSI and those obtained by leveling shows that the discrepancies between the two types of subsidence rates at all the LPs are within −5.1 to 4.2 mm/year. The mean and SD of the discrepancies are 0.1 and ±3.2 mm/year, respectively.

The results demonstrate that the PSI method based on high-resolution TSX imagery is useful in detecting road network subsidence. The subsidence measurements derived using the TSX PSI method have millimeter-level accuracy. The high-resolution and regional-scale characteristics of TSX PSI are helpful in detecting subsidence troughs that may damage the road network. The TSX PSI method is also advantageous in revealing the mechanism and origins of the subsidence phenomena.

Acknowledgments This work was jointly supported by the National Basic Research Program of China (973 Program) under Grant 2012CB719901, the National Natural Science Foundation of China under Grant 41074005, and the 2013 Doctoral Innovation Funds of Southwest Jiaotong University. We thank Infoterra GmbH and USGS for providing TerraSAR-X SAR images and SRTM DEM, respectively.

References

1. Rondinelli D, Berry M (2000) Multimodal transportation, logistics, and the environment: managing interactions in a global economy. Eur Manage J 18(4):398–410
2. Wu X, Hu SJ, Cui YP et al (2005) Study on the evaluation of harmonious development between railway transportation and economy. J China Railw Soc 27(3):20–25 (in Chinese)
3. Miller RD, Xia JH, Steeples DW (2009) Seismic reflection characteristics of naturally-induced subsidence affecting transportation. J Earth Sci 20(3):496–512
4. Hu RL, Yue ZQ, Wang LC, Wang SJ (2004) Review on current status and challenging issues of land subsidence in China. Eng Geol 76:65–77
5. Massonnet D, Feigl KL (1998) Radar interferometry and its application to changes in the Earth's surface. Rev Geophys 36(4):441–500
6. Helmut R (2009) Advances in interferometric synthetic aperture radar (InSAR) in earth system science. Prog Phys Geogr 33(6):769–791
7. Zebker HA, Villasenor J (1992) Decorrelation in interferometric radar echoes. IEEE Trans Geosci Remote Sens 30(5):950–959
8. Ding XL, Li ZW, Zhu JJ et al (2008) Atmospheric effects on InSAR measurements and their mitigation. Sensors 8:5426–5448
9. Ferretti A, Prati C, Rocca F (2001) Permanent scatterers in SAR interferometry. IEEE Trans Geosci Remote Sens 39(1):8–20
10. Hooper A, Zebker H, Segall P et al (2004) A new method for measuring deformation on volcanoes and other natural terrains using InSAR persistent scatterers. Geophys Res Lett 31:L23611
11. Mora O, Mallorqui JJ, Broquetas A (2003) Linear and nonlinear terrain deformation maps from a reduced set of interferometric SAR images. IEEE Trans Geosci Remote Sens 41(10):2243–2253
12. Liu GX, Buckley SM, Ding XL et al (2009) Estimating spatio-temporal ground deformation with improved persistent-scatterer radar interferometry. IEEE Trans Geosci Remote Sens 47(9):3209–3219
13. Liu GX, Jia HG, Zhang R et al (2011) Exploration of subsidence estimation by persistent scatterer InSAR on time series of high resolution TerraSAR-X images. IEEE J Sel Top Appl Earth Observ Remote Sens 4(1):159–169
14. Wegmüller U, Walter D, Spreckels V et al (2010) Nonuniform ground motion monitoring with TerraSAR-X persistent scatterer interferometry. IEEE Trans Geosci Remote Sens 48(2):895–904
15. U.S. Department of Transportation (2006) InSAR applications for highway transportation projects. Central Federal Lands Highway Division (FHWA-CFL/TD-06-002), 101. http://www.cflhd.gov/programs/techDevelopment/geotech/insar/documents/01_insar_entire_document.pdf. Accessed 21 Apr 2006
16. Shan W, Wang CJ, Hu Q (2012) Expressway and road area deformation monitoring research based on InSAR technology in isolated permafrost area. In: Proceedings of 2012 2nd International Conference on Remote Sensing, Environment and Transportation Engineering (RSETE), Nanjing, China, 5
17. Froese CR, Keegan T, Abbott B, Kooij MVD (2005) Managing risks associated with ground movement along railway corridors using spaceborne InSAR. The American Railway Engineering and Maintenance-of-Way Association, 16. http://www.arema.org/files/library/2005_Conference_Proceedings/00032.pdf. Accessed 26 Nov 2013
18. Ge LL, Li XJ, Chang HC (2010) Impact of ground subsidence on the Beijing–Tianjin high-speed railway as mapped by radar interferometry. Ann GIS 16(2):91–102
19. Enviro-Library (2009) Challenges and prospects of sustainable water management in Tianjin, 2008. Sustainable Groundwater Management in Asian Cities, 17. http://enviroscope.iges.or.jp/modules/envirolib/upload/981/attach/07_chapter3-4tianjin.pdf. Accessed 2 Jul 2009
20. Yu B, Liu GX, Li ZL et al (2013) Subsidence detection by TerraSAR-X interferometry on a network of natural persistent scatterers and artificial corner reflectors. Comput Geosci 58:126–136

12

The approach to calculate the aerodynamic drag of maglev train in the evacuated tube

Jiaqing Ma · Dajing Zhou · Lifeng Zhao · Yong Zhang · Yong Zhao

Abstract In order to study the relationships between the aerodynamic drag of maglev and other factors in the evacuated tube, the formula of aerodynamic drag was deduced based on the basic equations of aerodynamics and then the calculated result was confirmed at a low speed on an experimental system developed by Superconductivity and New Energy R&D Center of South Jiaotong University. With regard to this system a high temperature superconducting magnetic levitation vehicle was motivated by a linear induction motor (LIM) fixed on the permanent magnetic guideway. When the vehicle reached an expected speed, the LIM was stopped. Then the damped speed was recorded and used to calculate the experimental drag. The two results show the approximately same relationship between the aerodynamic drag on the maglev and the other factors such as the pressure in the tube, the velocity of the maglev and the blockage ratio. Thus, the pressure, the velocity, and the blockage ratio are viewed as the three important factors that contribute to the energy loss in the evacuated tube transportation.

Keywords Evacuated tube · Maglev train · Aerodynamic drag · Pressure in the tube

J. Ma · D. Zhou · L. Zhao · Y. Zhang · Y. Zhao
Superconductivity and New Energy R&D Center, Southwest Jiaotong University, Chengdu 610031, Sichuan, China

J. Ma (✉) · D. Zhou · L. Zhao · Y. Zhang · Y. Zhao
Key Laboratory of Magnetic Levitation Technologies and Maglev Trains (Ministry of Education of China), Southwest Jiaotong University, Chengdu 610031, Sichuan, China
e-mail: 357287962@qq.com

Y. Zhao
School of Materials Science and Engineering, University of New South Wales, Sydney, NSW 2052, Australia

1 Introduction

The speed of traditional trains is limited because of the dynamic friction between the wheels of the train and the fixed rail on the ground. When the trains are running at a low speed, most of the energy is consumed by friction. The trains can be levitated above the rail to avoid such friction with the technology of magnetic levitation [1]. There are three types of levitation technologies: electromagnetic suspension (EMS), electrodynamic suspension (EDS), hybrid electromagnetic suspension (HEMS) [2]. Even with these three methods, the velocity of trains could not be improved remarkably because of the aerodynamic drag. When the trains run at low speeds, this drag is not evident. At high speeds, however, the aerodynamic drag is too large to be neglected. Whatever the trains are levitated or not, the aerodynamic drag is the dominate part of drag when it runs at a high speed in the atmosphere near the earth's surface. At a speed range between 400 and 500 km/h, the aerodynamic drag accounts for 80 %–90 % of the total drag including the aerodynamic drag, the eddy resistance force, and the braking force [3]. The train speed is much lower than the airplane speed because airplanes flight in a circumstance of rarefied gas in the high altitude. In view of this fact, the evacuated tube transportation (ETT) was proposed to reduce the aerodynamic drag and improve the speed of the maglev train. Shen [4] and Yan [5] discussed the possibility, strategy, and the technical proposal for developing the ETT in China.

Theoretically speaking, when the inner part of the tube is in the condition of absolute vacuum, the aerodynamic drag for the levitation train inside the tube will be zero. However, this is very hard to realize. An alternative is to draw-off the gas partly and optimize the train shape. Therefore, the influence of the air pressure, the velocity, and the blockage ratio on the Maglev train in the evacuated tube system is a very interesting topic to study. Up to now, some research works have been done to explore what conditions are suitable for future ETT. Raghuathan and Kim [6] reviewed the state of the art on the aerodynamic and aeroacoustic problems of high-speed railway train and highlighted proper control strategies to alleviate undesirable aerodynamic problems of high-speed railway system. Various aspects of the dynamic characteristics were reviewed and aerodynamic loads were considered to study the aerodynamic drag [7]. Wu et al. [8] simulated the maglev train numerically with software STARCD based on the N–S equation of compressible viscosity fluid and k-ε turbulence model. The flow field, the pressure distribution and the aerodynamic drag coefficient were also analyzed to illustrate the relationship between the aerodynamic drag and the shape of the train in evacuated tube. The pressure distribution in the whole flow field and the relation between the aerodynamic drag and the basic parameters were derived in [9]. Shu et al. [10] simulated the flow field around the train based on the 3D compressible viscous fluid theory and draw the conclusion that its aerodynamic performance is relevant to the length of the streamlined nose. In [11–13], the simulated results showed that the speed, the pressure and the blockage ratio significantly affect the aerodynamic drag of the train in an evacuated tube.

In this paper, the experimental system model developed by Superconductivity and New Energy R&D Center of South Jiaotong University was used to study the aerodynamic drag in the tube. The basic mass conversation equation and the momentum conversation equation [14] were used at first to deduce the expression that describes the relationships between the drag and the main parameters such as the tube pressure, velocity, and the blockage ratio. Then the aerodynamic drag is calculated with that expression. Finally, the calculated results are compared with the experimental data to verify the validity of the deduced expression.

2 The mathematical model of the aerodynamic drag and the physical model of the ETT

In this paper, the N–S equation of compressible viscosity fluid and k-ε turbulence model are applied to calculating the flowing field of aerodynamics. Considering an infinitesimal part for any arbitrary circumstance, it is well known that every part follows the laws of mass conversation equation and the momentum conversation equation. These two equations [15, 16] are listed as:

$$\frac{\partial \rho}{\partial t} + \nabla \cdot (\rho v) = 0, \qquad (1)$$

$$\frac{\partial (\rho v_{\mathrm{i}})}{\partial t} + \nabla \cdot (\rho v_{\mathrm{i}} v) = \nabla \cdot (\mu \nabla v_{\mathrm{i}}) - \frac{\partial p}{\partial x} + F_{\mathrm{i}}, \qquad (2)$$

where

$$\rho = \rho\,(x, y, z, t)$$

ρ is density of the infinitesimal part, v velocity of the infinitesimal part, v_i each component of velocity (different when in different coordinates), F_i body force in each direction, μ dynamic viscosity, and p pressure of the infinitesimal part.

Eq. (1) means that the quality flowing into the infinitesimal part is equal to that out of this part. And Eq. (2) means that the rate of change of arbitrary mass's momentum is equal to the sum of the force acting upon it.

All of the following mathematical derivations in this paper are based on Eqs. (1) and (2). To study the evacuated tube transportation, an experimental model was developed.

This model called evacuated tube system for maglev train (ETSMT) includes three components: the evacuate tube, maglev, and propulsion system.

The tube is made of Perspex with the circumference of 10 m and the circular permanent magnetic guideway (PMG) is placed along the bottom of the tube. The positions of the tube, the train, and the PMG and the used rectangular coordinate system are shown in Fig. 1.

In Fig. 1, a_0 and b_0 stand for the width and the height of the tube respectively; a, b, and c stand for the width, the height, and the length of the maglev train, respectively. h_0 is the height of the PMG and h the levitated height.

As shown in Fig. 1, the train is levitated above the rail with height h and can only move in the x-direction due to the self-guiding characters of high-temperature-superconducting (HTS)-PMG system. For convenience, we suppose:

1) The tube is straight and the train runs along it only in the x-direction.
2) The magnetic flux of permanent magnetic rail is constant in the x-direction.

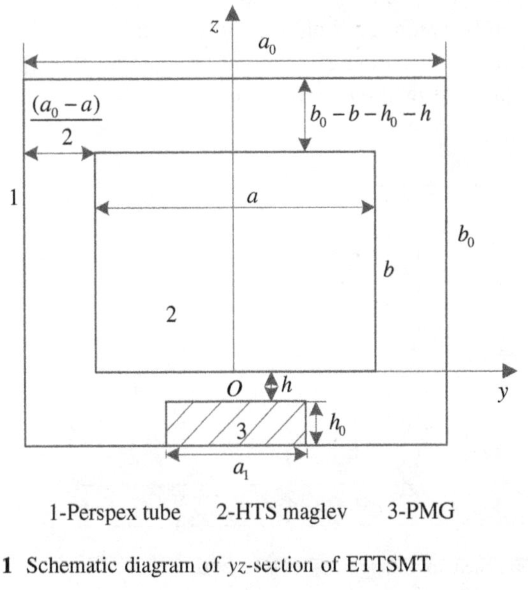

1-Perspex tube 2-HTS maglev 3-PMG

Fig. 1 Schematic diagram of yz-section of ETTSMT

3) The pressure of any part of inner tube is constant in the z-direction with neglecting the atmospheric molecular mass.
4) The train is in the center of the tube whatever the blockage rate is.

In the Cartesian rectangular coordinate system the velocity is expressed as:

$$v = v_x e_x + v_y e_y + v_z e_z, \tag{3}$$

where ρ is the atmospheric density, v_x, v_y, v_z are the components of velocity in x, y, z axis directions respectively, e_x, e_y, e_z are the unit vectors of each axis. When considering the assumptions of (1) and (2), v_y and v_z are zero. Then Eqs. (1) and (2) are modified as:

$$\frac{\partial \rho(x,t)}{\partial t} + \nabla \cdot (\rho v_x e_x) = 0, \tag{4}$$

$$\frac{\partial[\rho(x,t)v_x]}{\partial t} + \nabla \cdot [\rho(xt)v_x^2 \boldsymbol{e}_x] = \nabla \cdot (\mu \nabla v_x) - \frac{\partial \rho}{\partial x} + F_x. \tag{5}$$

To demonstrate the evident effect of the pressure in tube, the streamlined nose of the train is not adopted. The schematic diagram in moving direction of train is shown in Fig. 2.

As shown in Fig. 2, the maglev train is levitated above the PMG and there is no dynamic friction between the train and the rail. According to assumption (2), there is no vibration in the z-direction, which means that all of the kinetic energy of the free levitated running train is consumed because of the aerodynamic drag after the train gains the initial kinetic energy. In an ideal situation, when the tube pressure is zero, the aerodynamic drag is equal to zero. Since this condition is almost impossible to realize, the actual practice is to pull the air out of the tube to form a suitable pressure. The purpose of this work is to explore the relationship between the drag and the tube pressure.

The aerodynamic drag of a running maglev in this system is composed of three components:

(1) F_1: the force on windward side of the train due to the collision between air and the train,
(2) F_2: the air friction on four side faces of the train and
(3) F_3: the force caused by the different pressures of windward side and the tailstock side of the train

Each force will be discussed in following sections based on Eqs. (4) and (5).

2.1 The calculation of F_1

For simplification, a long section of the air ahead of the train is moving at the same velocity as the train because of the character of the air. An infinitesimal part of the air in area 1 is considered in Fig. 3. We suppose that the velocity of the thin layer of air is to vary after a tiny time dt after collision and the displacement of this layer is dx away from the windward side along x-axis within another tiny time dt.

The velocity of the infinitesimal air before collision is

$$v_1 = 0. \tag{6}$$

After collision, its velocity is equal to that of the train's. So the kinetic energy of this air is $\frac{1}{2}\rho dxdydz \cdot v_x^2$ and we have the equation:

$$\frac{1}{2}\rho dxdydz \cdot v_x^2 = dF_{1x}dx. \tag{7}$$

F_1 is equal to zero when the velocity of the train is smaller than sound velocity because the velocity of atmospheric molecule is equal to the sound velocity after the collision with the windward side of the train. When the velocity of train is larger than sound velocity, the air column with the length of $v_c \cdot dt$ is affected within the period of dt and the velocity of that air column is approximatively equal to train velocity. The force can be calculated by combining Eqs. (6) and (7) and the momentum conversation equation:

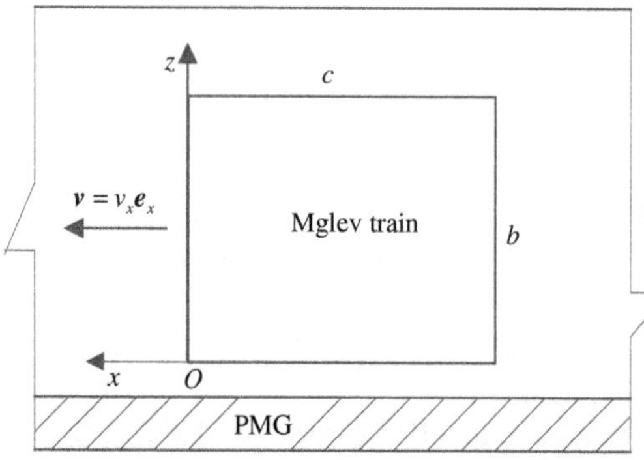

Fig. 2 The schematic diagram of xz-section of ETTSMT

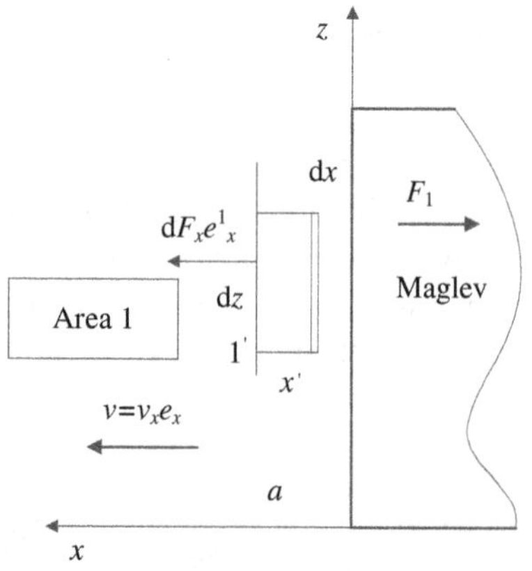

Fig. 3 Schematic diagram of the F_1

$$F_1 = -e_x \iint_{yz} dF_{2x} = -e_x \iint_{yz} \frac{1}{2}\rho_0 v_x^2 dydz = -e_x \frac{p}{2p_0}\rho_0 v_c \iint_{yz} v_x dydz \quad (8)$$

Thus,

$$F_1 = \begin{cases} 0, & v_x < v_c \\ -e_x v_c \iint_{yz} \frac{p}{2p_0}\rho_0 v_x dydz, & v_x \geq v_c \end{cases} \quad (9)$$

where $p_0 = 101,325$ Pa, and $\rho_0 = 1.293\ \text{kg/m}^3$.

Because of the Brownian movement of molecules, it is reasonable to neglect F_1 when the train runs at a low speed.

2.2 The calculation of F_2

In this EETSMT, there are four side faces where friction force generates, as shown in Fig. 4, where F_{2U}, F_{2B}, F_{2L} and F_{2R} represent the frictions on upper side, lower side, left side, and right side respectively. The left side is toward the inside of the paper and the right side is toward the outside.

On upper side, considering a infinitesimal volume of $dxdydz$, the area of contact between the infinitesimal and the air is $ds = dxdy$.

This component of the aerodynamic drag F_2 exits because of gas viscidity. The regularity of the fluid velocity distribution between the tube wall and train body side is described by the function of variable z [16]:

$$v'_x = f_{2U}(n) \quad (b \leq n \leq b_0), \quad (10)$$

where f_{2U} is velocity function of length n; b and b_0 are shown in Fig. 1.

The relationship between the fluid internal friction stress and the velocity gradient according to Newton's proposal is

$$\tau = \mu \frac{\partial f_{2U}(n)}{\partial n}, \quad (11)$$

where τ friction stress, μ viscosity, $\frac{\partial f_{2U}(n)}{\partial n}$ is the change rate of the velocity from train body side to the tube wall. The friction stress is the force on a unit area with the direction perpendicular to the velocity, and the viscosity μ is affected by the temperature instead of the air pressure. So the μ is constant when the temperature is unchanged. From the analysis above, the friction of infinitesimal $dxdydz$ is

$$dF_{2U} = \tau dydx \quad (12)$$

$$\boldsymbol{F}_{2U} = -\boldsymbol{e}_x \mu \iint_{S_{2U}} \frac{\partial f_U(n)}{\partial n} dxdy. \quad (13)$$

Likewise, F_{2B}, F_{2L} and F_{2R} can be deduced.

Then $F_2 = F_{2U} + F_{2B} + F_{2L} + F_{2R}$

$$= -e_x \mu \left[\iint_{S_{2U}} \frac{\partial f_U(n)}{\partial n} dxdy + \iint_{S_{2B}} \frac{\partial f_B(n)}{\partial n} dxdy + \iint_{S_{2L}} \frac{\partial f_L(n)}{\partial n} dxdz + \iint_{S_{2R}} \frac{\partial f_R(n)}{\partial n} dxdz \right] \quad (14)$$

2.3 The calculation of F_3

Figure 5 shows that F_3 is generated by pressure difference between the headstock and the tailstock side of the train. This force is larger when the velocity of the train is greater.

In Fig. 5, the pressure of the inner tube is p. The train windward side is x–z side with area S. For a small time interval dt, the train moves with distance dx. And there is no interpenetration of air between different areas 1, 2, and 3 within dt. So the velocity of infinitesimal gas is v_x when taking the train as a reference. According to the Bernoulli formula, the relationship between the pressure and the velocity at point A is:

$$p_{31} = p + \frac{\rho}{2} v_x^2, \quad (15)$$

And the pressure at point B is

$$p_{32} = p \quad (16)$$

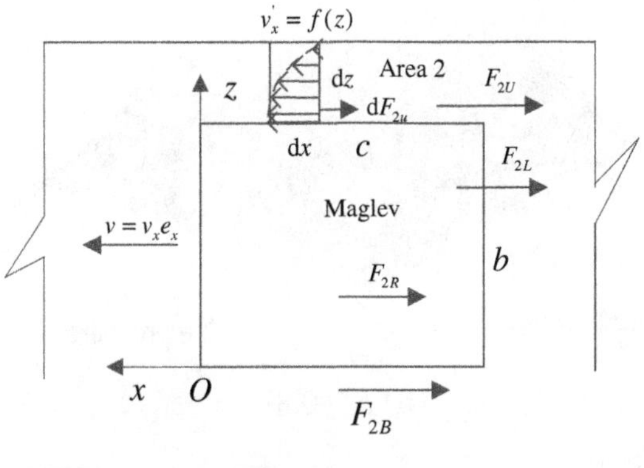

Fig. 4 Schematic diagram of the F_2

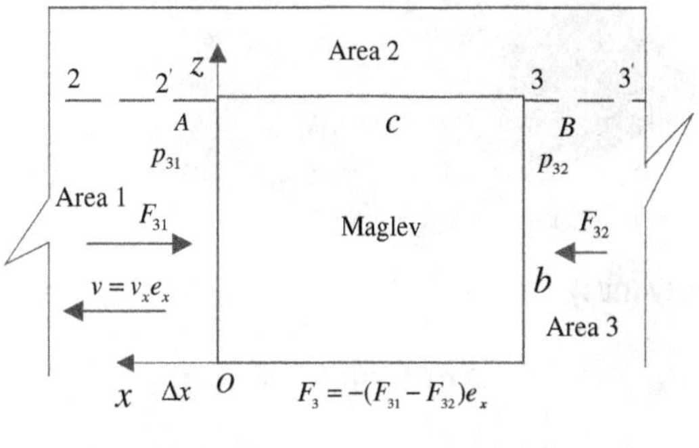

Fig. 5 Schematic diagram of the F_3

So the pressure difference is:

$$F_3 = -e_x \iint\limits_{yz} \frac{\rho}{2} v_x^2 \mathrm{d}y\mathrm{d}z. \tag{17}$$

2.4 Approximation of the total aerodynamic drag

The blockage rate is defined as

$$b_r = \frac{a \cdot b}{a_0 \cdot b_0} = \frac{S}{S_0}, \tag{18}$$

where a, b, a_0 and b_0 are also illustrated in Fig. 1.

When the train is running, the pressure, density, and flow velocity of arbitrary gas are functions of time and space. According to the assumptions and definitions mentioned above, Eqs. (9), (14), and (17) could be modified as:

$$F_1 = \begin{cases} 0, \\ -e_x v_c \iint\limits_{yz} \frac{p}{2p_0} \rho_0 v_x dydz \end{cases} = \begin{cases} 0, & v_x < v_c \\ -b_r S_0 \frac{p}{2p_0} \rho_0 v_c v_x e_x & v_x \geq v_c \end{cases} \tag{19}$$

$$\begin{aligned} F_2 &= -\boldsymbol{e}_x \left[\iint\limits_{S_{2U}} \mu \frac{\partial f_U(n)}{\partial n} dxdy + \iint\limits_{S_{2B}} \mu \frac{\partial f_B(n)}{\partial n} dxdy \right. \\ &\quad \left. + \iint\limits_{S_{2L}} \mu \frac{\partial f_L(n)}{\partial n} dxdz + \iint\limits_{S_{2R}} \mu \frac{\partial f_R(n)}{\partial n} dxdz \right] \\ &= -\boldsymbol{e}_x \mu c v_x \left(\frac{a}{b_0 - b - h_0 - h} + \frac{a}{h + h_0} + \frac{4b}{a_0 - a} \right) \\ &= -\boldsymbol{e}_x \mu c v_x \sqrt{S_0 b_r}. \end{aligned} \tag{20}$$

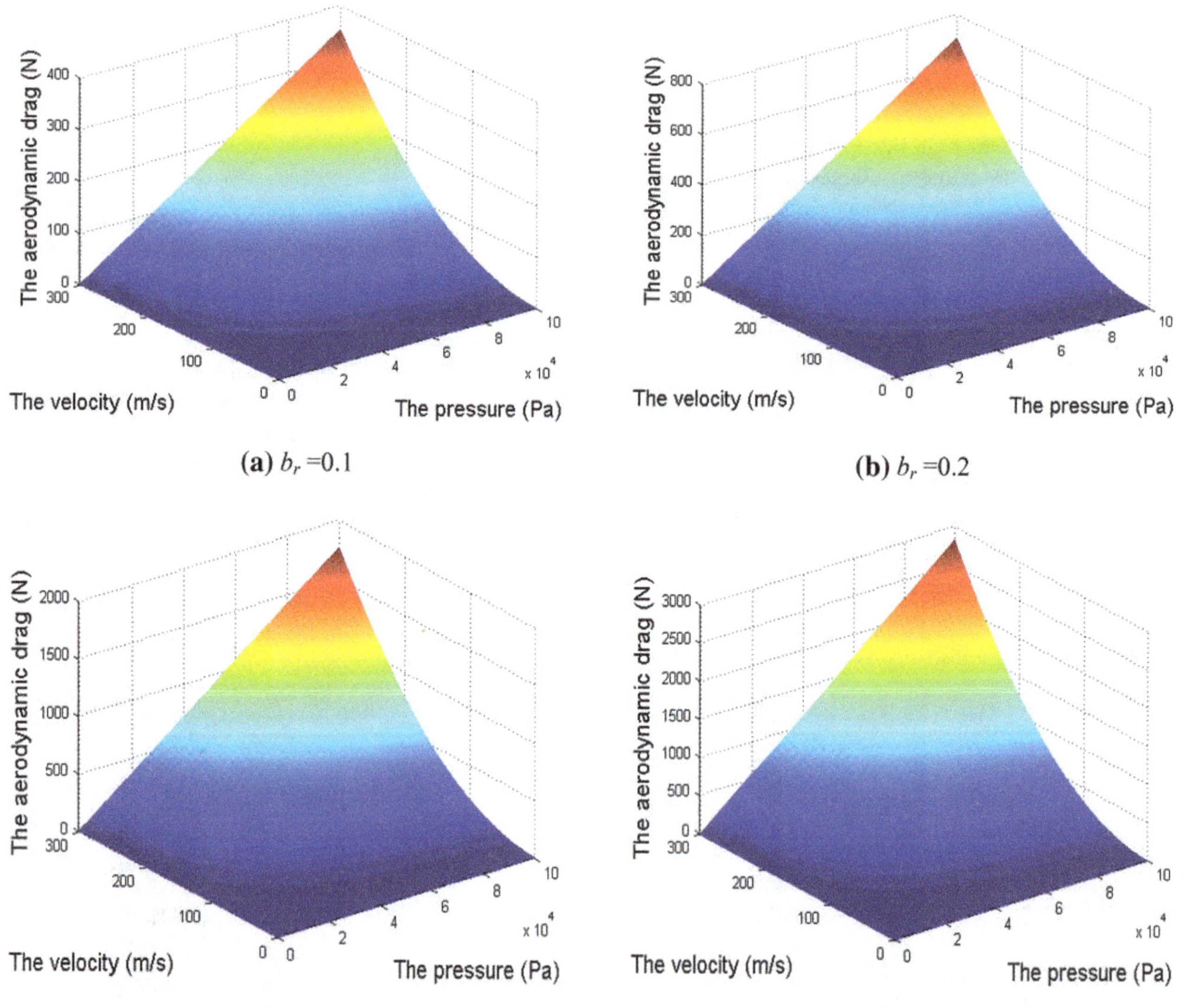

(a) b_r=0.1 (b) b_r=0.2 (c) b_r=0.5 (d) b_r=0.8

Fig. 6 The total calculated drag when blockage rate is 0.1, 0.2, 0.5, 0.8, respectively

$$F_3 = -e_x \iint\limits_{yz} (0 + \frac{\rho}{2} v_x^2) dydz = -e_x \int\limits_{-a/2}^{a/2} \int\limits_{0}^{b} (0 + \frac{p\rho_0}{2p_0} v_x^2) dydz = -e_x b_r S_0 \frac{p\rho_0}{2p_0} v_x^2 \quad (21)$$

The total aerodynamic drag is expressed as

$$F_x(b_r, v_x, p) = \begin{cases} b_r S_0 \frac{p}{2p_0} \rho_0 v_x^2 + \mu c v_x \sqrt{S_0 b_r}, & v_x < v_c, \\ b_r S_0 \frac{p}{2p_0} \rho_0 v_x (v_x + v_c) + \mu c v_x \sqrt{S_0 b_r}, & v_x \geq v_c. \end{cases} \quad (22)$$

According to Eq. (22), the relations between the total drag and the blockage ratio, the velocity and the pressure are shown in Fig. 6.

According to Fig. 6, we can calculate the total aerodynamic drag with Eq. (22) and the known parameters of blockage ratio b_r, velocity v, and pressure p, and easily obtain the relationship between them.

When $T = 288.15$ K, $\mu = 1.78 \times 10^{-5}$ kg/(m·s) and $v_c = 340$ m/s, the effect of the pressure variation could be neglected.

3 The experimental system of the evacuated tube

Figure 7 shows the ETSMT located in a tube made up of Perspex. It is vacuumized with a vacuum pump and the pressure inside the pipe can be detected by an instrument. We designed a control system to gain a fixed pressure ranging from 2,000 to 101,325 Pa. The experimental steps are listed:

1) The HTS maglev was fixed above the PMG with non-ferromagnetic material at some height such as 0.01 m and then the liquid nitrogen was poured into the train. After the train was levitated, the non-ferromagnetic material must be removed from the PMG.
2) The opening hole of the pipe was covered and then the vacuum pump was started with the control system to reach the design pressures such as 10,000, 8,000, and 5,000 Pa and etc.
3) The liner induction motor was started and then the train could be drove to move when the maglev train runs near the LIM. Thus, the train speed can be accelerated to a certain value such as 3 m/s.
4) The LIM was stopped when the train's speed reached an expected value. Then the time difference between the position check points A and B was recorded to gain the decreasing train velocity.
5) The velocity was calculated with necessary parameters.

All parameters of this experimental system in Fig. 1 are listed in Tables 1 and 2.

Table 1 Each fixed parameters in ETSMT

Parameters	Value(mm)
a_0	245
b_0	250
a_1	70
c	110
h_0	35
h	10

Table 2 Each experimental parameters in ETSMT

b_r	a (mm)	b (mm)
0.10	100	58.8
0.12	100	70.6
0.15	120	73.5
0.18	130	81.5
0.20	140	84

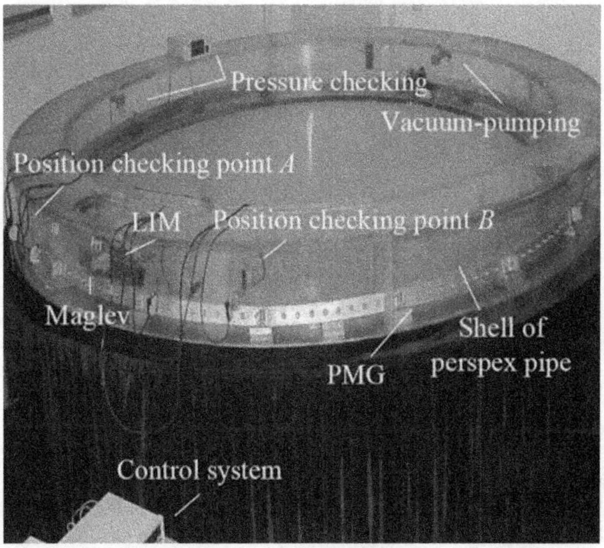

Fig. 7 The experimental system of ETT

4 The comparisons of theoretical and experimental results

The total aerodynamic drag of the running train cannot be measured directly because the train is freely levitated above the PMG. The average velocity between check points A and

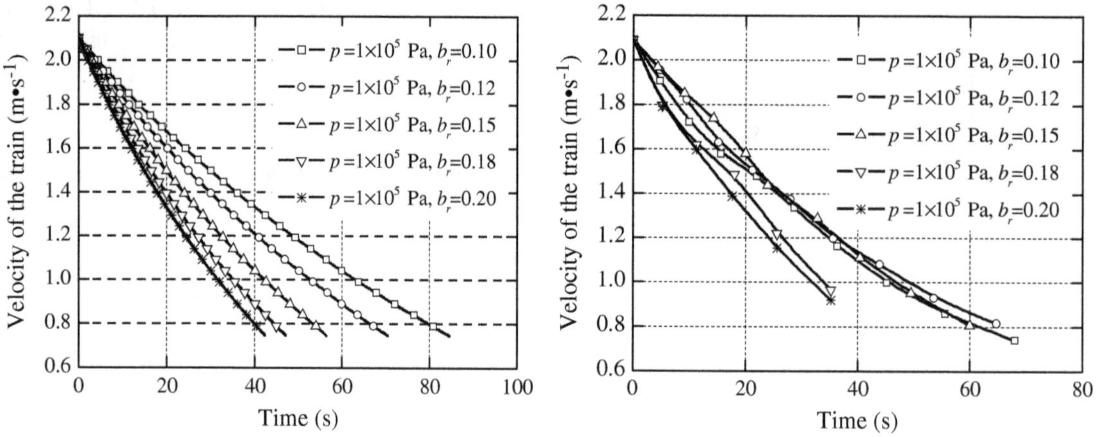

(a) The calculated results **(b)** The experimental results

Fig. 8 The relation between the velocity and the time when the pressure is 101,325 Pa

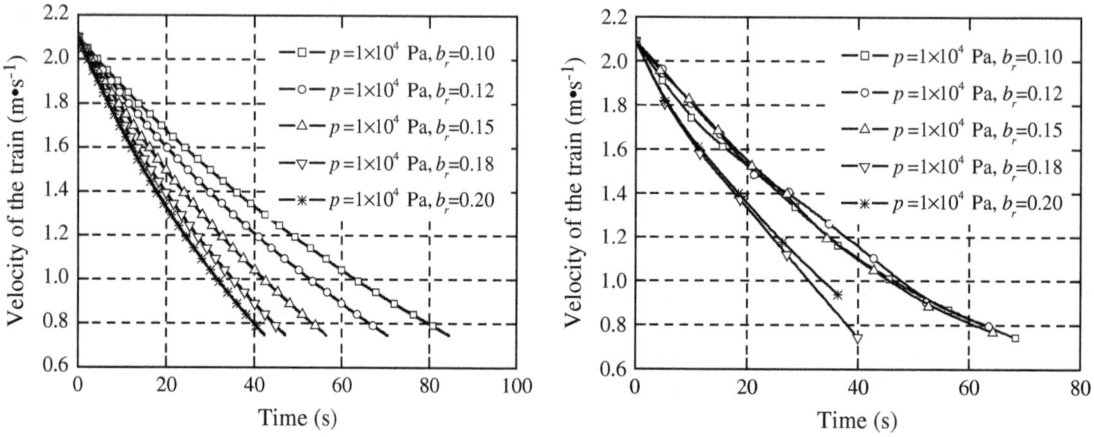

(a) The calculated results **(b)** The experimental results

Fig. 9 The relation between the velocity and the time when the pressure is 10,000 Pa

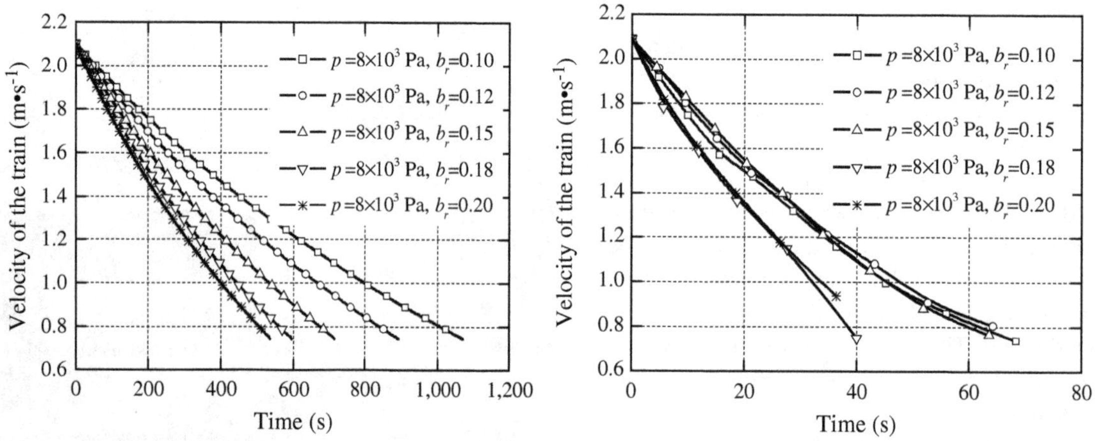

(a) The calculated results **(b)** The experimental results

Fig. 10 The relation between the velocity and the time when the pressure is 8,000 Pa

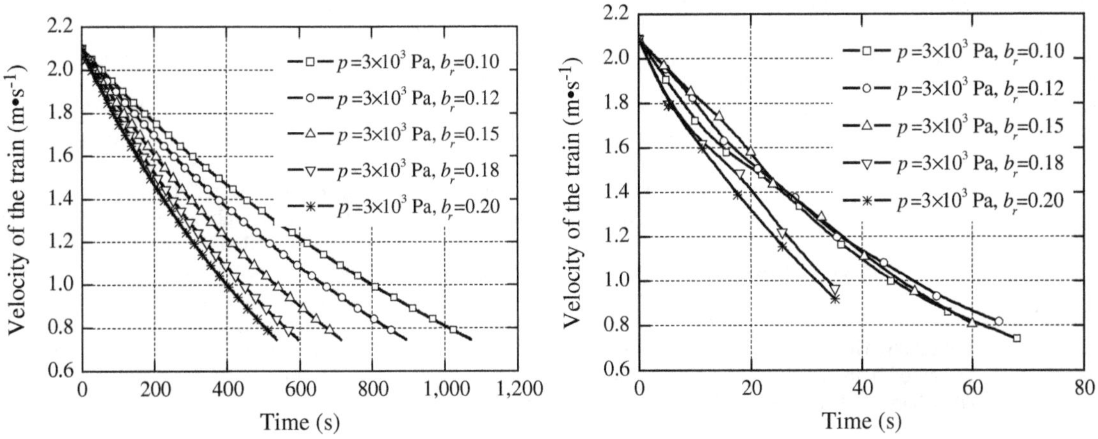

(a) The calculated results **(b)** The experimental results

Fig. 11 The relation between the velocity and the time when the pressure is 3,000 Pa

B in Fig. 7 can be calculated by the measured time difference and the length of arc $\overrightarrow{AB}$. The train velocity in experiment was speeded up to 2.2 m/s and then the linear motor was stopped. Figure 8a shows the relationship between the decreasing velocity and the time according to Eq. (22), while Fig. 8b shows the experimental result.

When the pressure is constant in the inner tube, the running time is less and the blockage ratio is larger. That is to say, if the blockage ratio is larger, so is the aerodynamic drag. Both the calculated and the experimental results show such a trend.

Figures 9, 10, and 11 illustrate the relation between the velocity and the time under different pressures. We can see if the pressure is decreased, the running time is longer because the negative acceleration is smaller. Let the velocity be 2.2 m/s and the blockage rate be 0.2 in Eq. (22). If the pressures are 101,325, 10,000, 8,000, and 3,000 Pa, the drags are 0.0,363, 0.0,036, 0.0,029, and 0.0,011 N respectively.

5 Conclusions

(1) When the pressure and the blockage ratio are constant, the aerodynamic drag is a quadratic function of the velocity. When the velocity of the train is bigger than the sound velocity, the formula of the aerodynamic drag becomes more complex.
(2) If the blockage ratio is smaller, the drag becomes smaller. In practice, the blockage ratio is impossible to be very small because of the limitation of the pipe's section size. So a suitable blockage ratio should be determined in the design of the ETT system.
(3) If the pressure in the tube is zero, the aerodynamic drag equals to zero no matter how the velocity and the blockage rate vary. This ideal condition is difficult to realize because of the technological limitation.

In this work, the comparison between the theoretical and experimental results was made when the velocity of the Maglev train is small. When the system runs in a lower pressure, more efforts must be made to solve more sophisticated technical problems. Thus, the speed, the pressure, and the blockage ratio each must have reasonable values for ETT. In such a case, the magnetic drag between the Maglev train and PMG may be negligible. In future study, we will consider the effect of magnetic drag between the Maglev train and PMG at high speeds.

Acknowledgments This paper was supported by the National Magnetic Confinement Fusion Science Program (No. 2011GB112001), the Program of International S&T Cooperation (No. S2013ZR0595), the Fundamental Research Funds for the Central Universities (Nos. SWJTU11ZT16, SWJTU11ZT31), the Science Foundation of Sichuan Province (No. 2011JY0031,2011JY0130).

References

1. Meins J, Miller L, Mayer WJ (1998) The high speed maglev transportation system transrapid. IEEE Transactions on Magn 24(2):808–811
2. Lee HW, K KC, Lee J (2006) Review of maglev train technologies. IEEE Trans Magn 42(7):1917–1925
3. Shen Z (2001) Dynamic interaction of high speed maglev train on girders and its comparison with the case in ordinary high speed railways. J Traffic Transp Eng 1(1):1–6 (in Chinese)
4. Shen Z (2005) On developing high-speed evacuated tube transportation in China. J Southwest Jiaotong Univ 40(2):133–137 (in Chinese)

5. Yan L (2006) Progress of the maglev transportation in China. IEEE Trans Appl Supercond 16(2):1138–1141
6. Raghuathan S, Kim HD, Setoguchi T (2002) Aerodynamics of high speed railway train. Prog Aerosp Sci 38(6):469–514
7. Cai YG, Chen SS (1997) Dynamic characteristics of magnetically levitated vehicle systems. Appl Mech Rev, ASME 50(11): 647–670
8. Wu Q, Yu H, Li H (2004) A study on numerical simulation of aerodynamics for the maglev train. Railw Locomot & CAR 24(2):18–20 (in Chinese)
9. Xu W, Liao H, Wang W (1998) Study on numerical simulation of aerodynamic drag of train in tunnel. J China Railw Soc 20(2):93–98 (in Chinese)
10. Shu X, Gu C, Liang X et al (2006) The Numerical simulation on the aerodynamic performance of high-speed maglev train with streamlined nose. J Shanghai Jiaotong Univ 40(6):1034–1037 (in Chinese)
11. Zhou X, Zhang D, Zhang Y (2008) Numerical simulation of blockage rate and aerodynamic drag of high-speed train in evacuated tube transportation. Chin J Vacuum Sci Technol 12(28):535–538 (in Chinese)
12. Chen X, Zhao L, MA J, Liu Y (2012) Aerodynamic simulation of evacuated tube maglev trains with different streamlined designs. J Mod Transp 20(2):115–120
13. Jiang J, Bai X, Wu L, Zhang Y (2012) Design consideration of a super-high speed high temperature superconductor maglev evacuated tube transport(I). J Mod Transp 20(2):108–114
14. Fletcher CAJ (1900) Computational Techniques for Fluid Dynamics(Vol.I and II). Springer-Verlag, Berlin
15. Launder BE, Spalding DB (1974) The numberical computation of turbulent flows [J]. Comput Methods Appl Mech Eng 3:269–289
16. Qian Y (2004) Aerodynamics (The first edition). Beihang University Press, Beijing (in Chinese)

13

Design and preliminary validation of a tool for the simulation of train braking performance

Luca Pugi · Monica Malvezzi · Susanna Papini · Gregorio Vettori

Abstract Train braking performance is important for the safety and reliability of railway systems. The availability of a tool that allows evaluating such performance on the basis of the main train features can be useful for train system designers to choose proper dimensions for and optimize train's subsystems. This paper presents a modular tool for the prediction of train braking performance, with a particular attention to the accurate prediction of stopping distances. The tool takes into account different loading and operating conditions, in order to verify the safety requirements prescribed by European technical specifications for interoperability of high-speed trains and the corresponding EN regulations. The numerical results given by the tool were verified and validated by comparison with experimental data, considering as benchmark case an Ansaldo EMU V250 train—a European high-speed train—currently developed for Belgium and Netherlands high-speed lines, on which technical information and experimental data directly recorded during the preliminary tests were available. An accurate identification of the influence of the braking pad friction factor on braking performances allowed obtaining reliable results.

Keywords Braking performances · Friction behavior of braking pads · Prediction tool

L. Pugi (✉) · S. Papini · G. Vettori
Department of Industrial Engineering, University of Florence, Via Santa Marta 3, 50139 Florence, Italy
e-mail: luca.pugi@unifi.it

M. Malvezzi
Department of Information Engineering and Mathematical Science, University of Siena, Via Roma 26, 53100 Siena, Italy

1 Introduction

Braking performance is a safety relevant issue in railway practice, impacting vehicle longitudinal dynamics, signaling, and traffic management, and its features and requirements are important also for interoperability issues [1].

EN 14531 regulation [2] provides indications concerning preliminary calculation of braking performance, giving a general workflow that can be adapted to different vehicle categories:

- Freight wagons,
- Mass transit,
- Passenger coaches,
- Locomotives, and
- High-speed trains.

The aim of the regulation [2] is to set a general method that should be shared among different industrial partners (industries, railway operators, safety assessors, etc.).

The availability of software tools aimed to simulate the performance of braking system is useful to speed up and optimize the design process [3]. Braking performance evaluation is also necessary to properly quantify the intervention curve of automatic train protection (ATP) systems [4, 5]. Some examples of train brake system simulators are available in the literature. In [6], David et al. presented a software tool for the evaluation of train stopping distance, developed in C language. In [7], the software TrainDy was presented; it was developed to reliably evaluate the longitudinal force distribution along a train during different operations. In [8], Kang described a hardware-in-the-loop (HIL) system for the braking system of the Korean high-speed train and analyzed the characteristics of the braking system via real-time simulations. In [9], many interrelationships between various factors and types of braking techniques were analyzed.

A simple but reliable tool able to simulate and predict the performances of braking system on the basis of a limited and often uncertain set of parameters could be useful and give interesting information to the designers on how to choose and optimize brake features, especially in the first phase of the design process of a new train.

In this work, the authors have developed a Matlab^{TM} tool called "TTBS01", which implements the method for the calculation of braking performances described in [2]. The tool has been validated on experimental results concerning AnsaldoBreda EMU V250. The results, which will be detailed through this paper, showed an acceptable agreement with experimental tests, and then confirmed the reliability of the proposed tool and its applicability to the prediction of stopping distance of different types of trains in various operative conditions, including degraded conditions and failure of some subsystems. The proposed tool can thus be adopted in the design phase to choose proper dimensions of the braking system components and to preliminarily evaluate their performance.

Since the detailed description of the calculation method is directly available on the reference regulation [2], in this work, the authors will give a more general description of the algorithm, focusing mainly on the considered test case, the numerical results, and the matters that have proven to be critical during the validation activities. A particular attention has been paid to some features that are originally not prescribed by the regulations in force, but could be considered to further increase result accuracy and reliability. In particular, some parameters, such as friction factor of braking pads, which should be slightly variable according to different operating conditions, were identified and tabulated.

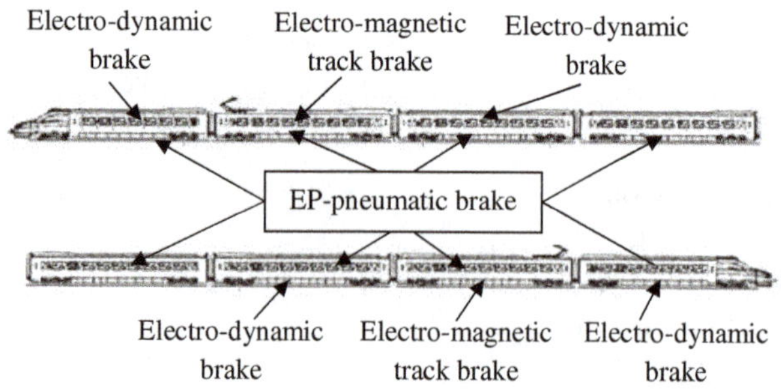

Fig. 1 EMU V250 vehicle composition and braking plant layout

2 The test case: the EMU V250 train

The simulation tool described in this paper, named "TTBS01", was tested and validated using the data obtained on an Ansaldo EMU V250 train: a high-speed electrical multiple unit for passenger transport with a maximum operating speed of 250 km/h (maximum test speed 275 km/h), composed of two train sets of eight coaches. The traction is distributed with alternating motor and trailer vehicles in the sequence "MTMTTMTM", where M indicates motorized coaches and T the trailer ones. The arrangement of each motorized wheelset is B0–B0. Train composition is shown in Fig. 1: the motorized coach traction motors can be used for electro-dynamic braking types, both regenerative and dissipative. The 2nd and the 7th coaches are equipped with an electro-magnetic track brake that should be adopted in emergency condition. The mandatory pneumatic braking system is implemented with the support of both direct and indirect electro-pneumatic (IEP) operating modes: the braking command can be directly transmitted by wire to the BCU (braking control unit) on each coach, or indirectly, by controlling the pressure of the pneumatic pipe, as seen in the simplified scheme shown in Fig. 2.

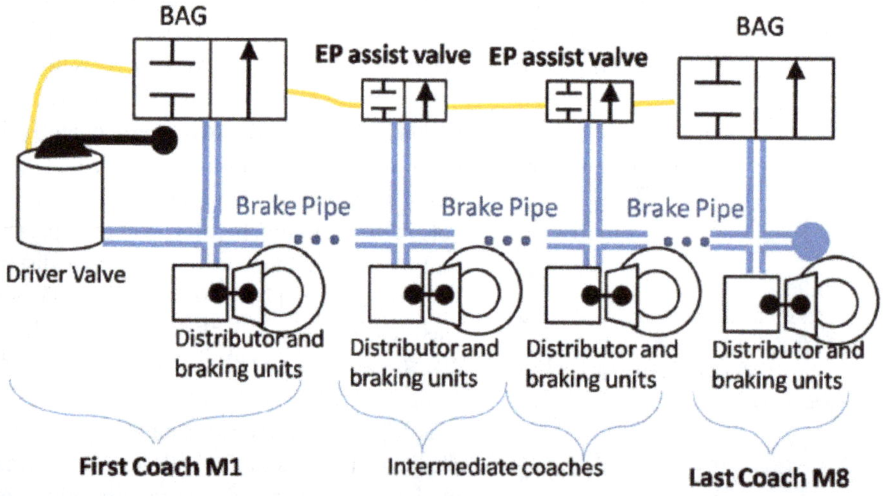

Fig. 2 Braking plant in the IEP mode

Fig. 3 Brake disks on trailer bogie

Finally, a backup mode where the brake plant is controlled as a standard pneumatic brake ensures interoperability with vehicles equipped with a standard UIC brake. Each axle is equipped with three brake disks for trailing axles (as in Fig. 3), and two for the motorized ones, where electric braking is available, too. In this configuration, the magnetic track brake should be available, since a pressure switch commanded using the brake pipe controls the track lowering (threshold at 3 bar absolute).

The corresponding configuration of the pneumatic brake plant and the inertia values used for calculations are described in Tables 1 and 2.

2.1 Further controls: double pressure stage and load sensing

The pressure applied to brake cylinders and consequently the clamping and braking forces are regulated as a function of train mass (load sensing) and speed (double pressure stage). Load sensing allows optimizing braking performance with respect to vehicle inertia and weight. Double pressure stage allows protecting friction components against excessive thermal loads (double pressure stage). Both the systems allow preventing over-braking: according to the regulations [1] and [10], braking forces applied to wheels have to be limited, in order to prevent over-braking, defined as "brake application exceeding the available wheel/rail adhesion".

In particular, the braking forces are usually regulated, e.g. on freight trains, using a load-sensing pressure relay, simplified scheme of which is represented in Fig. 4. A sensing device mounted on the primary suspension stage produces a pressure load signal that is approximately proportional to the axle load. The reference pilot pressure command, produced by the brake distributor, is amplified by the relay in order to feed brake cylinders, using the leverage schematically represented in Fig. 4. The systems work as a servo pneumatic amplifier with a pneumo-mechanic closed-loop regulation, aiming to adapt the pneumatic impedance of the distributor output to the flow requirements of the controlled plant. The gain is adjustable since the pivot of the leverage, and consequently, the amplification ratio is regulated by the pressure load signal.

Table 1 Main parameters of the braking plant [5, 6]

Coach	Bogie	Wheel diameter (new) (mm)	Wheel diameter (worn) (mm)	Brake radius (mm)	Number of disks/ axle	Dynamic pad friction level	Brake actuator piston surface (cm^2)	Spring counter force/actuator (N)	Caliper efficiency	Ratio of the caliper
M1	1	920	850	299	2	0.42	506,7	1,300	0.95	2.82
	2	920	850	299	2	0.42	506,7	1,300	0.95	2.82
T2	3	920	850	243	3	0.42	506,7	1,300	0.95	2.69
	4	920	850	243	3	0.42	506,7	1,300	0.95	2.69
M3	5	920	850	299	2	0.42	506,7	1,300	0.95	2.82
	6	920	850	299	2	0.42	506,7	1,300	0.95	2.82
T4	7	920	850	243	3	0.42	506,7	1,300	0.95	2.69
	8	920	850	243	3	0.42	506,7	1,300	0.95	2.69
T5	9	920	850	243	3	0.42	506,7	1,300	0.95	2.69
	10	920	850	243	3	0.42	506,7	1,300	0.95	2.69
M6	11	920	850	299	2	0.42	506,7	1,300	0.95	2.82
	12	920	850	299	2	0.42	506,7	1,300	0.95	2.82
T7	13	920	850	243	3	0.42	506,7	1,300	0.95	2.69
	14	920	850	243	3	0.42	506,7	1,300	0.95	2.69
M8	15	920	850	299	2	0.42	506,7	1,300	0.95	2.82
	16	920	850	299	2	0.42	506,7	1,300	0.95	2.82

Table 2 Vehicle loading conditions and inertia values for braking plant calculation [5, 6]

Coach	Bogie	VOM load (Tare) (t)	TSI load (t)	CN load (normal) (t)	CE load (exceptional) (t)	Bogie mass (t)	Rotating mass/axle (t)
M1	1	15.9	16.7	17	17.6	9.93	1.5
	2	13.9	15	15.4	16.3	9.81	1.5
T2	3	13.9	15	15.3	16.6	7.85	0.6
	4	14	15.1	15.4	16.5	7.85	0.6
M3	5	13.6	14,8	15.2	16.1	9.81	1.5
	6	14.1	15.5	15.9	16.8	9.81	1.5
T4	7	11.2	12.8	13.3	14.2	7.85	0.6
	8	12.1	13.7	14.2	15	7.85	0.6
T5	9	12	13.6	14.1	14.9	7.85	0.6
	10	11.3	12.8	13.2	14.1	7.85	0.6
M6	11	14.1	15.7	16.2	17	9.81	1.5
	12	13.8	15.3	15.8	16.7	9.81	1.5
T7	13	14	15.6	16.1	16.9	7.85	0.6
	14	14.1	15.6	16.1	17	7.85	0.6
M8	15	13.7	15.2	15.7	16.5	9.81	1.5
	16	15.9	16.9	17.2	17.8	9.93	1.5
Train mass (t)		435.2	478.6	492.2	520		
Train rotating mass (t)							33.6

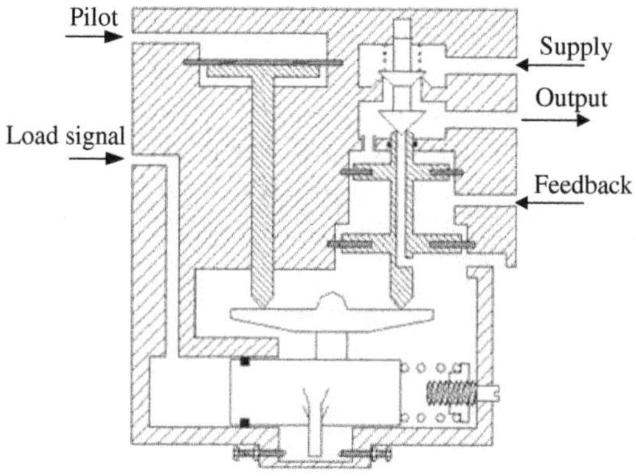

Fig. 4 Pressure relay/load-sensing device

On freight trains, where the difference between the tare and fully loaded vehicle masses could be in the order of 300 % (from 20 to 30 t/vehicle for the empty wagon to 90 t/vehicle for the fully loaded one), load sensing is very important. For high-speed trains, such as EMU V250, the difference between VOM and CE loading conditions, as visible in Table 2, is not in general lower than 10 %–20 %.

As a consequence, the corresponding variation in terms of deceleration and dissipated power on disks is often numerically not much relevant and is partially tolerated by regulations in force [10] for high-speed trains with more than 20 axles, in emergency braking condition or in other backup mode, where the full functionality of the plant should not be completely available.

For the reasons of safety, the correct implementation of the double stage pressure ensuring that lower pressure is applied on cylinders for traveling speed of over 170 km/h is much more important. This is important because the energy dissipated during a stop braking increases approximately with the square of train traveling speed and, as a consequence, a reduction of disk clamping forces may be fundamental to avoid the risk of excessive thermal loads. Furthermore, the adhesion limits imposed by [10] prescribe a linear reduction of the braking forces between 200 and 350 km/h, according to a linear law which corresponds to a reduction of the braking power of about one-third in the above-cited speed range.

2.2 Electrical braking and blending

Electrical or electro-dynamical brakes are a mandatory trend for a modern high-speed train. Most of the more modern EMUs have the traction power distributed over a high number of axles. On EMU V250 train, nearly 50 % of the axles is motorized and nearly 55 % of the total train weight is supported by motorized bogies.

As a consequence, a considerable amount of the total brake effort should be distributed to traction motors, by performing regenerative or dissipative braking, according to the capability of the overhead line for managing the corresponding recovered power. In particular, not only regenerative but also dissipative electric braking is quite attractive, considering the corresponding reduction of wear of friction braking components such as pads and disks. Since electric braking is applied in parallel with the conventional pneumatic one, an optimized mixing strategy in the usage of both systems, usually called blending, has to be performed.

In Fig. 5, the electric braking effort available on a motorized coach as a function of the train traveling speed and of the electrification standard of the overhead line is shown. Three different operating conditions can be recognized:

- Maximum pneumatic braking force: under a certain traveling speed, the corresponding operating frequencies of the traction system are too low. On the other hand, also the demanded braking power is quite low, and so it can be completely managed by means of the pneumatic braking system.
- Minimum pneumatic braking: in this region, the electric braking effort is limited to a maximum value, often related to the motor currents. If a higher braking effort is required, then the pneumatic brake is activated to supply the difference.
- Pneumatic braking increases to supply insufficient electric power: as speed increases, the performances of the motor drive system are insufficient to manage the corresponding power requirements, limiting the maximum braking effort to the associated iso-power curve. As a consequence, the contribution of the pneumatic braking power tends to increase with speed.

3 Summary of the European standards for brake calculation

The EN 14531 (first draft 2003) describes the fundamental algorithms and calculations for the design of brake equipment for railway vehicles. The procedure provides the calculation of various aspects related to the performance: stopping or slowing distances, dissipated energy, force calculations, and immobilization braking. For the purposes of this work, the Part 6 of the regulation: "Application to high-speed trains" is of interest. The general algorithm to calculate braking distances is described in the regulation: the input data consist of train and brake characteristics, and the method to estimate the deceleration as a combination of different braking forces acting on the train is suggested as a function of the initial speed [1]. Moreover, the criteria for the technical and operational compatibility between the infrastructures and the rolling stock are defined in L.245/402 technical specification for interoperability (TSI) published in the *Official Journal of the European Communities* in 2002. The essential requirements for trans-European high-speed rail systems are related to safety, reliability, availability, health, environmental protection, and technical compatibility. Notably, the brake system requirements for high-speed rail systems are established; i.e., the minimum braking performance is defined as the minimum deceleration and evaluated as a function of speed [2]. On the other hand, the European norm UIC544-1 (4th edition, October 2004) defines the method for computation of the braking power through the braked mass and determination of the deceleration [3].

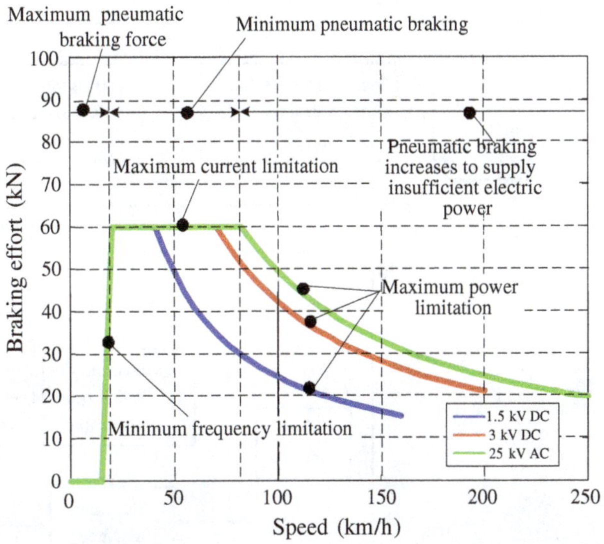

Fig. 5 Typical behaviors of electric and pneumatic braking efforts on motorized bogies

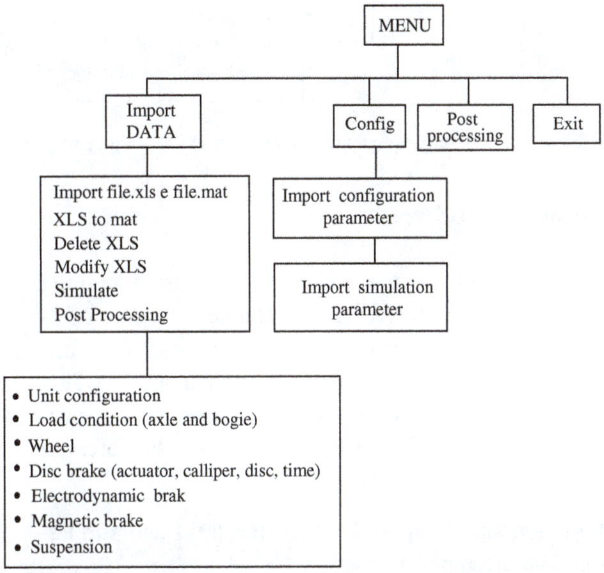

Fig. 6 Interface structure of the TTBS01 tool

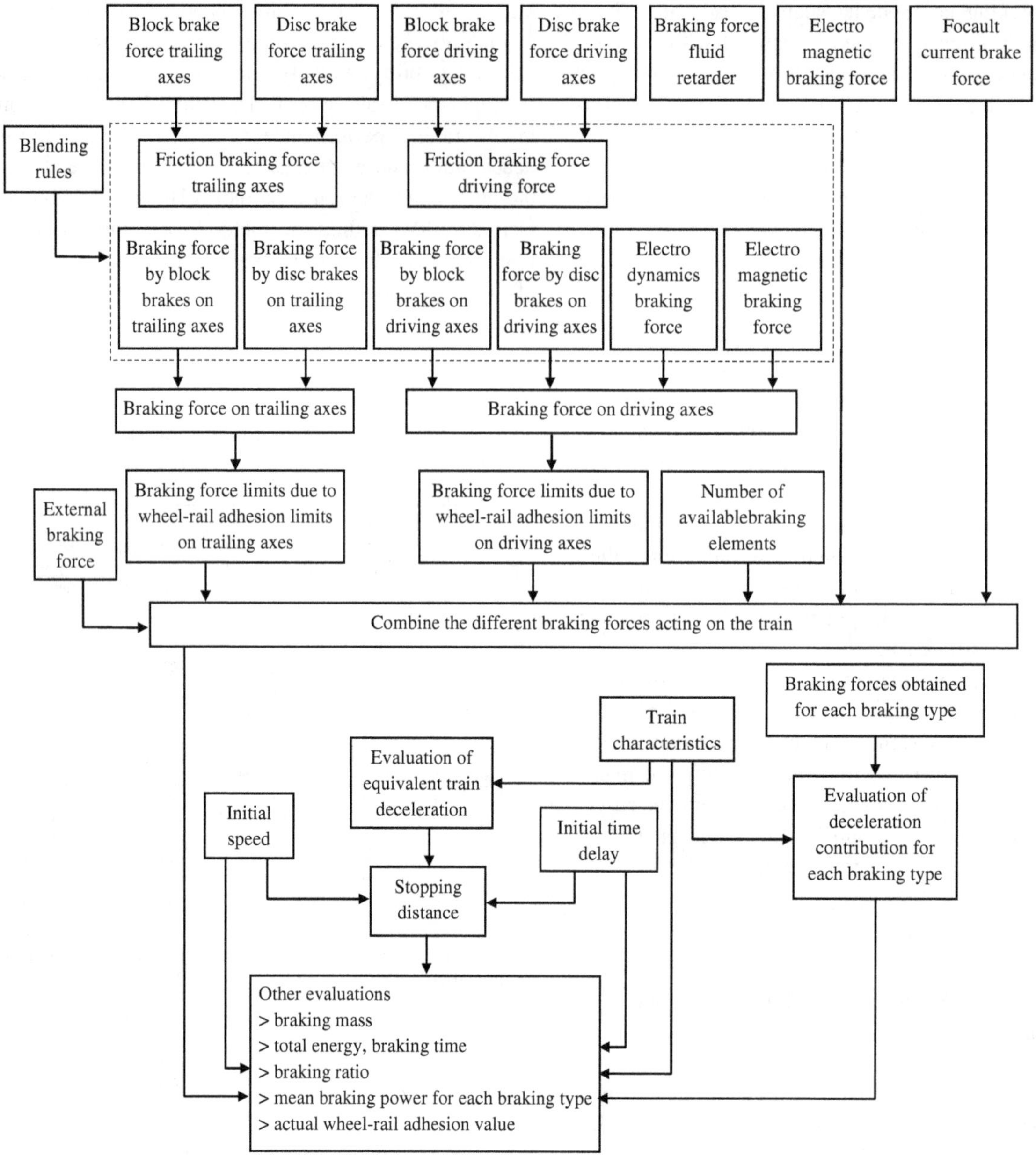

Fig. 7 Flow chart of braking calculations performed according to [2]

4 Software—TTBS01

The software tool for the computation of train braking systems, named TTBS01, has been implemented in Matlab^{TM}. The algorithm provides a graphical user interface (GUI) to help the user to insert and modify input data. It is organized in different windows and grouped in four sections, as shown in the scheme of Fig. 6 and the software user interface in Fig. 8.

- Pre-processing (Import DATA): the train and simulation data are input by user.
- Configuration (Config.): data are saved and stored in files.
- Calculation: braking system calculation is performed according to [2], and the corresponding flowchart is shown in Fig. 7.
- Post-processing: the user can view the representative brake output in several charts.

5 Tool validation

The validation of tool results was carried out by comparing simulation results with test data [11, 12].

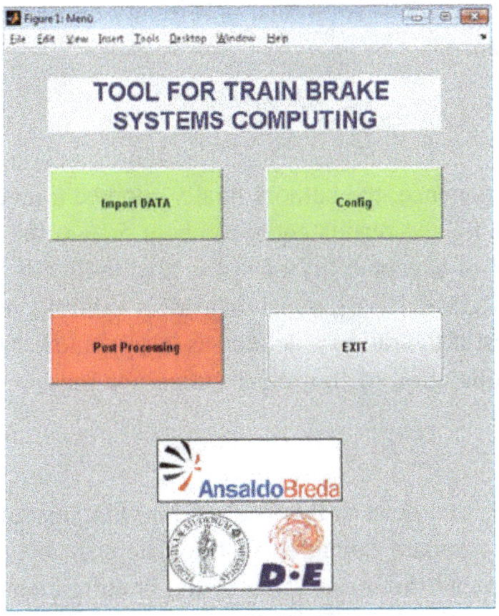

Fig. 8 Main menu window of TTBS01

Totally, a population of about 50 braking test runs was investigated, which were performed on a train equipped with the sensor layout described in Table 3.

The brake performance test concerns the emergency and service braking at several initial speeds, considering the different working and operating conditions of the braking system (direct electro-pneumatic, indirect electro-pneumatic, pneumatic, etc.). The test runs were finished in normal adhesion condition, where the wheel slide protection (WSP) system did not work. The test runs were performed on a complete V250 unit, coaches of which had passed all the single-coach tests, with a fully working braking system (all other subsystems involved in the braking functionality).

The braking runs for the test procedure were performed in three different load conditions: VOM, TSI, and CE, as defined in [1]:

- VOM load condition, defined as mass empty, ready for departure;
- TSI load condition, corresponding to mass normal load; and
- CE load condition, defined as mass exceptional load.

5.1 Acceptance criteria

In order to verify and validate the TTBS01 simulation tool, the relative error e_s between the simulated stopping distance s_{simul} and the experimental one s_{test} is defined as (1), and the corresponding speed and acceleration profiles have been evaluated.

$$e_s = \frac{s_{test} - s_{simul}}{s_{simul}}. \tag{1}$$

According to [13–16], the repeatability of braking performances in terms of mean deceleration has to satisfy the requirements summarized in Table 4, where the probability of degraded braking performances is shown. The relative error on stopping–braking distance s, for an assigned initial speed v_0, is approximately proportional to the mean deceleration, as stated by (2):

$$s = \frac{v_0^2}{2a} \Rightarrow \frac{\partial s}{\partial a} = \frac{v_0^2}{2a} \Rightarrow \frac{\partial s}{s} = -\frac{\partial a}{a}. \tag{2}$$

Table 3 Sensor layout adopted for experimental test runs on EMU V250 [5, 6]

	Pressure transducer	Radar Doppler sensor	Servo-acelerometer	Thermocouples
Accuracy	0.5 % respect to full range	±1 km/h	0.1 % respect to full range	K type
Range	0–12 bar	0–500 km/h	1 g	thermocouples
Quantity and layout	8 pressure transducer on brake plant	1/on a coach carbody	1/on a coach carbody	4/on disks

Table 4 Statistic distribution of degraded braking performances according to [7, 8]

Probability (no. of tests)	10^{-1} (10^1)	10^{-2} (10^2)	10^{-3} (10^3)	10^{-4} (10^4)	10^{-5} (10^5)
Mean deceleration	0.969	0.945	0.926	0.905	0.849
Nominal deceleration	(−3.1 %)	(−5.5 %)	(−7.4 %)	(−9.5 %)	(−15.1 %)

Table 5 Calculated longitudinal eigenfrequencies of EMU V250 according to [17] (Hz)

Compostion	First eigenfrequency	Second eigenfrequency	Third eigenfrequency	Fourth eigenfrequency	Fifth eigenfrequency
Standard (8 coaches)	2.4	4.7	6.9	6.9	8.8
Doubled (16 coaches)	1.2	2.4	3.6	4.8	5.9

Considering a population of 50 test runs, a 4 % error between simulation and test results was considered as acceptable.

The statistical distribution of the degraded braking performances defined according to [13, 14] is summarized in Table 4, which is referred to as a homogenous population of braking tests. Since in the campaign on EMU V250, each test was performed with different boundary and operating variables, a higher variability with respect to the expected simulation results should be expected.

In addition, some further considerations have to be made concerning longitudinal train oscillations. During the tests, a 1–2-Hz longitudinal mode was observed by both speed and acceleration sensors, which accorded with the results of a previous modal analysis [17] as shown in Table 5, and more generally with the typical longitudinal eigenfrequencies of train formations [18, 19]. In particular, the phenomenon is clearly recognizable from the acceleration profiles depicted in Fig. 9, while a qualitative comparison between experimental and simulation speed profiles, with respect to the linear regression curve built on experimental data, is shown in Fig. 10.

This phenomenon causes a variability of about 1–2 km/h on the measured speed with respect to the mean value (about 1 %–1.5 % with respect to the launching speed). The sensitivity of error on braking distance to the correct evaluation of the launching speed, as shown in (3), produces about 2–3 % additional uncertainty on estimated braking distance.

$$s = \frac{v_0^2}{2a} \Rightarrow \frac{\partial s}{\partial v_0} = \frac{v_0}{a} \Rightarrow \frac{\partial s}{s} = \frac{2\partial v_0}{v_0}. \qquad (3)$$

As a consequence, the authors finally adopted a level of acceptability for the results equal to about 5 %–6 %.

This level of acceptability of test is also indirectly prescribed by UIC544-1 [20], which considers valid the result of braking test if the ratio σ_r, defined as in (4), is lower than 0.03 for a population of four consecutive test runs.

$$\sigma_r = \frac{\sigma}{s_{mean}}, \qquad (4)$$

where s_{mean} is the mean of the measured braking distances, and σ is the standard deviation of the difference between the measured and the mean value of the braking distance.

Considering the definition of mean error and standard deviation, the condition (4) corresponds to an admissible relative error on the measured braking distances of about 6 %–6.5 %, which is thus larger than the one adopted for the TTBS01 validation procedure.

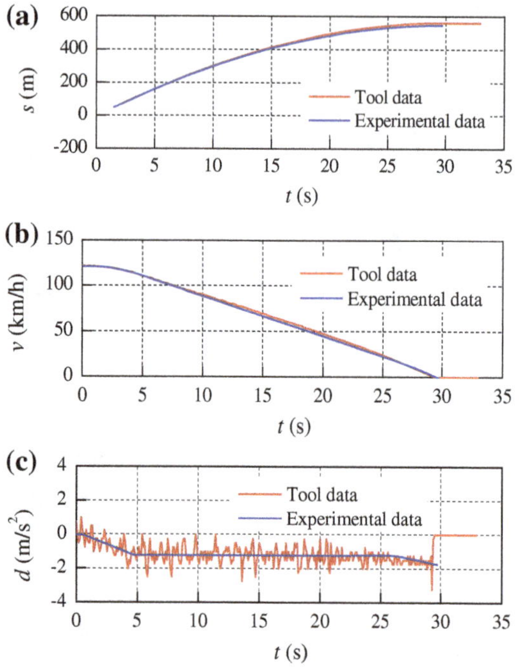

Fig. 9 Space (**a**), speed (**b**), and deceleration (**c**) profiles measured and calculated during a braking maneuver

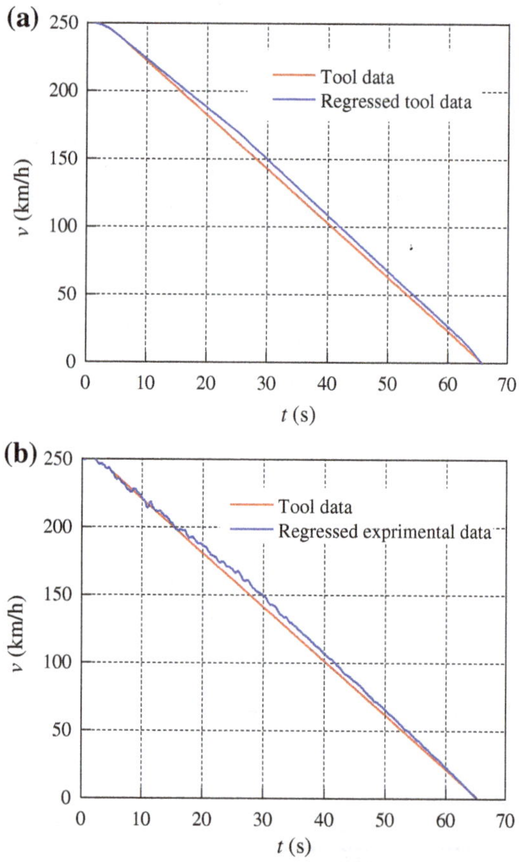

Fig. 10 Comparison between simulated (**a**) and experimental (**b**) speed profiles with linear regressed curves

5.2 Identification of brake pad friction factor and preliminary validation of the tool

Applying the TTBS01 procedure with the calculation described in [2] to the cases covered by the experimental data led to unsatisfactory results in terms of statistical distribution of the error e_s, as shown in Fig. 11: only 60 % of the simulated test runs were able to satisfy the requirements, even when considering a 5.5 % admissible value for e_s.

Taking the real behavior of a friction brake pad as the example of Fig. 12 [12, 14], the following considerations arise: the brake pad friction factor is clearly dependent on three parameters: the speed, the dissipated energy that mainly depends on clamping forces and starting speed, and the clamping forces applied to the pad. As a consequence, by adopting the measured data of the friction [19] and using a narrower population of tests on the train (four braking tests over a population of 50), we identified a feasible behavior of the pad friction factor as a function of the traveling speed and the loading condition of the train (Fig. 13). In fact, the clamping forces of the brakes are self-regulated according to the vehicle weight and the traveling speed, once the mean values of the clamping forces with respect to the dissipated power is fixed.

By modifying the software TTBS01 according to the proposed brake pad behavior, we obtained the results satisfying the criteria for the software validation, with an acceptable value of e_s lower than 5.5 % (exactly 5.35 %) as shown in Fig. 14. It is also worthy to point out that after the modification, the number of elements under the threshold of 2 %–4 % is more than doubled.

Finally, the first ten braking test simulations are compared with the experimental results in Figs. 15 and 16. One can see that a good-fitting agreement in terms of shape of speed profiles is evident. In particular, the results in Figs. 15 and 16 refer to emergency braking maneuvers

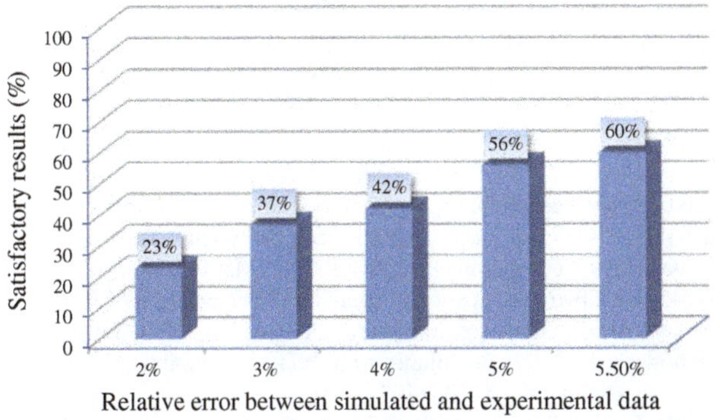

Fig. 11 Number of satisfactory simulated results as a function of the admissible value of e_s (constant brake pad friction faction)

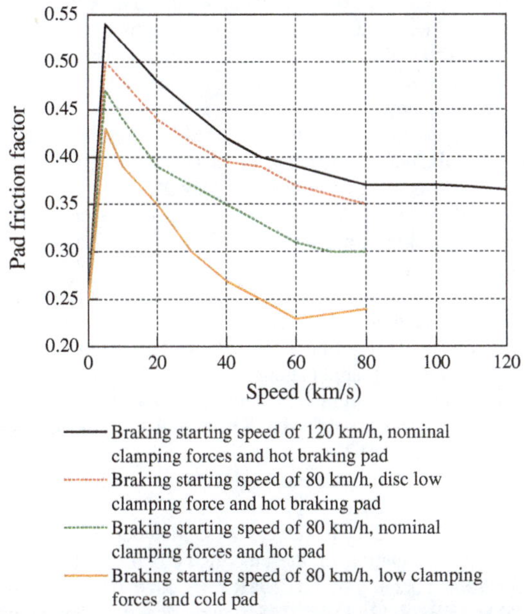

Fig. 12 Measured behavior of brake pad friction factor [12, 14], test performed on test rig [23] according UIC test program [22]

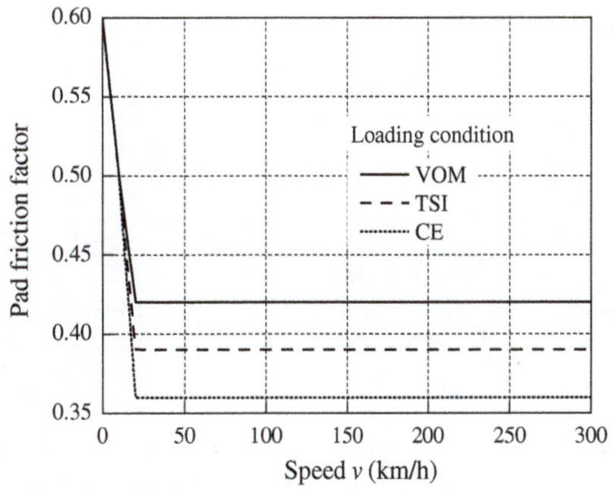

Fig. 13 Variable braking pad friction factor implemented on TTBS01 for the validation on EMU V250

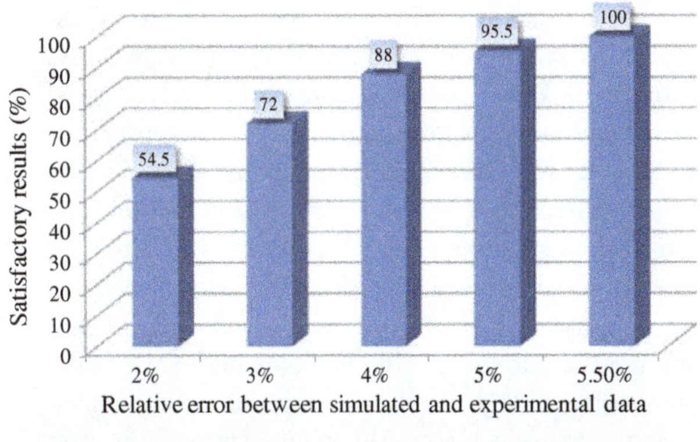

Fig. 14 Number of satisfactory simulation results as a function of the admissible value of e_s (variable pad friction factor is implemented)

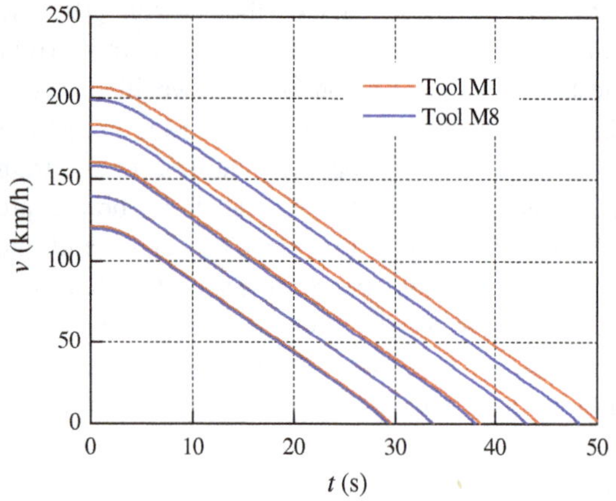

Fig. 15 Simulated test runs (different launching speed and motion sense) with emergency braking

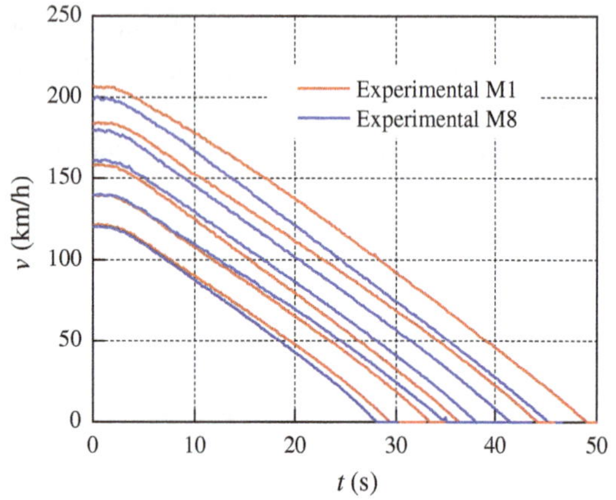

Fig. 16 Experimental speed profiles measured on ten braking test runs (different launching speed and motion sense) with emergency braking

performed in the VOM loading condition (vehicle tare), repeated twice in both the sense of motion over the line.

6 Conclusion

Preliminary validation of TTBS01 tool on EMU V250 experimental data has provided an encouraging feedback. As a consequence, TTBS01 should be considered both as a good tool for the preliminary simulation of braking systems and a base to build up real-time code for the monitoring of brake system performances. It is worthy to mention that the calculation method suggested by EN regulations in force [2] could be not reliable, since the typical behavior of braking forces, as influenced by braking pads, is not taken into account. For the purpose of UIC homologation [21], brake pads have to be widely tested, and even more complicated testing activities are performed by manufacturers. For each approved pad, a huge documentation concerning the variability of the friction factor with respect to speed and load conditions can be easily found. Therefore, the proposed method that calculates train braking performances by taking into account the variability of brake pad friction factors has a high feasibility. It is highly recommendable that the implementation of this feature in standard calculation methods is prescribed by regulations in force. Moreover, the use of reliability statistical methods proposed by ERRI documents should be further investigated.

Acknowledgments The authors wish to thank Ansaldo Breda for their competence and their practical and cooperative approach to problems, which greatly helped in realizing the positive conclusion of this research activity.

References

1. Technical specification for interoperability relating to the rolling stock subsystem of the trans-European high-speed rail system referred to in Article 6(1) of Council Directive 96/48/EC, 30 May 2002
2. EN14531 Railway applications—methods for calculation of stopping and slowing distances and immobilisation edn, 15 Sept 2009
3. Piechowiak T (2009) Pneumatic train brake simulation method. Veh Syst Dyn 47(12):1473–1492
4. Vincze B, Geza T (2011) Development and analysis of train brake curve calculation methods with complex simulation. Adv Electr Electron Eng 5(1-2):174–177
5. Yasunobu S, Shoji M (1985) Automatic train operation system by predictive fuzzy control. In: Sugeno Michio (ed) Industrial applications of fuzzy control. North Holland, Amsterdam, pp 1–18
6. David B, Haley D, Nikandros G (2001) Calculating train braking distance. In: Proceedings of the Sixth Australian workshop on Safety critical systems and software vol 3, Australian Computer Society, Inc., Sydney
7. Cantone L, Karbstein R, Müller L, Negretti D, Tione R, Geißler HJ (2008) TrainDynamic simulation—a new approach. In: 8th World Congress on Railway Research, May 2008
8. Kang Chul-Goo (2007) Analysis of the braking system of the Korean high-speed train using real-time simulations. J Mech Sci Technol 21(7):1048–1057
9. Wilkinson DT (1985) Electric braking performance of multiple-unit trains—Proceedings of the Institution of Mechanical Engineers, Part D. J Automob Eng 199(4):309–316
10. EN 15734-1 Railway applications—braking systems of high speed trains—part 1: requirements and definitions, Nov 2010
11. OBVT50 Brake performance test—vehicle type test procedure—EMUV250, 14 May 2010
12. OBVT50 Brake performance test—vehicle type test procedure—Test Report, 22 Nov 2010
13. UIC B 126/DT 414 UIC B 126/DT 414, Methodology for the safety margin calculation of the emergency brake intervention curve for trains operated by ETCS/ERTMS, June 2006
14. ERRI 2004 ERRI B 126/DT 407, Safety margins for continuous speed control systems on existing lines and migration strategies for ETCS/ERTMS, Nov 2004 (3rd draft)

15. Malvezzi M, Presciani P, Allotta B, Toni P (2003) Probabilistic analysis of braking performance in railways. In: Proc. of the IMechE, J Rail Rapid Transit, vol 217 part F, pp 149–165
16. Malvezzi M, Papini R. Cheli S, Presciani P (2003) Analisi probabilistica delle prestazioni frenanti dei treni per la determinazione dei coefficienti di sicurezza da utilizzare nei modelli di frenatura dei sistemi ATC In: Atti del Congresso CIFI, Ricerca e Sviluppo nei Sistemi Ferroviari, Napoli, pp 8–9 Maggio 2003
17. Pugi L, Conti L (2009). Braking simulations of Ansaldo Breda EMU V250. In: Proceeding of IAVSD Congress 2009
18. Pugi L, Rindi A, Ercole A, Palazzolo A, Auciello J, Fioravanti D, Ignesti M (2011) Preliminary studies concerning the application of different braking arrangements on Italian freights trains. Veh Syst Dyn 8:1339–1365 ISSN: 0042-3114
19. Pugi L, Rindi A, Ercole A, Palazzolo A, Auciello J, Fioravanti D, Ignesti M (2009) Attività di studio e simulazione per l'introduzione del regime di locomotiva lunga. Ingegneria Ferroviaria 10:833–852 ISSN: 0020-0956
20. UIC 544-1 Freins—performance de freinage, 4th edn, Oct 2004
21. UIC 541-3 (2010) Brakes–Disc Brakes and their application, General Conditions for the approval of Brake Pads, 7th edn, July 2010
22. Approval tests with disc brake pads of the type Becorit BM 46 according UIC 541-3 VE (6th edn November 2006), Test Report 14 Dec 2007
23. Pugi L, Rinchi M (2002) A test rig for train brakes. In: AITC-3rd AIMETA International Tribology Conference, Salerno, p 18–20 Sept 2002

14

Optimal relay allocation strategy based on maximum rate in two-user multi-relay system

Ji Liu · Xia Lei · Haibo Cao · Maozhu Jin

Abstract Cooperative communication can enhance the performance of wireless networks via relays, and so how to allocate the relay to obtain the optimal performance of system is a key issue. In this article, we consider a cooperative system where two users communicate with the destination via relays, and these relays connect with the destination by cable. Through the theoretical derivation and analysis, we obtain the optimal relay allocation strategy based on the maximum rate under the condition of relays setting forwarding thresholds. The result shows that the system has the maximum transmission rate when the relays are allocated equivalently between users. Moreover, compared with the single-user system, the results prove that diversity gain has a decisive effect on the performance in low SNR. However, with the SNR increasing, the impact of diversity gain on system rate will be reduced. In high SNR, spatial freedom degree of the channel of multiple users is brought to enhance the performance instead of diversity gain. Numerical results demonstrate the validity of the allocation strategy and conclusion arrived in this study, but we do not find the similar arguments in the literature so far, for comparison with the conclusion of this study.

Keywords Cooperative communication · Relay allocation · Total rate

J. Liu · X. Lei (✉) · H. Cao
National Key Laboratory of Communications,
University of Electronic Science and Technology of China,
Chengdu 610054, China
e-mail: leixia@uestc.edu.cn

M. Jin
Business School of Sichuan University, Chengdu 610065, China

1 Introduction

Recently, the cooperative communication has been explored more and more as an impressive technique to combat the effects of channel fading and to reduce the energy consumption in wireless networks [1, 2]. In addition, cooperative communication efficiently reduces the probability of outage occurrence and solves the problems of intermittent communication and the signal detection in low signal-to-noise ratio (SNR) [3]. Moreover, the purposes of cooperative communication are to make full use of network nodes resources and to improve the transmission rate and reliability of wireless networks.

In cooperative systems, to utilize the wireless resources efficiently, relay allocations are often considered [4]. Bletsas et al. [5, 6] proposed an opportunistic relay which selects the relay with best channel condition cooperating source node for data transmission. The scheme does not require the use of complex space–time coding, but simultaneously reduce the resource consumption and simplify the information interaction between relays and destination. Hwang et al. [7] modified the basic opportunistic approach [5] by reducing the number of channel estimations. It defines a predefined SNR threshold, and the relay can take part in only if it satisfies such threshold. In [8] and [9], the sources include their residual power level on request-to-send (RTS) packets, allowing all overhearing nodes to estimate channel state information (CSI), thus making optimal power allocation. The relay selection decision depends upon the relay transmission power and CSI, as well as the residual power of source and relay nodes. In [10], the author developed a distributed relay selection for decode-and-forward (DF) cooperative networks, which uses an error-probability-minimizing criterion to select relay and which can achieve full diversity without requiring the relay to know whether it decodes correctly. In the case of opportunistic relay selection in cooperative networks with secrecy constraints, where an eavesdropper node tries to overhear the source message, two new opportunistic relay selection techniques, which incorporate the quality of the relay–eavesdropper links and take into account secrecy-related issues, were investigated by Krikidis [11]. Meanwhile, Zhang et al. [12] proposed an energy-efficient cooperative relay selection scheme that utilized the transmission power more efficiently in cooperative relay systems. Based on a suboptimal solution, the energy-efficient relay selection scheme was given, which selected the relay stations with the best energy efficiency and decided the optimal number of cooperative relay stations. In addition, threshold-based approaches which rely on a certain threshold to reduce the number of competing relays were addressed in [13] and [14].

The existing researches in the literature are based on the traditional cooperative model, relays of which primarily receive the information from the source and forward to the destination in the next stage [15–17]. Previous researches of relay allocations have obtained certain achievements. However, there are many studies [5–12] which require relays to know instantaneous CSI, and [13, 14] did not consider the performance of multiuser system. Compared with the traditional model, the cooperative model in which relays use the cable to receive signal instead of wireless medium has not been paid more attention, but still has important significance.

In this article, on the basis of the previous researches, we study the relay allocation strategy of two-user multi-relay system based on maximum rate by setting a forwarding threshold for relay without need to know the instantaneous CSI. With the proposed optimal relay allocation strategy, the performance of the two-user multi-relay system is compared with that of the single-user system, and the impacts that affect the rate of system are analyzed under different power transmission.

2 System model

In Fig. 1, we consider a wireless network composed of two users and M relays, and the destination is a central node that receives the information from relays by cable. The relay nodes, which consist of large base stations, can communicate with each other through the central node in order to get the state information of the other relays. In this article, we mainly solve the problem of the relay allocation between users.

Orthogonal transmissions are used for simultaneous transmissions among different users by using different

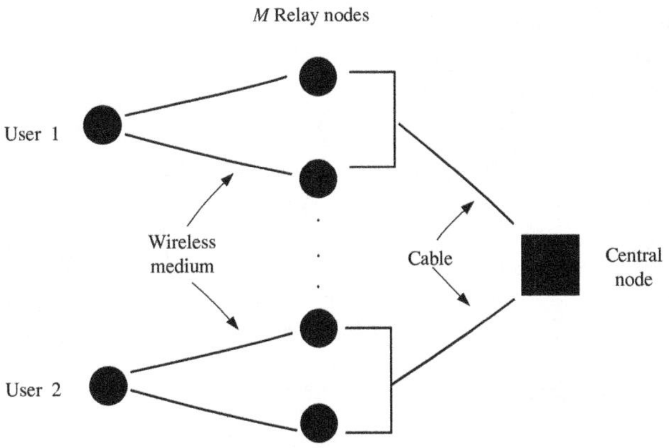

Fig. 1 The diagram of two-user M-relay system

channels (e.g., with different frequency bands) and time division multiplexing is employed for each user [18]. Let $h_{i,j}$, where $i = 1,2$ and $j = 1,2,\ldots,M$, denote the channel between the ith user and the jth relay node. $|h_{i,j}|$ obeys Rayleigh distribution.

R_i, the transmission rate of the ith user, is defined as [19]

$$R_i = \log_2\left(1 + \sum_{j \in S_i} |h_{i,j}|^2 \cdot r_0\right), \quad (1)$$

where r_0 represents the SNR with the noise environment alone, and S_i is defined as the set of relays allocated to ith user. The maximum-ratio-combining (MRC) is employed at each relay set, and the combined signals are transmitted to the central node by cable.

3 The relay allocation strategy of two-user M-relay system

In this section, we consider the single-user and the two-user systems.

According to (1), R_{1-M} is given by

$$R_{1-M} = \log_2\left(1 + \sum_{j=1}^{M} |h_j|^2 \cdot r_0\right) = \log_2(1 + r), \quad (2)$$

where r represents the receiving SNR of the combined signals from relay sets. If the channel power is normalized, then $|h_j|^2$ obeys an exponential distribution with $\lambda = 1$.

The probability density function (PDF) of the sum of N random variables with the same exponential distribution is given by [20]

$$f_{X_1+X_2+\cdots+X_n}(x) = \frac{\lambda^n}{(n-1)!} x^{n-1} \exp(-\lambda x). \quad (3)$$

The PDF of r is written as

$$f_r(x) = \frac{1}{r_0 \cdot (M-1)!}\left(\frac{x}{r_0}\right)^{M-1} \exp\left(-\frac{x}{r_0}\right). \tag{4}$$

The distribution function of R_{1-M} is given by

$$F_{R_{1-M}}(y) = P(Y<y) = P(\log_2(1+r)<y) = \int_0^{2^y-1} f_r(x)\mathrm{d}x. \tag{5}$$

The PDF of R_{1-M} is given by

$$f_{R_{1-M}}(y) = F'_{R_{1-M}}(y) = \frac{2^y \cdot \ln 2}{r_0 \cdot (M-1)!}\left(\frac{2^y-1}{r_0}\right)^{M-1} \cdot \exp\left(-\frac{2^y-1}{r_0}\right). \tag{6}$$

The rate of single-user M-relay system is

$$R_{1-M} = E_{1-M}(Y) = \int_0^{\infty} y \cdot f_{R_{1-M}}(y)\mathrm{d}y. \tag{7}$$

In addition, for the two-user M-relay system, assume that the power of transmission is normalized. Referring to (1), R_{2-M} is given as follows:

$$R_{2-M} = \log_2\left(1+\sum_{i\in S_1}|h_i|^2 \cdot \frac{1}{2}r_0\right) + \log_2\left(1+\sum_{j\in S_2}|h_j|^2 \cdot \frac{1}{2}r_0\right) = \log_2(1+r_1)+\log_2(1+r_2) = R_1+R_2, \tag{8}$$

where user 1 is allocated $|S_1| = a$ relays and user 2 is allocated $|S_2| = M-a$ relays. Assuming that the channels between each user and relays are independent, r_1 and r_2 are independent, identical distributed.

In a similar way, according to (6), the PDFs of R_1 and R_2 are expressed as follows:

$$f_{R_1}(y) = \frac{2^y 2^a \ln 2}{r_0(a-1)!} \cdot \left(\frac{2^y-1}{r_0}\right)^{a-1} \times \exp\left(-\frac{2(2^y-1)}{r_0}\right), \tag{9}$$

$$f_{R_2}(y) = \frac{2^y 2^{M-a} \ln 2}{r_0(M-a-1)!} \cdot \left(\frac{2^y-1}{r_0}\right)^{M-a-1} \times \exp\left(-\frac{2(2^y-1)}{r_0}\right). \tag{10}$$

Then, the PDF of $R_{2-M} = R_1 + R_2$ is given by [20]

$$f_{R_{2-M}}(y) = \int_{-\infty}^{+\infty} f_{R_1}(\tau) \cdot f_{R_2}(y-\tau)\mathrm{d}\tau = \int_0^{y} f_{R_1}(\tau) \cdot f_{R_2}(y-\tau)\mathrm{d}\tau. \tag{11}$$

Thus, (8) is also expressed as

$$R_{2-M} = E_{2-M}(Y) = \int_0^{\infty} y \cdot f_{R_{2-M}}(y)\mathrm{d}y = \int_0^{\infty} y \cdot \int_0^{y} \frac{2^M 2^y (\ln 2)^2}{r_0^2(a-1)!(M-a-1)!} \times \left(\frac{2^\tau-1}{r_0}\right)^{a-1} \times \left(\frac{2^{y-\tau}-1}{r_0}\right)^{M-a-1} \times \exp\left(-\frac{2(2^\tau+2^{y-\tau}-2)}{r_0}\right)\mathrm{d}\tau\mathrm{d}y. \tag{12}$$

Using numerical integration in (12), we can obtain the rates R_{2-M} of different relay allocation strategies in definite r_0. Furthermore, considering M as even number, we obtain the maximum rate $R_{2-M}^{\max}$ at $a = M/2$.

When we compare the single-user system with the two-user system, (2) and (8) will be analyzed. Let

$$A = \sum_{j=1}^{M} |h_j|^2, \quad B = \sum_{i\in S_1} |h_i|^2 \quad \text{and} \quad C = \sum_{j\in S_2} |h_j|^2,$$

and let us assume that the power of each channel is normalized; then we get $A = B + C$. If the relays are allocated equivalently between users, then $B = C = A/2$.

Let

$$1 + Ar_0 = \left(1+\frac{1}{2}Br_0\right)\left(1+\frac{1}{2}Cr_0\right).$$

We get a cross point $r'_0 = \frac{2A}{BC}$. Then, if the number of relays allocated between users is equal, $r'_0 = \frac{8}{A}$. Therefore, if SNR $r_0 < r'_0$, the rate of single-user system is higher; on the contrary, it is the two-user system.

4 The relay allocation strategy of relay node with a forwarding threshold

When relays set the forwarding thresholds, it means that if the SNR of the combined signal is greater than rth, the relays will receive the information and forward to the central node; otherwise, the information will be discarded. In this way, the receiving complexity of system can be reduced.

Similar to the above analysis, for the single-user M-relay system, after relays set the forwarding thresholds, the PDF of R is expressed as

$$f_R^{\text{th}}(y) = \begin{cases} \int_0^{\log_2(1+r_{\text{th}})} f_R(x)\mathrm{d}x, & y = 0, \\ f_R(y), & y > \log_2(1+r_{\text{th}}). \end{cases} \tag{13}$$

Then, the rate of the single-user system is given by

$$R^{\text{th}}_{1-M} = E^{\text{th}}_{1-M}(Y) = \int_{\log_2(1+r_{\text{th}})}^{\infty} y f^{\text{th}}_R(y)\mathrm{d}y. \tag{14}$$

For the two-user M-relay system, referring to (9) and (10), after relays set the forwarding thresholds, let $f^{\text{th}}_{R_1}(y)$ denote the PDF of rate at user 1 to whom a relays are allocated, and $f^{\text{th}}_{R_2}(y)$ denote the PDF of rate at user 2 to whom $M - a$ relays are allocated. Then, the PDF of the total rate R is written as

$$f^{\text{th}}_{R_{2-M}}(y) = f^{\text{th}}_{R_1}(y) * f^{\text{th}}_{R_2}(y) = \begin{cases} a \cdot b, & y=0, \\ \int_{\log_2(1+r_{\text{th}})}^{y-\log_2(1+r_{\text{th}})} f_{R_1}(\tau) f_{R_2}(y-\tau)\mathrm{d}\tau \\ \quad + a \cdot f_{R_1}(y) + b \cdot f_{R_2}(y), & y > \log_2(1+r_{\text{th}}), \end{cases} \tag{15}$$

where

$$a = \int_0^{\log_2(1+r_{\text{th}})} f_{R_2}(x)\mathrm{d}x, \quad \text{and}$$

$$b = \int_0^{\log_2(1+r_{\text{th}})} f_{R_1}(x)\mathrm{d}x.$$

Therefore, the total transmission rate of the system is given as follows:

$$R^{\text{th}} = E(Y) = \int_{\log_2(1+r_{\text{th}})}^{\infty} y \cdot f^{\text{th}}_{R_{2-M}}(y)\mathrm{d}y. \tag{16}$$

In a similar way to Sect. 3, after relays set the forwarding thresholds, we also reach the cross point $r'_{0\,\text{th}} = \frac{2A}{BC}$. Owing to the relays discard, the SNR signals of which are less than the threshold, the receiving SNR will decrease for the equal transmission power in low SNR. Thus, the user needs more power to reach the same transmission rate, $r'_{0\,\text{th}} > r'_0$. With the increase of threshold r_{th}, $r'_{0\,\text{th}}$ will be larger. Likewise, with the increasing number of relays, diversity gain will be enhanced. In the case when other conditions remain unchanged, the required power to reach the same transmission rate will be lower, and $r'_{0\,\text{th}}$ will decrease.

5 Simulation results

In this section, simulations are performed to demonstrate the above theory. In the Rayleigh channel, by setting different thresholds for relays, we compared different relay allocation strategies between theory and simulation. The simulation parameters are shown in Table 1.

Table 1 Simulation parameters

Channel	Rayleigh
Number of users	1, 2
Number of relay nodes	4, 6
Forwarding threshold of relay	0, 8, 10 dB
Transmitted total power	Normalization

Figures 2 and 3 illustrate the total transmission rates of different relay allocation strategies without the forwarding threshold of relay. For the two-user M-relay system, the allocation scheme in which the relays are allocated equivalently between users can get the maximum rate. Comparing the two-user system with the single-user system, we can find the cross point SNR r'_0 in the figures. If

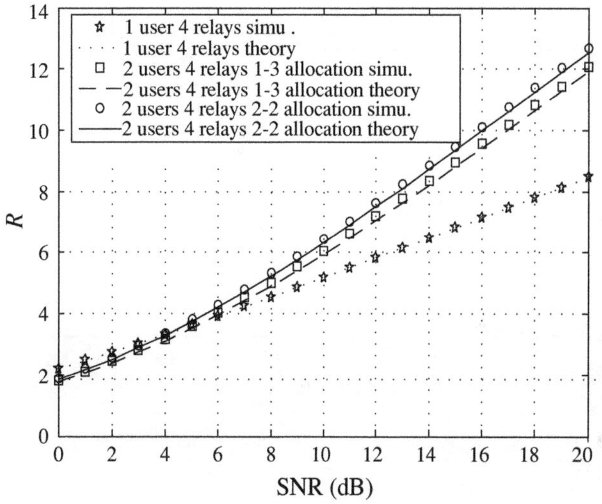

Fig. 2 Total transmission rates of 4-relay systems without threshold

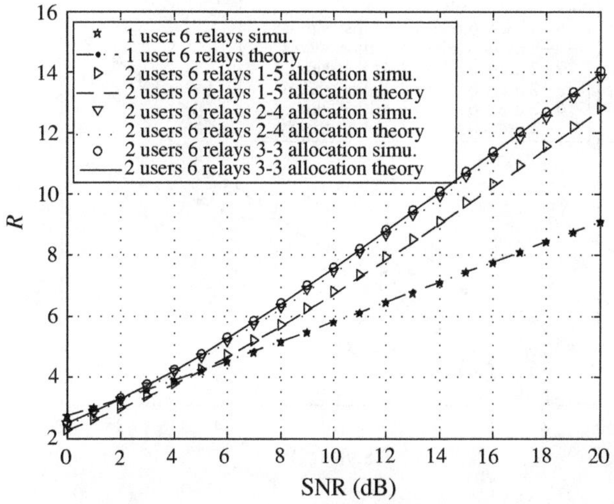

Fig. 3 Total transmission rates of 6-relay systems without threshold

SNR $r_0 < r'_0$, the rate of the single-user system is higher; otherwise, the two-user system has a higher rate. In addition, with the number of relays increasing, the system can get more diversity gain. This means that the required power to reach the same rate will be lower, and $r'_{0\text{th}}$ will be greater.

Figures 4 and 5 illustrate the total transmission rates of different relay allocation strategies with 8 dB forwarding thresholds of relays. Compared with Figs. 2 and 3, the combined signals less than the threshold are discarded. The total rate of system is decreased in low SNR, and the required power to reach the same rate will be increased in contrast to the threshold-free case. Therefore, when the threshold is set, the cross point $r'_{0\text{th}}$ will be increased.

Figures 6 and 7 illustrate the total transmission rates of different relay allocation strategies with increasing forwarding threshold. Compared with Figs. 4 and 5, the number of the discarded signals is enlarged. Under the unchanged channel condition, the acquired power to reach the same rate will increase in low SNR. Therefore, with the thresholds increasing, the cross point $r'_{0\text{th}}$ will increase.

By analyzing the single-user and the two-user systems with the same number of relays in the simulations, we found that diversity gain decided the system performance

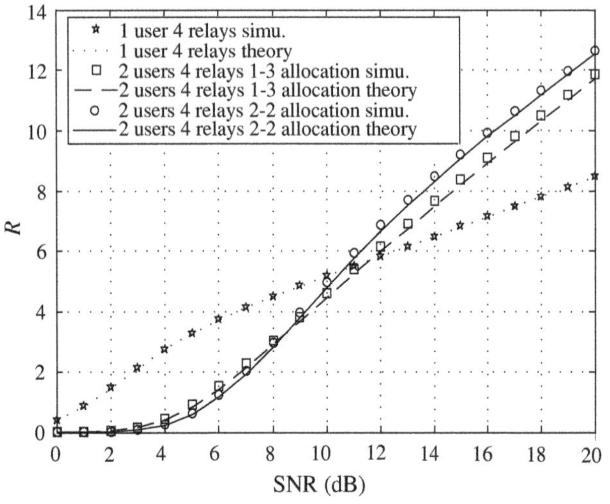

Fig. 4 Total transmission rates of 4-relay systems with 8 dB threshold

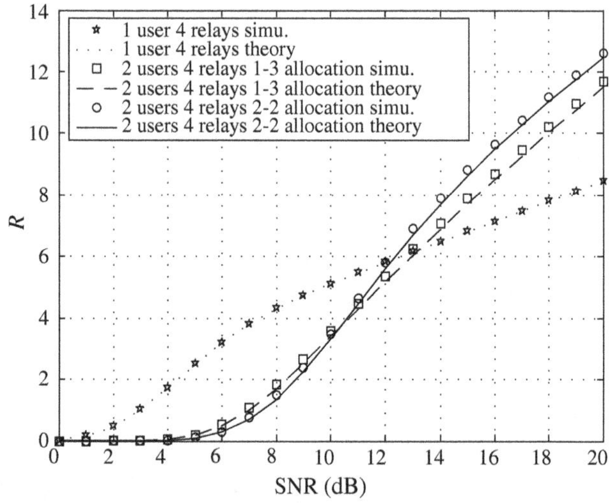

Fig. 6 Total transmission rates of 4-relay systems with 10 dB threshold

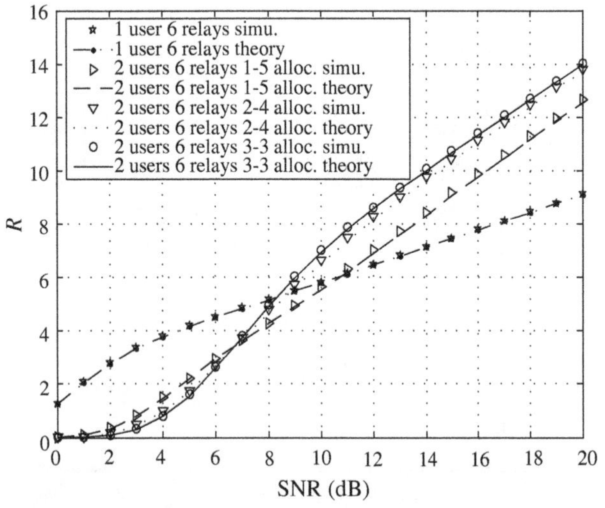

Fig. 5 Total transmission rates of 6-relay systems with 8 dB threshold

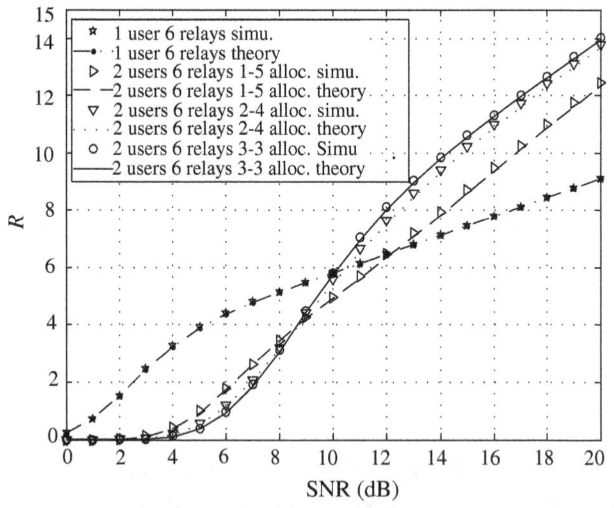

Fig. 7 Total transmission rates of 6-relay systems with 10 dB threshold

in low SNR. Moreover, the transmission rate of the single-user system is higher than that of the two-user system, because the diversity gain of the single-user system is larger. In high SNR, the impact of diversity gain on system rate decreased. With the SNR increasing, the impact can be ignored, and the spatial freedom degree of the channel is brought to enhance the performance in multiple users instead of diversity gain.

6 Conclusion

This article mainly investigates how to allocate the relays to obtain the optimal performance of system. Under the conditions of relays setting forwarding thresholds, a two-user multi-relay system was analyzed. The results show that the system can obtain the maximum transmission rate when the relays are allocated equivalently between users. In addition, comparison between the two-user system with the single-user system proves that diversity gain has a decisive effect on the performance in low SNR. Moreover, the single-user system can obtain the maximum transmission rate because of the larger diversity gain. However, with the SNR increasing to a certain value, the impact of diversity gain on system rate will reduce. In high SNR, the spatial freedom degree of the channel of multiple users instead of the diversity gain will improve the performance effectively.

Acknowledgments This study was supported by the National Natural Science Foundation of China under Grant numbers 61032002, 61101090, and 60902026, and the Chinese Important National Science & Technology Specific Projects under Grant number 2011ZX03001-007-01.

References

1. Laneman JN, Tse DNC, Wornell GW (2004) Cooperative diversity in wireless networks: efficient protocols and outage behavior. IEEE Trans Inform Theory 50(12):3062–3080
2. Jamal T, Mendes P (2010) Relay selection approaches for wireless cooperative networks, In: 2010 IEEE 6th international conference on wireless and mobile computing, networking and communications, Niagara Falls, Ontario, 11–13 Oct 2010, pp 661–668
3. Morillo-Pazo J, Trullols , Barcelo JM et al (2008) A cooperative ARQ for delay-tolerant vehicular networks, In: Proceedings of IEEE ICDCS, Beijing, June 2008, pp 192–197
4. Wei Y, Yu FR, Song M (2010) Distributed optimal relay selection in wireless cooperative networks with finite-state Markov channels. IEEE Trans Veh Technol 59(5):2149–2158
5. Bletsas A, Khisti A, Reed D et al (2006) A simple cooperative diversity method based on network path selection. IEEE J Sel Areas Commun 24(3):659–672
6. Bletsas A, Lippnian A, Reed DP (2005) A simple distributed method for relay selection in cooperative diversity wireless networks, based on reciprocity and channel measurements, In: Proceeding of IEEE 61st vehicular technology conference, Stockholm, May 30–June 1 2005, pp 1484–1488
7. Hwang KS, Ko YC (2007) An efficient relay selection algorithm for cooperative networks, In: Proceedings of IEEE VTC, Baltimore, Sept 30–Oct 3 2007, pp 81–85
8. Chen Y, Yu G, Qiu P et al (2006) Power-aware cooperative relay selection strategies in wireless ad hoc networks, In: Proceedings of IEEE PIMRC, Helsinki, Sep 2006, pp 1–5
9. Adam H, Bettstetter C, Senouci SM (2008) Adaptive relay selection in cooperative wireless networks, In: Proceedings of IEEE PIMRC, Cannes, Sept 2008, pp 1–5
10. Liu J, Lu K, Cai X (2010) Distributed error-probability—minimizing relay selection for cooperative wireless networks, In: 2010 44th annual conference on information sciences and systems, Princeton, 17–19 Mar 2010, pp 1–6
11. Krikidis I (2010) Opportunistic relay selection for cooperative networks with secrecy constraints. Commun IET 4(15):1787–1797
12. Zhang T, Zhao S, Cuthbert L et al (2010) Energy-efficient cooperative relay selection scheme in MIMO relay cellular networks. In: IEEE international conference on communication systems (ICCS), Singapore, 17–19 Nov 2010, pp 269–273
13. Huang KS, Ko YC (2007) An efficient relay selection algorithm for cooperative networks. In: Proceedings of IEEE VTC, Baltimore, Sept 30–Oct 3 2007, pp 81–85
14. Bali Z, Ajib W, Boujemaa H (2010) Distributed relay selection strategy based on source-relay channel. In: 2010 IEEE 17th international conference on telecommunications, Doha, 4–7 Apr 2010, pp 138–142
15. Zhang J, Zhuang H, Li T et al (2009) A novel relay selection strategy for multi-user cooperative relaying networks. In: IEEE 69th vehicular technology conference, Barcelona, 26–29 Apr 2009, pp 1–5
16. Hosseini K, Adve R (2010) Comprehensive node selection and power allocation in multi-source cooperative mesh network. In: 2010 44th annual conference on information sciences and systems, Princeton, 17–19 Mar 2010, pp 1–6
17. Baidas MW, Mackenzie AB (2012) An auction mechanism for power allocation in multi-source multi-relay cooperative wireless networks. IEEE Trans Wirel Commun 11(9):3250–3260
18. Phan KT, Nguyen DHN, Le-Ngoc T (2009) Joint power allocation and relay selection in cooperative networks, In: Proceedings of IEEE global telecommunications conference (GLOBECOM), Honolulu, Nov 30–Dec 4 2009, pp 1–5
19. Phan KT, Le-Ngoc T, Vorobyov SA (2009) Power allocation in wireless multi-user relay networks. IEEE Trans Wirel Commun 8(5):2535–2545
20. Rice JA (1995) Mathematical statistics and data analysis, 2nd edn. Duxbury Press, Belmont

15

Wave–current interaction with a vertical square cylinder at different Reynolds numbers

Azhen Kang · Bing Zhu

Abstract Large eddy simulation is performed to study three-dimensional wave–current interaction with a square cylinder at different Reynolds numbers, ranging from 1,000 to 600,000. The Keulegan–Carpenter number is relevantly a constant of 0.6 for all cases. The Strouhal number, the mean and the RMS values of the effective drag coefficient in the streamwise and transverse directions are computed for various Reynolds numbers, and the velocity of a representative point in the turbulent zone is simulated to find the turbulent feature. It is found that the wave–current interaction should be considered as three-dimensional flow when the Reynolds number is high; under wave–current effect, there exists a critical Reynolds number, and when the Reynolds number is smaller than the critical one, current effect on wave can be nearly neglected; conversely, with the Reynolds number increasing, wave–current–structure interaction is sensitive to the Reynolds number.

Keywords Large eddy simulation (LES) · Wave–current–structure interaction · Drag coefficient · Vortex shedding · Reynolds number

1 Introduction

In recent years, the highway and road systems have gone through a rapid expansion in China, resulting in the construction of many sea bridges. The piers of these bridges deep into sea must endure large wave forces and tidal actions. The high ocean waves and turbulent currents often cause the large vibration or deformation of bridges. The design for sea bridge piers relies on the accurate prediction of wave–current forces and vortex-shedding frequency. Accordingly, wave–current–structure interaction is a focus in the studies of coastal and offshore bridges.

Park et al. [1] used the linear potential theory to investigate the fully nonlinear wave–current–body interaction in terms of three-dimensional numerical tank. His study was based on weak current and no flow separation. However, in reality, flow separation will always occur even though the current is weak. Thus, in order to deal with the physical problems of wave–current–structure interactions, the turbulence closure model was developed. Deardorff [2] established a large eddy simulation (LES) model which solves the large-scale eddy motions and modeled the small-scale turbulent fluctuations to solve the turbulent flows with large Reynolds numbers. Sohankar [3] applied a LES model to studying flow interaction with a bluff body from moderate-to-high Reynolds numbers by employing two different sub-grid scale models, namely, the Smagorinsky and a dynamic one-equation models. Koo and Kim [4] investigated nonlinear wave–current interactions with fixed or freely floating bodies using a two-dimensional fully-nonlinear numerical wave tank (NWT). Li and Lin [5] developed a two-dimensional numerical tank to simulate the coaction processes based on the Reynolds-averaged Navier–Stokes equations. Cheng et al. [6] used the lattice Boltzmann method to simulate a two-dimensional incompressible linear shear flow over a square cylinder and investigated the effect of shear rate on the frequency of vortex shedding as well as the drag force. The complexities of flow and turbulence patterns as well as the pressure fields induced by the wave–current attacks require a three-dimensional model to predict the deformation of piers of bridge structures. Vengadesan and Nakayama [7] evaluated

A. Kang (✉) · B. Zhu
School of Civil Engineering, Southwest Jiaotong University, Chengdu 610031, China
e-mail: xiaokang_198610@163.com

turbulent flow over a square cylinder by employing three subgrid-scale SGS stress closure LES models. Lin and Li [8–11] used LES to study wave–current–body interaction, and obtained lots of significant revelations of nonlinear wave–current–body interactions. Tan [12, 13] applied an LES model to simulate three-dimensional wave interaction with structures and studied wave–current–body interactions.

Based on the Navier–Stokes equation, the wave-generation method of defining inlet boundary conditions is applied in this article to build an LES model. The model is then used to study wave–current interaction with a vertical square cylinder for various Reynolds numbers. The drag force and vortex feature caused by the nonlinear wave–current–structure interaction are, respectively, numerically simulated with various high Reynolds numbers. The mean, RMS of drag coefficient and the Strouhal number which represent the vortex shedding frequency are calculated and compared.

2 Theoretical model

2.1 Model description

The governing equations for spatially averaged mean flow are as follows, which are obtained by filtering the classical Navier–Stokes equations [2]:

$$\overline{\frac{\partial u_i}{\partial x_i}} = 0, \tag{1}$$

$$\frac{\partial \overline{u_i}}{\partial t} + \frac{\partial}{\partial x_j}\left(\overline{u_i u_j}\right) = -\frac{1}{\rho}\frac{\partial \overline{p}}{\partial x_i} + \overline{g_i} + \frac{1}{\rho}\frac{\partial \overline{\tau}_{ij}}{\partial x_j}, \tag{2}$$

where $i = j = 1, 2, 3$ represent the three directions of three-dimensional fluid particle, the variables with overbars are spatially averaged quantities, ρ is the fluid density, g_i is the gravitational acceleration in the ith component, $\overline{u}_i$ is the mean velocity in the ith component, $\overline{p}$ is the filtered pressure, and $\overline{\tau}_{ij}$ is the viscous stress of filtered velocity field. The sub-grid scale tensors are defined by the difference of $\overline{u_i u_j}$ and $\overline{u_i}\,\overline{u_j}$ produced from the filtering, which can be described by means of sub-grid Reynolds stress:

$$\overline{\tau_{ij}^R} = -\rho\left(\overline{u_i u_j} - \overline{u_i}\,\overline{u_j}\right). \tag{3}$$

The sub-grid model is solved based on the eddy viscosity model hypothesis; thus, the isotropic residual stress tensor is defined as follows:

$$\overline{\tau_{ij}^r} = \overline{\tau_{ij}^R} - \frac{2}{3}k_r\delta_{ij}, \tag{4}$$

where $k_r = \frac{1}{2}\tau_{ij}^R$ is residual kinetic energy, and δ_{ij} is Kronecker delta function. From Eq. (4), the isotropic residual stress tensor terms can be absorbed into the filtered pressure terms, i.e.,

$$\overline{P} = \overline{p} + \frac{2}{3}k_r \tag{5}$$

By Substituting Eqs. (3)–(5) into Eq. (2), we transform the filtered momentum equation into

$$\frac{\partial \overline{u_i}}{\partial t} + \frac{\partial}{\partial x_j}\left(\overline{u_i u_j}\right) = -\frac{1}{\rho}\frac{\partial \overline{P}}{\partial x_i} + \overline{g_i} + \frac{1}{\rho}\frac{\partial \overline{\tau}_{ij}}{\partial x_j} + \frac{1}{\rho}\frac{\partial \overline{\tau_{ij}^r}}{\partial x_j}. \tag{6}$$

The Smagorinsky sub-grid scale model [14] is applied to calculate $\overline{\tau_{ij}^r}$, that is,

$$\overline{\tau_{ij}^r} = 2\rho\gamma_t\overline{S}_{ij} = \rho\gamma_t\left(\frac{\partial \overline{u}_i}{\partial x_j} + \frac{\partial \overline{u}_j}{\partial x_i}\right), \tag{7}$$

where eddy viscosity coefficient $\gamma_t = L_s^2\sqrt{2\overline{S}_{ij}\overline{S}_{ij}}$, Smagorinsky length scale $L_s = C_s W$, in which C_s is the Smagorinsky model dimensionless coefficients ranging from 0.1 to 0.2 due to various calculating fluids. In this study, C_s is taken to be 0.15, W is the length scale of minimum vortex, and $W = (\Delta_x + \Delta_y + \Delta_z)^{1/3}$, where Δ_x, Δ_y, Δ_z are the grid spacings of x, y, and z directions. Thus, the LES momentum equations which are obtained by means of Smagorinsky sub-grid model without the filtered signs can be described as

$$\frac{\partial u_i}{\partial t} + \frac{\partial}{\partial x_j}\left(u_i u_j\right) = -\frac{1}{\rho}\frac{\partial P}{\partial x_i} + g_i + \frac{1}{\rho}\frac{\partial}{\partial x_j}\left[\rho(\gamma + \gamma_t)\left(\frac{\partial \overline{u}_i}{\partial x_j} + \frac{\partial \overline{u}_j}{\partial x_i}\right)\right], \tag{8}$$

$$\frac{\partial u_i}{\partial x_i} = 0, \tag{9}$$

where γ is the molecular viscous coefficient.

To trace the three-dimensional free surface transformation, the so-called σ-coordinate transformation [15] is applied to map the irregular physical domain into a cube where the free surface and bottom boundary condition can be set precisely. In this study, operator-splitting method [16] is used to solve Eqs. (8) and (9).

2.2 Model validation

To validate the numerical model mentioned above, a three-dimensional model is set up. The square cylinder is vertically located in a numerical wave–current basin with the dimension of 30 m × 10 m × 1 m. The still water depth is 1 m. The side length of square cylinder is 1 m × 1 m, and its height is 1 m. The center of the cylinder is located at the centerline in the y direction and 10 m away from the left boundary. A uniform and undisturbed current with the speed of 0.22 m/s is set on the inflow boundary conditions;

the corresponding Reynolds number is $Re = u_0 \times L/v = 2.2 \times 10^4$. A nonuniform mesh system is used on the horizontal plan with the grids of 130 × 80. Near the square cylinder, the finest grids $\Delta x = \Delta y = 0.005$ m is deployed and coarser grids father away. In the vertical direction, the uniform grids are used, 20 grids in total. A timestep of $\Delta t = 0.001$ s is used in the computation. The computation time t is a dimensionless value by parameter l/u_0, and force is a dimensionless value by parameter ρu_0^2. Totally 200 dimensionless t are calculated.

The time history curves of drag and lift coefficient are given in Fig. 1. The mean force coefficients (C_d), mean-square deviation of drag and lift coefficients (C_d' and C_l'), and normalized shedding frequencies (Strouhal number St) are given in Table 1.

The data in Table 1 are analyzed from 100t to 200t (where $t = t^*u_0/l$) during which the turbulence has achieved full development. The corresponding numerical results are compared with Lyn's experiment results [17]. It can be seen from the comparison that the numerical result agrees with the experiment result well.

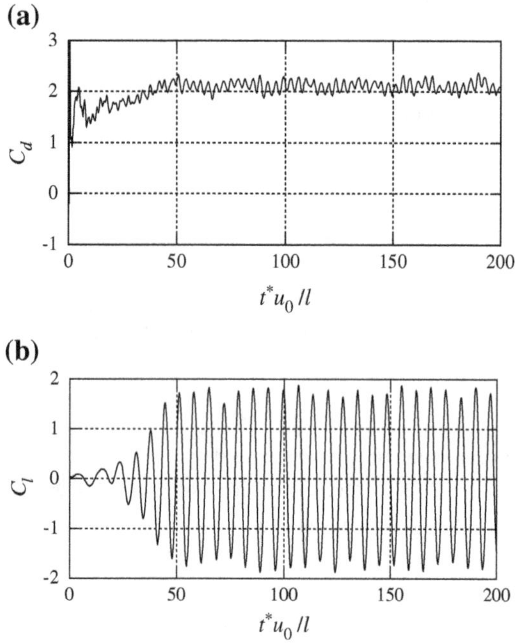

Fig. 1 The time history of drag and lift coefficients. **a** C_d, **b** C_l

Table 1 Calculated mean force coefficient (C_d), mean-square deviation of drag and lift coefficients (C_l' and C_d'), and normalized shedding frequencies (Strouhal number St)

Item	C_l'	C_d	C_d'	St
This article	1.2475	2.1094	0.1892	0.139
Ref. [17], $Re = 22{,}000$	–	2.05–2.23		0.135
Vickery, 1 [18]	1.32	–	0.17	0.12
Vickery, 2 [18]	1.27	–	0.17	–
Lee, $Re = 176{,}000$ [19]	1.22	2.05	0.22	–

2.3 Model conditions

In this study, the three-dimensional wave–current interaction with a vertical square cylinder at various Reynolds numbers is investigated. The numerical model is the same as that of Sect. 2.2 (see Fig. 2a). Inflow boundary condition is generally set on the left side of computational domain where both free surface and velocities are provided based on either laboratory measurements or theoretical expression. Thus, a linear wave train with a wave period of 4 s and a wave height of 0.05 m together with the current is sent from the left boundary. The corresponding Reynolds number ranges from 1.0×10^4 to 6.0×10^5 and the current speeds from 0.001 to 0.6 m/s. The value of $KC = u_pT/L \approx 0.6$. The right side of computational domain is set by radiation boundary that can absorb the wave and flow energy. On free surface, the zero normal and tangential stresses are enforced with zero pressure. On bottom or solid wall boundary, no-slip boundary is applied. A nonuniform mesh system is used on the horizontal plan with the grids of 130 × 80. Near the square cylinder, the finest grids $\Delta x = \Delta y = 0.05$ m are deployed and coarser grids father away. In the vertical direction, the uniform grids are used, 20 grids in total.

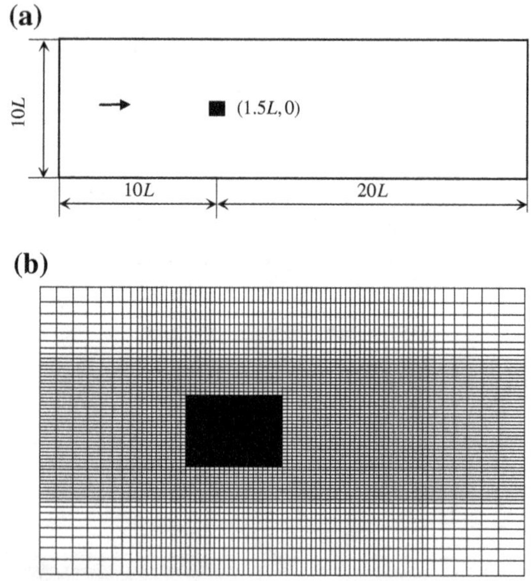

Fig. 2 Mesh arrangement near the *square cylinder* on the *horizontal* plane. **a** Computational domain arrangement, **b** Mesh arrangement near the square cylinder

3 Results and discussions

In Fig. 3, the calculated vorticity on the middle elevation at the same time moment $t = 100$ s for different Reynolds numbers is plotted as a gray scale color map. Two extreme cases of wave-only case ($KC = 0.6$) and current-only case ($Re = 6.0 \times 10^5$) are also discussed.

Figure 3a–c show that the current is so weak that its effect on wave can be neglected when Re is less than 1.5×10^5. The flow is essentially symmetric about the

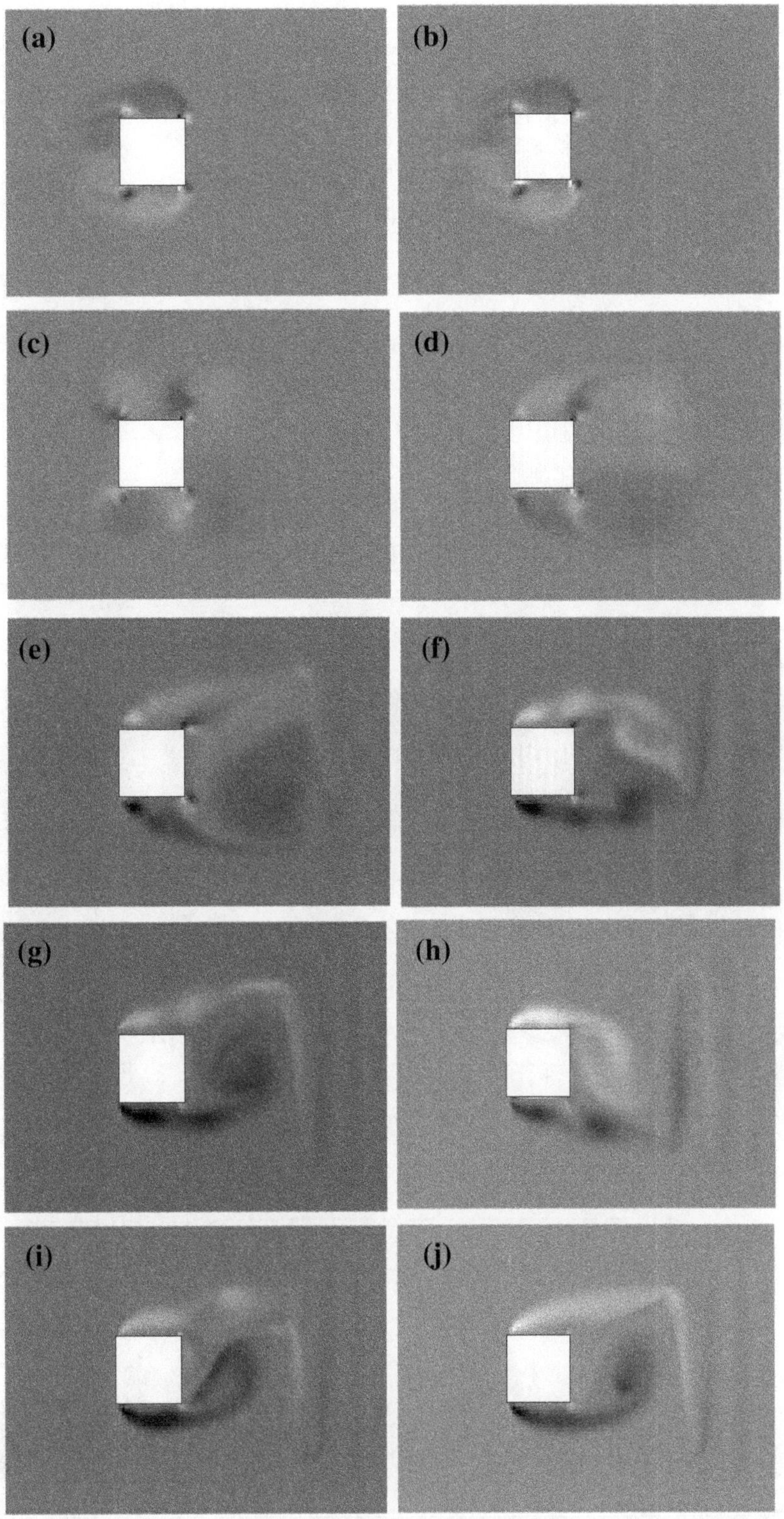

Fig. 3 Calculated vorticity on the middle elevation at $t = 100$ s for different Reynolds numbers. **a** Wave only, **b** $KC = 0.6$, $Re = 10{,}000$, **c** $KC = 0.6$, $Re = 100{,}000$, **d** $KC = 0.6$, $Re = 150{,}000$, **e** $KC = 0.6$, $Re = 200{,}000$, **f** $KC = 0.6$, $Re = 300{,}000$, **g** $KC = 0.6$, $Re = 400{,}000$, **h** $KC = 0.6$, $Re = 500{,}000$, **i** $KC = 0.6$, $Re = 600{,}000$, **j** $KC = 0$, $Re = 600{,}000$

centerline in the y direction during all time history under constant KC number and relatively weak current effect. Flow separations from corners exist but the vortices are attached to the structure. Under wave-only effect of wave height $H = 0.05$ m as well as wave period $T = 4$ s, the value of $KC = u_pT/L \approx 0.6$, where u_p is the maximum wave-induced fluid particle velocity, $u_p \approx 0.15$ m/s [10]. Thus, when the Reynolds number magnitude is smaller

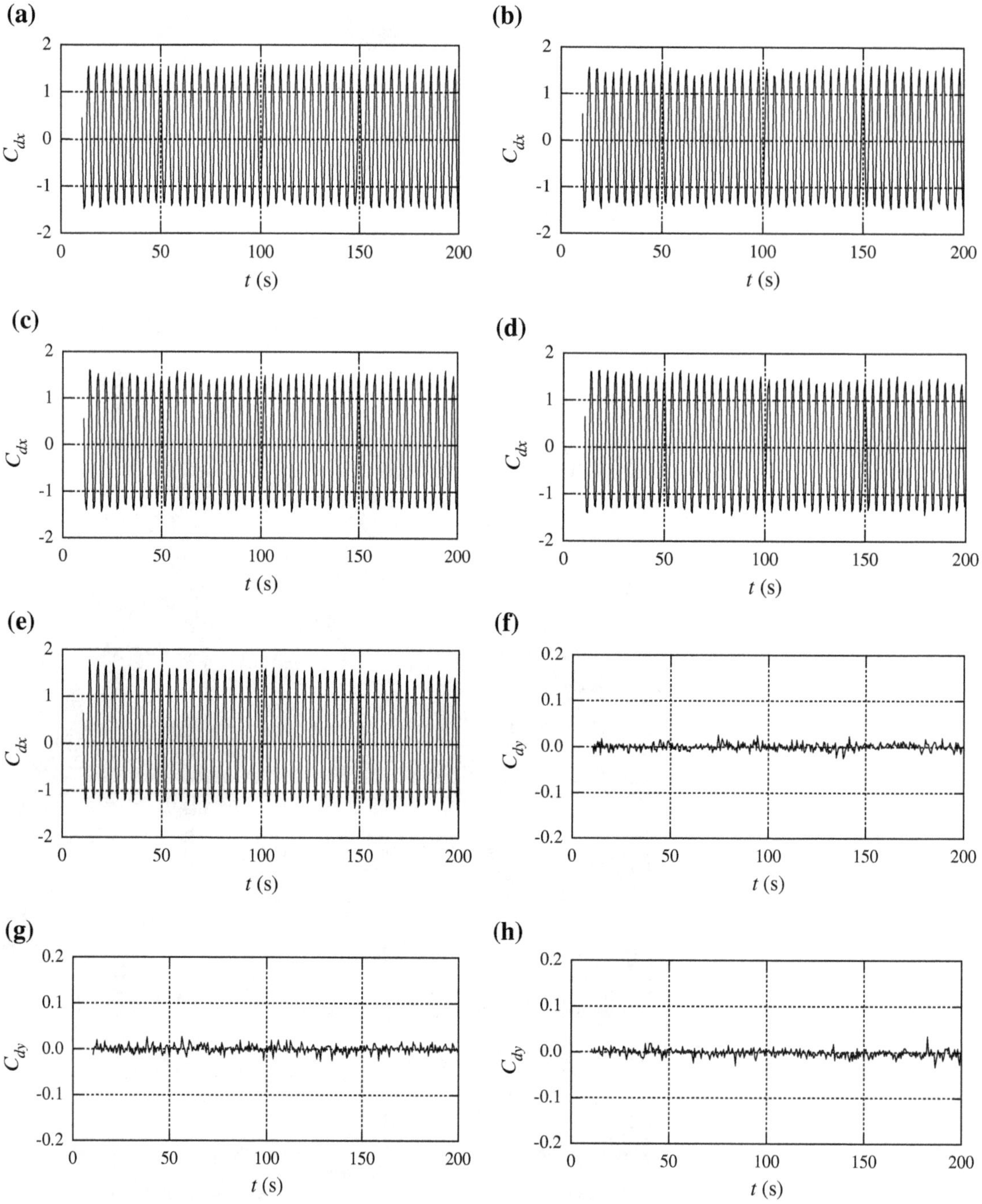

Fig. 4 Effective drag coefficient and corresponding energy spectra for Reynolds numbers (1.0×10^4 to 2.0×10^5). **a** Wave only, **b** $KC = 0.6$, $Re = 10{,}000$, **c** $KC = 0.6$, $Re = 100{,}000$, **d** $KC = 0.6$, $Re = 150{,}000$, **e** $KC = 0.6$, $Re = 200{,}000$, **f** Wave only, **g** $KC = 0.6$, $Re = 10{,}000$, **h** $KC = 0.6$, $Re = 100{,}000$, **i** $KC = 0.6$, $Re = 150{,}000$, **j** $KC = 0.6$, $Re = 200{,}000$, **k** Wave only, **l** $KC = 0.6$, $Re = 10{,}000$, **m** $KC = 0.6$, $Re = 100{,}000$, **n** $KC = 0.6$, $Re = 150{,}000$, **o** $KC = 0.6$, $Re = 200{,}000$

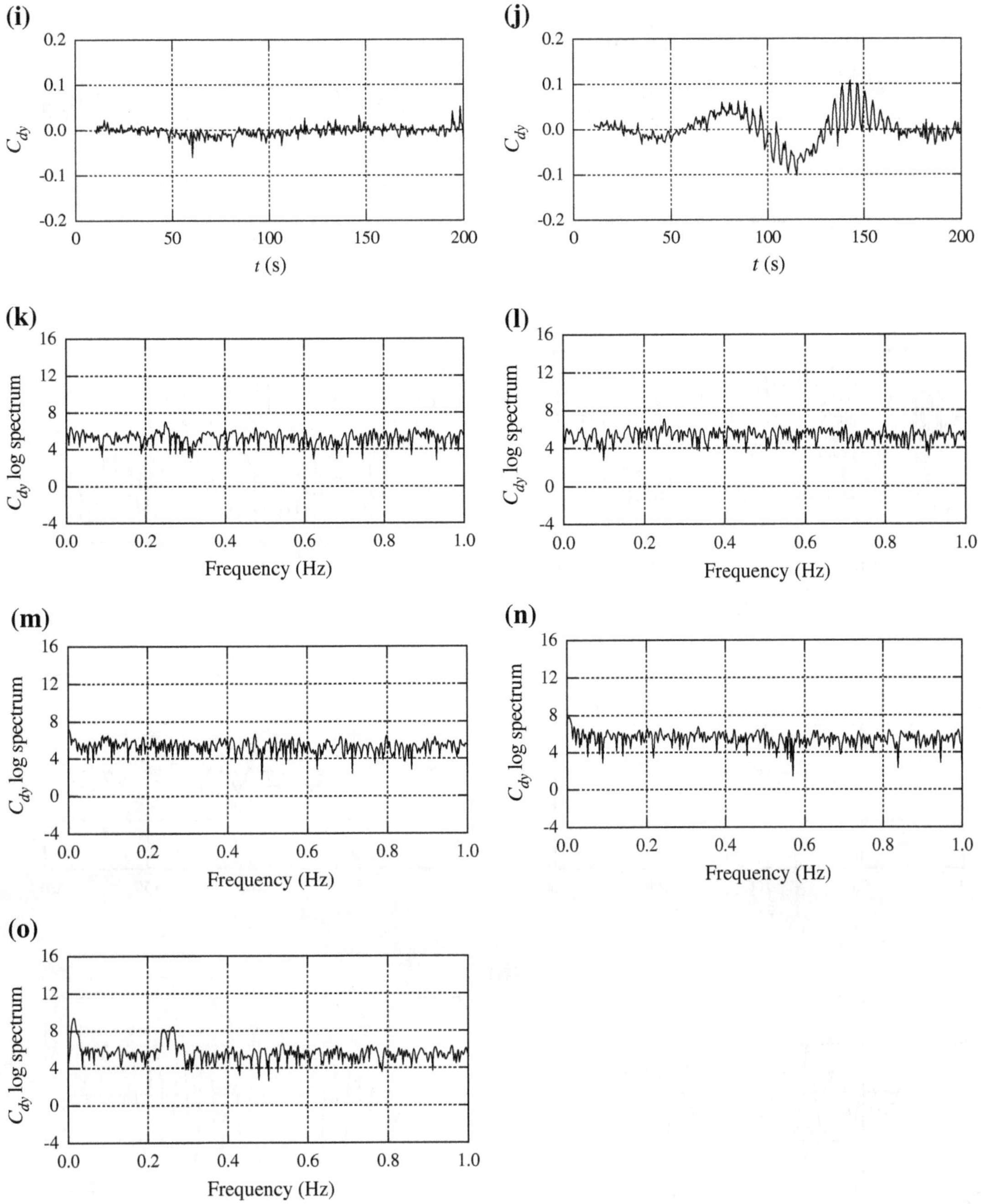

Fig. 4 continued

than $Re_{\text{critical}} = u_p L/\nu = 1.5 \times 10^5$, the effect of Reynolds number on wave is almost very little. The drag force and vortex form caused by wave–current interaction with $KC = 0.6$ as well as weak current are consistent with that of wave action only. Conversely, when the Reynolds number magnitude is the same as Re_{critical}, the wave–current nonlinear interaction is obvious. Although with the same KC number, the trailing vortex forms differ from one another. With increasing the Reynolds number from 1.5×10^5 to 6.0×10^5, the vortex form becomes more complicated; however, the separation point remains at the upstream corners at all times for different Reynolds numbers. From Fig. 3i–j, it is expected that the vortex feature under wave–current effect becomes much more

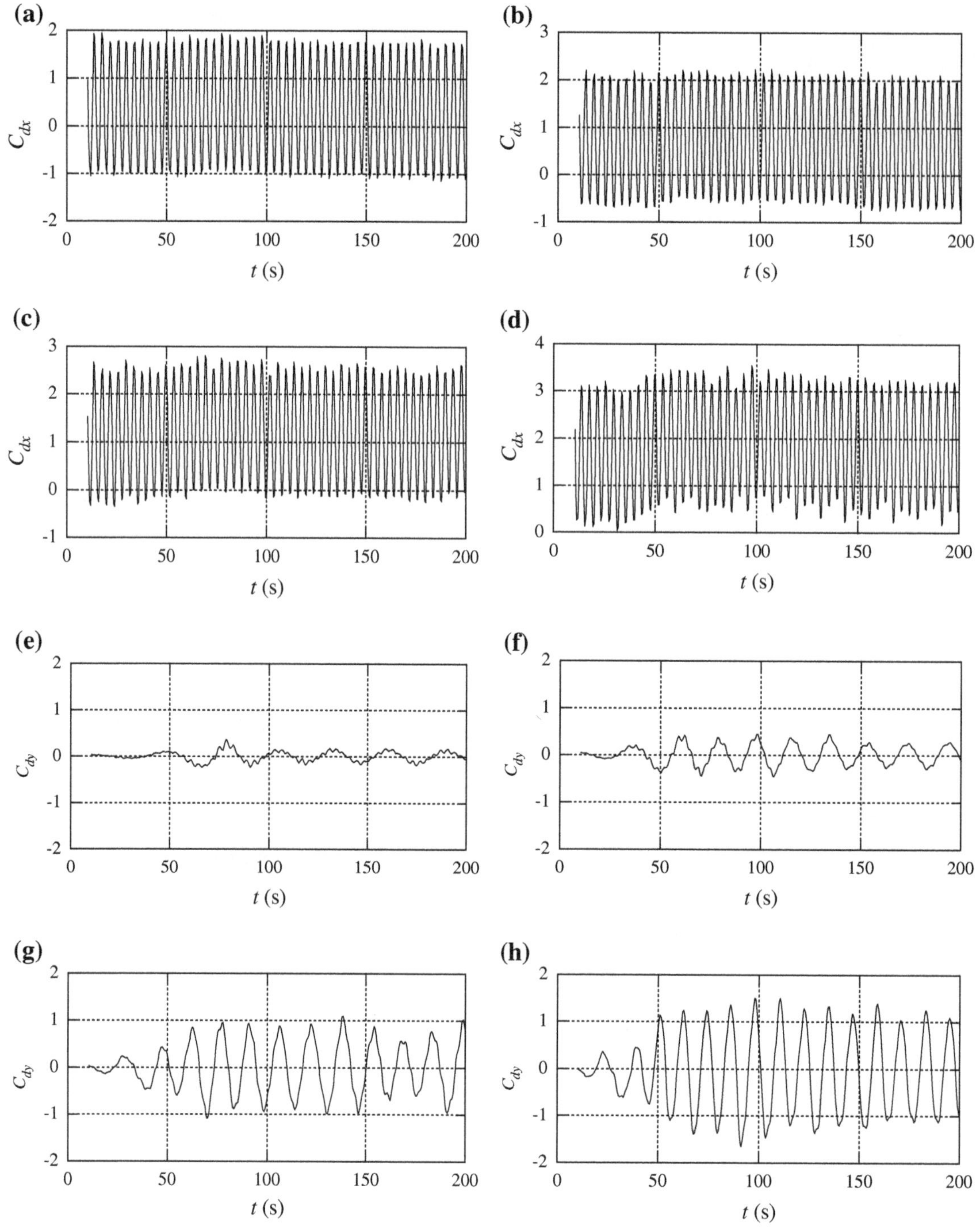

Fig. 5 Effective drag coefficient and corresponding energy spectra for Reynolds numbers(3.0×10^5 to 6.0×10^5). **a** $KC = 0.6$, $Re = 300{,}000$, **b** $KC = 0.6$, $Re = 400{,}000$, **c** $KC = 0.6$, $Re = 500{,}000$, **d** $KC = 0.6$, $Re = 600{,}000$, **e** $KC = 0.6$, $Re = 300{,}000$, **f** $KC = 0.6$, $Re = 400{,}000$, **g** $KC = 0.6$, $Re = 300{,}000$, **h** $KC = 0.6$, $Re = 600{,}000$ **i** $KC = 0.6$, $Re = 300{,}000$, **j** $KC = 0.6$, $Re = 400{,}000$, **k** $KC = 0.6$, $Re = 500{,}000$, **l** $KC = 0.6$, $Re = 600{,}000$

complicated in comparison to that of the current-only effect due to the presence of wave oscillating motion.

In the presence of wave, there also exists inertial force caused by the unsteady oscillating motion besides the drag force. Thus, to simplify the analysis and facilitate the comparisons among different cases, all forces are normalized by $\rho u^2(Ld)/2$, in this study u in all cases is taken as 0.6 m/s so that the effective force coefficient in the x and y directions, e.g., C_{dx} and C_{dy}, are obtained. The time history curve of effective drag coefficient and

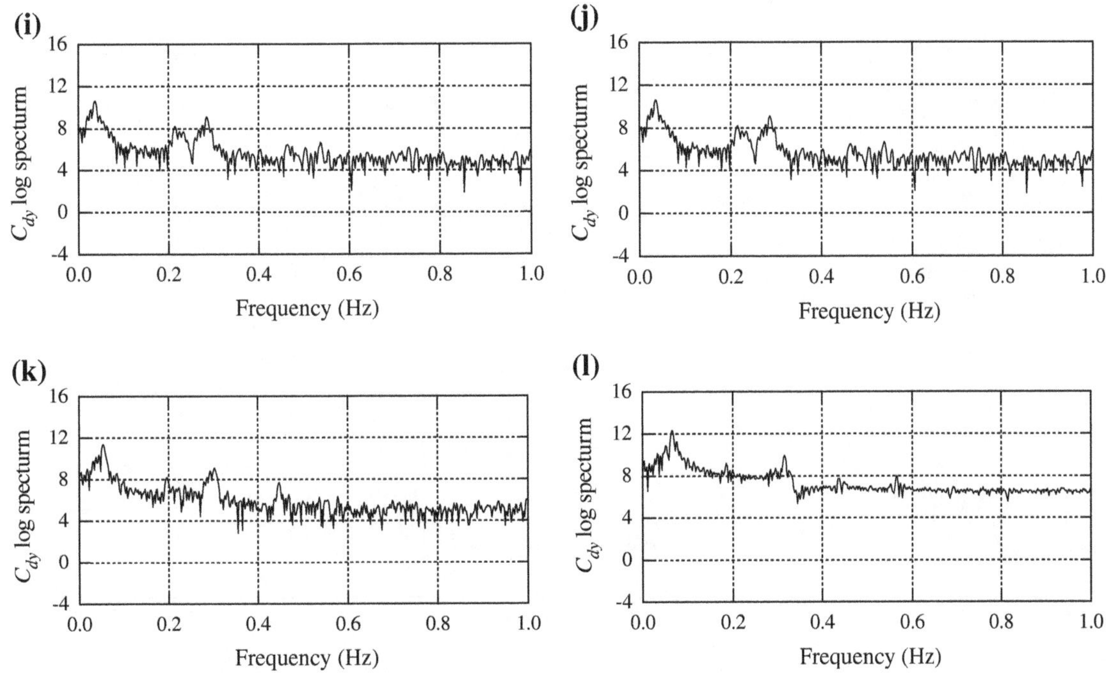

Fig. 5 continued

corresponding energy spectra with different Reynolds numbers have been presented in Figs. 4 and 5. From Fig. 4, when Reynolds number is smaller than Re_{critical}, the drag coefficient, lift coefficient and corresponding energy spectra are the same as those of wave-only case. This again proves that the effect of current on wave is so little that it can be neglected when the corresponding Reynolds number achieves a critical value. However, when Reynolds number is bigger than Re_{critical}, it is found that $C_{\text{d}x}$ and C_{dy} have been increased with the Reynolds number increasing (see Fig. 5). The vortex shedding frequency increased as the Reynolds number increased. This fact can be verified from Fig. 5i–l which is obtained from the logarithmic spectrum of C_{dy}.

The streamwise velocity U, transverse velocity V, vertical velocity W at one representative point on the horizontal plane of middle elevation, namely, point I at $x = 1.5$ m and $y = 0.0$ m (as shown in Fig. 1 for its position) are depicted in Fig. 6. It gives the time histories of resolved velocity at this point for various Reynolds numbers. Reynolds numbers ranging from 3.0×10^5 to 6.0×10^5 are plotted to discover the patterns obviously in Fig. 6. From Fig. 6a–d, it is found that in the first ten periods, the turbulent forms of all cases are complicated. The higher the Reynolds number, the more significant the velocity fluctuation. From Fig. 6e–h, it is seen that vortex shedding period is decreased as Reynolds numbers increased. Another feature is that the vertical velocity (W) magnitude is increased because of the increasing of Reynolds number. Thus, vertical velocity W should be considered to wave–current interaction with high corresponding Reynolds numbers. It is of great importance to consider the wave–current as three-dimensional flow.

The variations of Strouhal number and effective drag coefficient under different Reynolds numbers are shown in Fig. 7 and Table 2. The vortex shedding frequency and the mean wave–current force are nearly close to zero for the case of Reynolds number smaller than 1.0×10^5, which is the same as the wave-only case. As the Reynolds number increases, the Strouhal number, the mean force coefficient and the RMS of coefficients increase, and they all show the same variation tendency. It is expected that the mean ones, the RMS ones, and Strouhal number are all sensitive to various Reynolds numbers under wave–current effect. From Fig. 7d, it is found that the Strouhal number increases slightly for the higher Reynolds numbers ranging from 3.0×10^5 to 6.0×10^5.

4 FEM simulation

Wave–current interaction with a vertical square cylinder is investigated numerically in this study. The results show as follows:

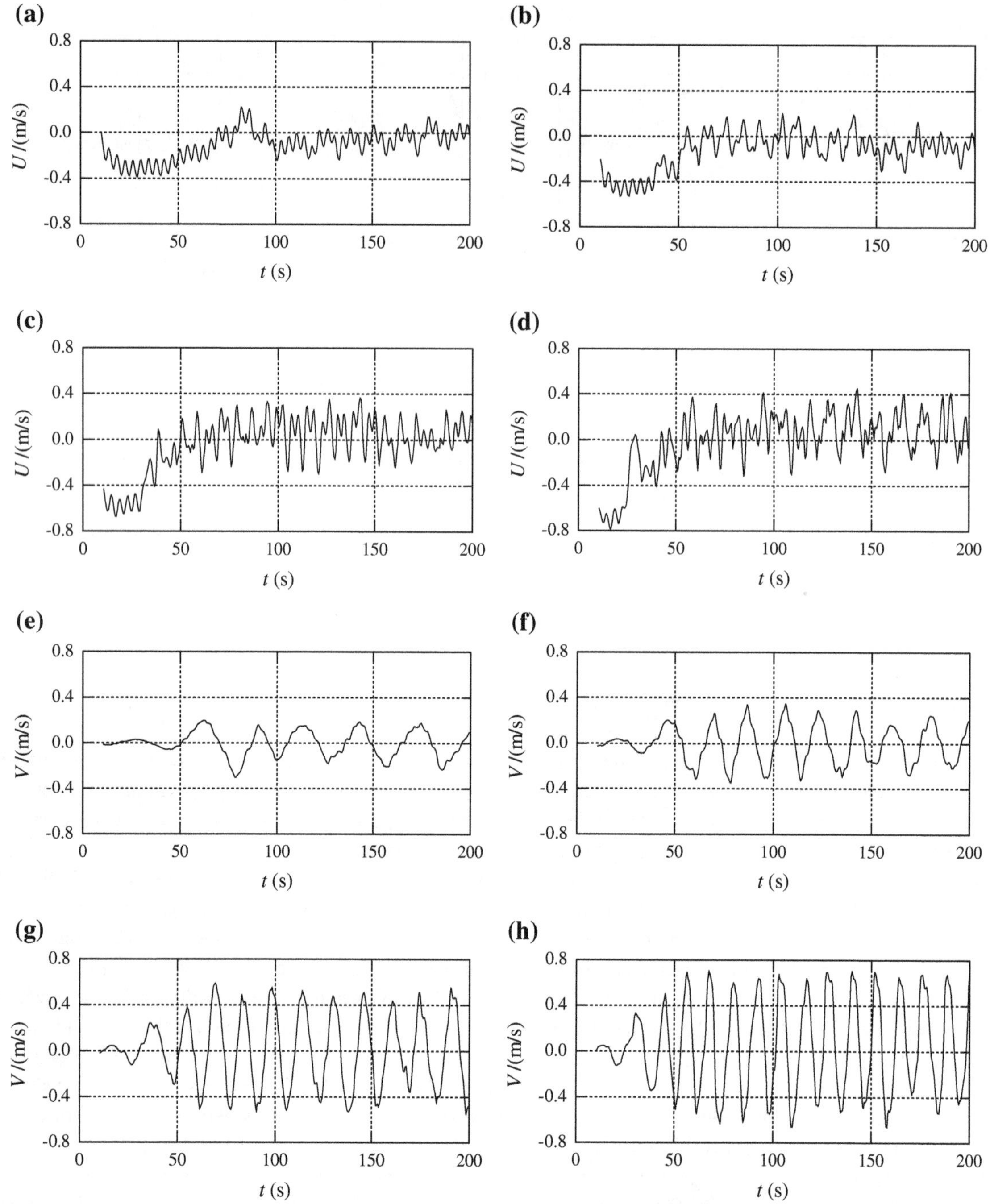

Fig. 6 Velocity in three directions at point(x = 1.5 m and y = 0.0 m) for different Reynolds numbers. **a** $KC = 0.6$, $Re = 300{,}000$, **b** $KC = 0.6$, $Re = 400{,}000$, **c** $KC = 0.6$, $Re = 500{,}000$, **d** $KC = 0.6$, $Re = 600{,}000$, **e** $KC = 0.6$, $Re = 300{,}000$, **f** $KC = 0.6$, $Re = 400{,}000$, **g** $KC = 0.6$, $Re = 500{,}000$, **h** $KC = 0.6$, $Re = 600{,}000$ **i** $KC = 0.6$, $Re = 300{,}000$, **j** $KC = 0.6$, $Re = 400{,}000$, **k** $KC = 0.6$, $Re = 500{,}000$, **l** $KC = 0.6$, $Re = 600{,}000$

(1) The vortex shedding frequency has been reduced because of wave–current nonlinear interaction.

(2) When the corresponding Reynolds number is smaller than a critical one, current effect on wave can be nearly neglected.

(3) With the Reynolds number increasing, however, wave–current–structure interaction is sensitive to the Reynolds number; in this case, the effect of Reynolds number on the global quantities, the mean value,

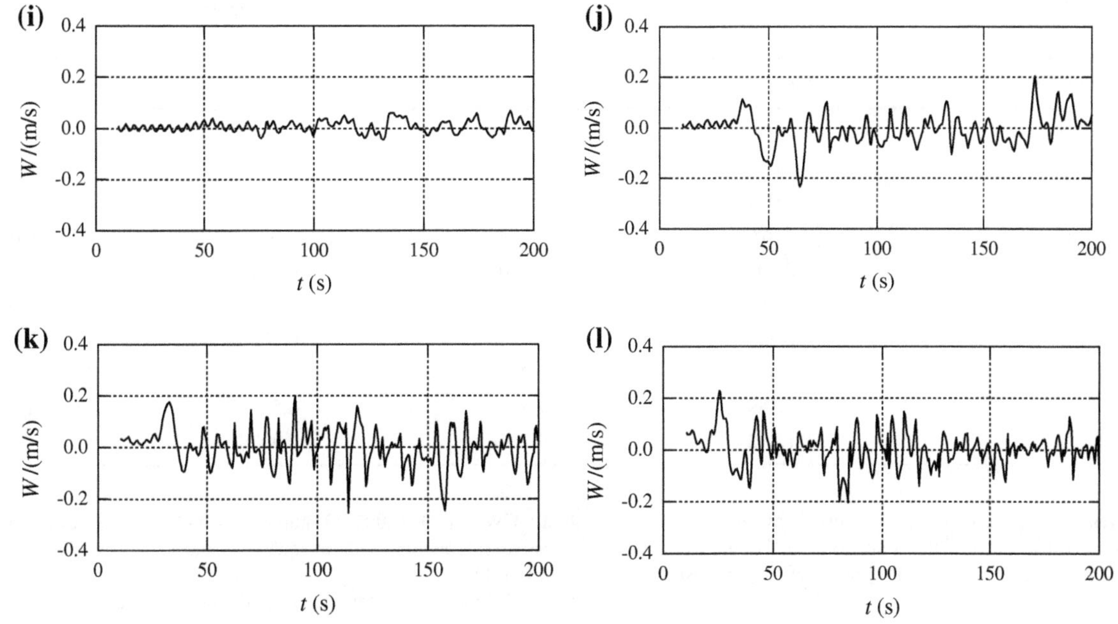

Fig. 6 continued

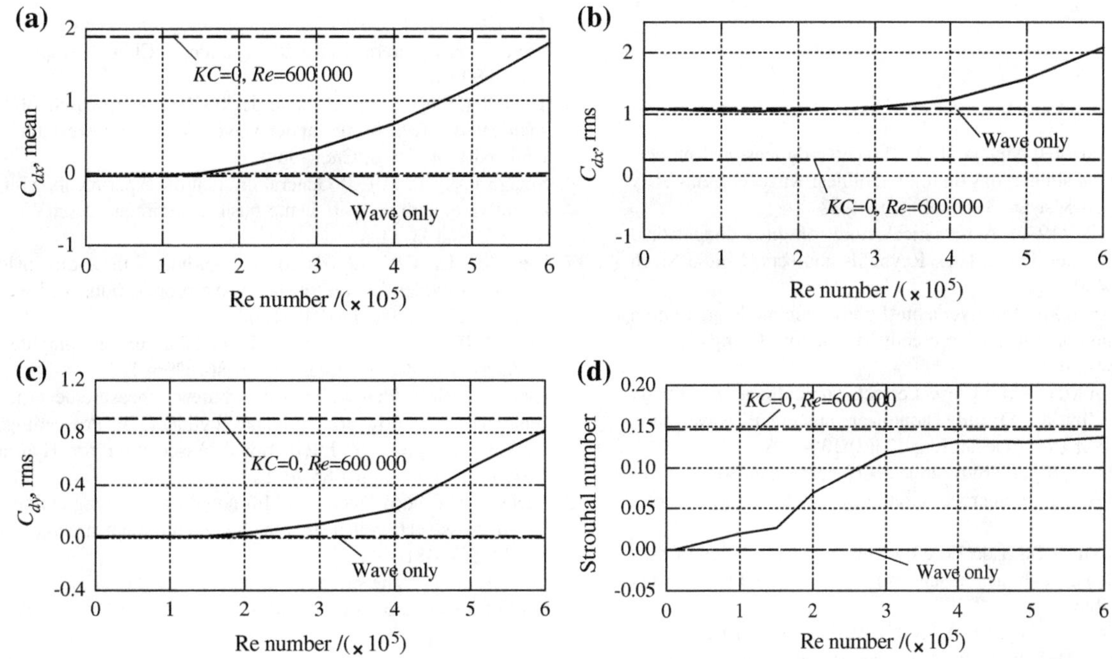

Fig. 7 The mean and RMS drag coefficient as well as Strouhal number (St) versus different Reynolds numbers. **a** C_{dx} (mean) , **b** C_{dx} (RMS), **c** C_{dy} (RMS), **d** St number

RMS value of drag force coefficient and turbulent feature are much more obvious and unignorable.

(4) The vertical velocity is increased with the Reynolds muber increasing, and the magnitude of it weighs against the streamwise velocity. Thus, it is of great importance to consider the wave–current interaction as three-dimensional flow.

Table 2 Calculated mean force coefficients, RMS of fluctuation of coefficients, and normalized shedding frequencies (Strouhal number St) for various Reynolds numbers

Wave–current parameters	$\bar{C}_{dx}$	$\bar{C}'_{dx}$	$\bar{C}'_{dy}$	St
$KC = 0.6$, $Re = 1.0 \times 10^3$	−0.0373	1.0877	0.0074	–
$KC = 0.6$, $Re = 1.0 \times 10^4$	−0.0374	1.0803	0.0075	–
$KC = 0.6$, $Re = 1.0 \times 10^5$	−0.0349	1.0815	0.0084	–
$KC = 0.6$, $Re = 1.0 \times 10^5$	−0.0259	1.0551	0.0081	0.019531
$KC = 0.6$, $Re = 1.0 \times 10^5$	−0.0014	1.0505	0.0111	0.026042
$KC = 0.6$, $Re = 1.0 \times 10^5$	0.0849	1.062	0.0336	0.0685
$KC = 0.6$, $Re = 1.0 \times 10^5$	0.338	1.1086	0.1058	0.117333
$KC = 0.6$, $Re = 1.0 \times 10^5$	0.687	1.2319	0.2145	0.13175
$KC = 0.6$, $Re = 1.0 \times 10^5$	1.1892	1.5709	0.5352	0.131
$KC = 0.0$, $Re = 1.0 \times 10^5$	1.8927	0.2568	0.9080	0.1465

Acknowledgments The research is supported by the National Natural Science Foundation of China (No. 51178397), and Technological Research and Development Programs of the Ministry of Railways (No. 2010G004-L).

References

1. Park JC, Kim MH, Miyata H (2001) Three-dimensional numerical wave tank simulations on fully nonlinear wave–current–body interactions. J Mar Sci Technol 6(2):70–82
2. Deardorff JW (1970) A numerical study of three-dimensional turbulent channel flow at large Reynolds numbers. J Fluid Mech 41(2):453–480
3. Sohankar A (2006) Flow over a bluff body from moderate to high Reynolds numbers using large eddy simulation. Comput Fluids 35(10):1154–1168
4. Koo W, Kim MH (2007) Current effects on nonlinear wave-body interactions by a 2D fully nonlinear numerical wave tank. J Waterw Port Coast Ocean Eng 133(2):136–146
5. Li Y, Lin M (2010) Hydrodynamic coefficients induced by waves and currents for submerged circular cylinder. Procedia Eng 4:253–261
6. Cheng M, Whyte DS, Lou J (2007) Numerical simulation of flow around a square cylinder in uniform-shear flow. J Fluids Struct 23(2):207–226
7. Vengadesan S, Nakayama A (2005) Evaluation of LES models for flow over bluff body from engineering application perspective. Sadhana 30(1):11–20
8. Li CW, Lin P (2001) A numerical study of three-dimensional wave interaction with a square cylinder. Ocean Eng 28(12): 1545–1555
9. Lin P (2004) A numerical study of solitary wave interaction with rectangular obstacles, Coast Eng 51(1):35–51
10. Lin P, Li CW (2003) Wave–current interaction with a vertical square cylinder. Ocean Eng 30(7):855–876
11. Lin P (2006) A multiple-layer sigma-coordinate model for simulation of wave–structure interaction. Comput Fluids 35(2): 147–167
12. Tan CJ, Zhu B (2010) Numerical study of three-dimensional wave–current interaction with cylinders. Chin J Appl Mech 27(4):680–686
13. Tan CJ (2009) Coupled cable-deck vibration research of long span cable stayed bridge under wave effect. Southwest Jiaotong University of China, Chengdu
14. Smagorinsky JS (1963) General circulation experiments with the primitive equations, part I: the basic experiment. Mon Weather Rev 91:99–163
15. Lin PZ, Li CW (2002) A σ-coordinate three-dimensional numerical model for free surface wave propagation. Int J Numer Methods Fluids 38(11):1045–1068
16. Benque JP (1982) A new method for tidal current computation. J Waterw Port Coast Ocean Div 108(3):396–417
17. Lyn DA (1989) Phase-averaged turbulence measurements in the separated shear flow around square cylinder, In: Proceedings of the 23rd congress of International Association for Hydraulic Research, Ottawa, pp A85-A92
18. Vickery BJ (1966) Fluctuating lift and drag on a long cylinder of square cross-section in a smooth and in a turbulent stream. J Fluid Mech 25(3):481–494
19. Lee BE (1975) The effect of turbulence on the surface pressure field of a square prism. J Fluid Mech 69(2):263–282

16

An overview of a unified theory of dynamics of vehicle–pavement interaction under moving and stochastic load

Lu Sun

Abstract This article lays out a unified theory for dynamics of vehicle–pavement interaction under moving and stochastic loads. It covers three major aspects of the subject: pavement surface, tire–pavement contact forces, and response of continuum media under moving and stochastic vehicular loads. Under the subject of pavement surface, the spectrum of thermal joints is analyzed using Fourier analysis of periodic function. One-dimensional and two-dimensional random field models of pavement surface are discussed given three different assumptions. Under the subject of tire–pavement contact forces, a vehicle is modeled as a linear system. At a constant speed of travel, random field of pavement surface serves as a stationary stochastic process exciting vehicle vibration, which, in turn, generates contact force at the interface of tire and pavement. The contact forces are analyzed in the time domain and the frequency domains using random vibration theory. It is shown that the contact force can be treated as a nonzero mean stationary process with a normal distribution. Power spectral density of the contact force of a vehicle with walking-beam suspension is simulated as an illustration. Under the subject of response of continuum media under moving and stochastic vehicular loads, both time-domain and frequency-domain analyses are presented for analytic treatment of moving load problem. It is shown that stochastic response of linear continuum media subject to a moving stationary load is a nonstationary process. Such a nonstationary stochastic process can be converted to a stationary stochastic process in a follow-up moving coordinate.

L. Sun (✉)
Department of Civil Engineering, The Catholic University of America, Washington, DC 20064, USA
e-mail: sunl@cua.edu

L. Sun
International Institute of Safe, Intelligent and Sustainable Transportation & Infrastructure, Southeast University, Nanjing 210096, China

Keywords Vehicle–pavement interaction · Random field · Continuum medium · Spectral analysis · Green's function · Linear system

List of Symbols

$\mathbf{F}(\mathbf{x}, t)$	Moving source
$\delta(\cdot)$	Dirac delta function
v	Source velocity
$P(t)$	Source magnitude
$\mathbf{h}[\cdot]$	Impulse response function
$\mathbf{H}[\cdot]$	Frequency response function
$S(\cdot)$	Power spectral density
$R(\cdot)$	Correlation function
$E[\cdot]$	Expectation
σ^2	Variance
ψ	Standard deviation
$\mathbf{G}[\cdot]$	Green's function
$\mathbf{u}[\cdot]$	Response function of continuum media

1 Introduction

The investment of the United States in the nation's transportation infrastructure alone (highways, bridges, railways, and airports) amounted to $7 trillion by 1999. To preserve infrastructure longevity in a cost-effective manner, the research in pavement design and infrastructure management has been growing rapidly in recent years. From a pavement design point of view, pavement response, damage, and performance are essentially the result of long-term vehicle–pavement interaction. When vehicle speed is low, the dynamic effect of vehicular loads on pavements is insignificant. However, with the promotion of high-speed surface transportation in the world, this dynamic effect must be taken into account to develop more rational pavement design methods. For instance, real causative mechanisms that lead to fatigue damage of pavement material might be frequency dependent. From an infrastructure management point of view, vehicle–pavement interaction has a profound impact on the way that existing technologies of structural health monitoring, environmental vibration mitigation, nondestructive testing and evaluation, and vehicle weight-in-motion are to be improved, innovated and implemented. For instance, modern high-speed surface transportation systems are normally accompanied by rises in levels of noise and vibration that may cause a significant detrimental effect to the ecology. Vehicle–pavement interaction-induced structure-borne and ground-borne vibrations emit and propagate toward some extent. Residents may experience hardship from uncomfortable vibration, and high-precision equipment may suffer from malfunctioning to irreparable damage. To mitigate noise and vibration in surrounding areas of roadway, it is necessary to investigate predominant frequencies of vehicle–pavement interaction, the source of vibration, so as to develop effective noise and vibration countermeasures. The study of vehicle–pavement interaction also plays a critical role in developing better inversion algorithms for nondestructive testing and evaluation of transportation infrastructure. In addition, taking into account dynamic effects of vehicle vibration caused by rough surface may considerably improve accuracy and reliability of weigh-in-motion systems, which measure a vehicle's weight as it travels at a normal speed.

Vehicle–pavement interaction is not only a central problem to pavement design, but also has a profound impact on infrastructure management, vehicle suspension design, and transportation economy. Figure 1 shows a central role of vehicle–pavement interaction played in a wide variety of applications. From a vehicle-design point of view, a designer needs to consider vehicle vibration and controllability, which affect ride quality and vehicle maneuver. Vehicle–pavement interaction also has a huge economic impact. As pavement performance deteriorates and roadway surface gets rougher, both operation costs (fuel, tire wear, and routine maintenance) and roadway maintenance cost of the vehicle increase dramatically, accompanied by decreasing transportation productivity. There is no doubt that there is a great and urgent need for fundamental research on vehicle–pavement interaction due to rapid deterioration of huge highway infrastructure nationwide, tight maintenance budget, and the key role played by vehicle–pavement interaction.

Since the American Association of State Highway Officials (AASHO) road test, the fourth power law has been widely used by pavement engineers to design highway and airfield pavement and to predict the remained life and cumulated damage of pavements [1–3]. Besides the damage caused by static loads, dynamic loads may lead to additional pavement damage. A consequence of a high power in the damage law is that any fluctuation of pavement loading may cause a significant increase in the damage suffered by pavement structures. A number of recent filed measurements and theoretical investigations showed that vehicle vibration-induced pavement loads are moving stochastic loads [4–8]. The researchers concluded that vibrations of vehicles were related primarily to pavement surface roughness, vehicle velocity, and suspension types [9–13].

Estimation of pavement damage caused by dynamic loads varies anywhere from 20 % to 400 % [14]. The smaller estimates are based on the assumption that peak dynamic loads (and hence the resulting pavement damage) are distributed randomly over the pavement surface. The larger estimates are based on the assumption that vehicles consistently apply their peak wheel loads in the same areas of the pavement. Theoretical studies by Cole [15] and Hardy and Cebon [16] confirmed that for typical highway traffic, certain areas of the pavement surface always suffered the largest wheel forces, even when the vehicles had a wide range of different suspension systems, payloads, and speeds.

Complicated relationships exist among vehicle suspension, dynamic wheel loads, pavement response, and damage [17–19]. On one hand, it has been known for years on how to manufacture automobiles operating properly on a variety of pavement surfaces. On the other hand, however, the effect of vehicle design on pavement has not been thoroughly studied. For instance, Orr [20] stated in his study that comparatively little was known about the influence of suspension design on pavement in the automobile industry yet.

A number of obstacles exist in revealing vehicle–pavement interaction. A theoretical foundation universally applicable to the involved specific problems is no doubt very attractive. It will not only provide a guide for experimental study and validation, but will also enable better design and maintenance of vehicles and pavements. As such, this article provides an overview of a unified theory for dynamics of vehicle–pavement interaction under moving and stochastic loads. This article covers three major aspects of the subject: pavement surface, tire–pavement contact forces, and response of continuum media under moving and stochastic vehicular loads developed by Sun and his associates (Fig. 2). The remainder of this article is organized as follows. Section 2 addresses mathematical description of pavement surface roughness. Section 3 studies contact force generated due to vehicle–pavement interaction, which is the source to both vehicle dynamics and pavement dynamics. Section 4 investigates pavement response under moving stochastic loads. Section 5 discusses difficulties and deficiencies in the existing research, and projects further research and applications. Section 6 makes concluding remarks.

2 Pavement surface

Irregularities in pavement surface are often called roughness. All pavement surfaces have some irregularities. The National Cooperative Highway Research Program defines "roughness" as "the deviations of a pavement surface from a true planar surface with characteristic dimensions that affect vehicle dynamics, ride quality, and dynamic pavement loads [21]." Factors contributing to roughness, in a general sense, include vertical alignment, cracks, joints, potholes, patches and other surface distresses.

Newly constructed pavements may be poorly finished or have design features such as construction joints and thermal expansion joints, which can be main sources of vehicle vibration [22, 23]. Measurements show that pavement roughness can be modeled as a random field consisting of different wavelength. Considerable effort has been devoted to describing pavement surface roughness [24–26]. The peak-to-valley measurements, the average deviation from a straight edge, and the cockpit acceleration are several distinct approaches that have been suggested for pavement surface characterization [27]. Practical limitations of these three approaches could be found in the study of Hsueh and Penzien [28].

Spectrum of a deterministic function referring to its Fourier transform reveals sinusoid components that a deterministic periodic or nonperiodic function is composed of. Spectrum analysis of a deterministic function $f(t)$ can be obtained if and only if this deterministic function satisfies Dirichlet condition and absolute integrability condition [29]:

$$\int_{-\infty}^{\infty} |f(t)|\ \mathrm{d}t < +\infty. \tag{1}$$

Pavement surface as a random field of elevation does not decay as spatial coordinates extend to infinity; therefore, it does not satisfy inequality (1). For this reason, taking Fourier analysis to a sampled pavement surface does not make too much sense in theory. However, the correlation vector of random field does decay as spatial coordinates extend to infinity. Hence, applying Fourier transform to correlation vector of pavement surface, commonly known as power spectral density, does serve the purpose of spectrum analysis in theory [30]. As such, mathematical description of pavement surface takes place under the theoretical framework of spectral analysis of stochastic process.

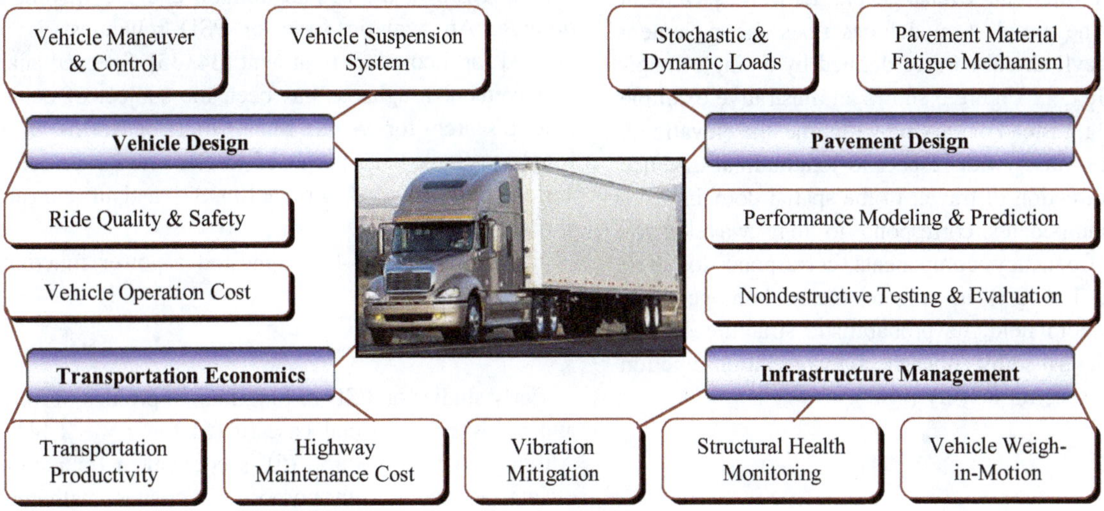

Fig. 1 A central role of vehicle–pavement interaction played in various applications

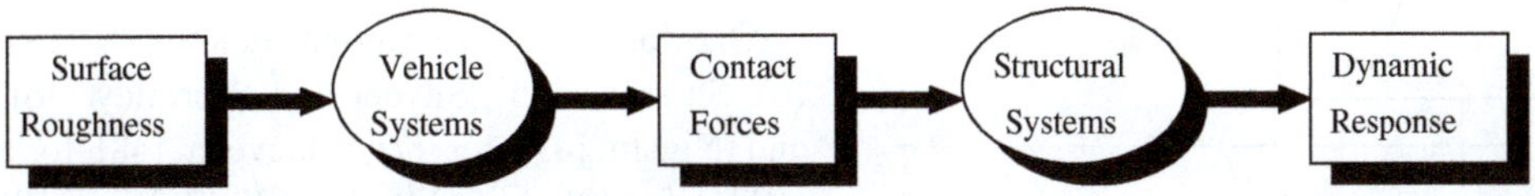

Fig. 2 Dynamics of vehicle–pavement interaction

The following subsections consist of a general mathematical framework for describing pavement surface roughness, a review of statistical description of pavement surface roughness, a description of periodic joints of cement concrete pavement, and three different hypotheses when projecting a one-dimensional road profile to a two-dimensional pavement surface.

2.1 One-dimensional description

Given the manifest complexity exhibited in pavement surface as a random field, making some assumptions becomes indispensable to simplify mathematical description of pavement surface. Commonly used assumptions on surface roughness are that surface roughness is an ergodic and homogeneous random field with elevation obeying Gaussian distribution [31–33]. The assumption of ergodicity makes sure that the temporal average of a sample of stochastic process equals to the statistical mean of stochastic process, which enable one to obtain the statistical characteristics of a stochastic process by measuring only a few samples. The assumption of homogeneity ensures random property of surface roughness is independent to the sites measured. When a vehicle travels at a constant speed, a homogeneous random filed is converted to a stationary stochastic process. The assumption of Gaussian distribution ensures that transformation of such randomness through a linear system is still Gaussian.

Road profile along longitudinal direction (i.e., direction of travel) is often an exchangeable term of pavement roughness. The simplest model describes the pavement surface as a cylindrical surface defined by a single longitudinal profile $\xi(x)$. Figure 3 shows an illustrative example of road profile. Instead of varying with time, the elevation ξ of the surface varies with respect to longitudinal distance x along the direction of travel. In the spatial domain, low-frequency components correspond to long wavelengths while high frequency components correspond to short wavelengths. Let $\xi(x)$ be a zero mean, homogeneous, Gaussian random field. Its probabilistic structure can be completely described by either the autocorrelation function or the power spectral density (PSD).

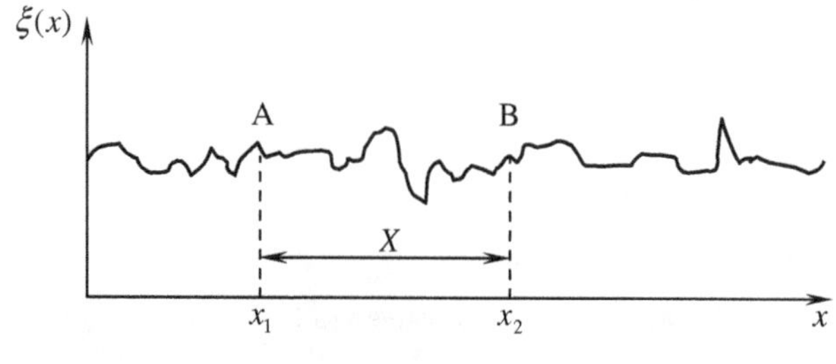

Fig. 3 One-dimensional road profile

According to the Winner–Khintchine theory [31], the following expressions constitute a pair of Fourier transform:

$$S_{\xi\xi}(\Omega) = (2\pi)^{-1}\int_{-\infty}^{\infty} R_{\xi\xi}(X)\mathrm{e}^{-i\Omega X}\mathrm{d}X, \tag{2a}$$

$$R_{\xi\xi}(X) = \int_{-\infty}^{\infty} S_{\xi\xi}(\Omega)\mathrm{e}^{i\Omega X}\mathrm{d}\Omega, \tag{2b}$$

where X represents the distance between any two points along the road. Wavenumber spectrum, $S_{\xi\xi}(\Omega)$, is the direct PSD function of wavenumber Ω , which represents spatial frequency defined by $\Omega = 2\pi/\lambda$ where λ is the wavelength of roughness. Under the assumption of homogeneity, spatial autocorrelation function $R_{\xi}(X)$ is defined by

$$R_{\xi\xi}(X) = E[\xi(x_1)\xi(x_2)] = E[\xi(x_1)\xi(x_1+X)] \tag{3}$$

for any x_1 and X. Here $E[\cdot]$ is the expectation operator and can be calculated by

$$E[\xi(x)] = \lim_{X\to\infty}\frac{1}{2X}\int_{-X}^{X}\xi(x)\mathrm{d}x. \tag{4}$$

This expectation, differing from statistical mean of a random process, is temporal average of a sample of stochastic process.

2.1.1 PSD roughness

As a one-dimensional random field, roughness can be characterized in the spatial domain and the wavenumber domain. An analytic form of PSD roughness is often desired for theoretical treatment [34, 35]. Spectral analysis of pavement roughness has been the subject of considerable research for years. Many function forms of PSD roughness have been proposed to characterize bridge deck surface [36], rail-track surface [33, 37], and airfield runway surface [24, 38].

Sayers et al. [39, 40] suggested a power function for PSD roughness:

$$S_{\xi\xi}(\Omega) = A\Omega^{\alpha}. \tag{5}$$

Early studies on PSD of longitudinal profiles of runways and roads are a special case of (5) for $\alpha = -2$ [41–44]. Although power function (5) is convenient for parameter estimation and design purpose, it creates mathematical difficulties at $\Omega = 0$, where $S_{\xi\xi}(\Omega)$ becomes infinite. For this reason, two distinct function forms have been proposed in later studies.

One form is to use rational functions: Sussman [45] for (6), Sussman [45], Snyder and Wormley [46], and Yadav and Nigam [47] for (7), Macvean [48] for (8), Bolotin (1984) for (9), and Gillespie [25, 26] for (10), to name a few.

$$S_{\xi\xi}(\Omega) = \frac{S_0(\Omega^2 + \alpha^2 + \beta^2)}{(\Omega^2 - \alpha^2 + \beta^2)^2 + 4\Omega^2\alpha^2}, \quad (6)$$

$$S_{\xi\xi}(\Omega) = \frac{S_0}{\Omega^2 + \alpha^2}, \quad (7)$$

$$S_{\xi\xi}(\Omega) = \frac{S_0}{(\Omega^2 + \alpha^2)^2}, \quad (8)$$

$$S_{\xi\xi}(\Omega) = \frac{4\sigma^2}{\pi} \frac{\alpha\Omega_0^2}{(\Omega^2 - \Omega_0^2) + 4\alpha^2\Omega^2}, \quad (9)$$

$$S_{\xi\xi}(\Omega) = S_0[1 + \Omega_0^2\Omega^{-2}](2\pi\Omega)^{-2}. \quad (10)$$

Another form is to use piecewise functions. For instance, (11) was suggested by the International Organization for Standardization (ISO) to cover different frequency ranges [49–52].

$$S_{\xi\xi}(\Omega) = \begin{cases} C_{sp}\Omega_a^{-w_1} & \text{for } 0 \le \Omega \le \Omega_a \\ C_{sp}\Omega^{-w_2} & \text{for } \Omega_a \le \Omega \le \Omega_b, \\ 0 & \text{for } \Omega_b < \Omega \end{cases} \quad (11)$$

where Ω_a represents reference frequency, and Ω_b represents cut-off frequency. Iyengar and Jaiswal [33] provided a similar split power law in the form of (11) for rail-track in which $C_{sp} = 0.001(\text{m}^{-2}$ cycles/m) and $w_1 = w_2 = 3.2$. Marcondes et al. [22, 23] proposed an exponential piecewise function:

$$S_{\xi\xi}(\Omega) = \begin{cases} A_1 \exp(-k\Omega^p) & \text{for } \Omega \le \Omega_1, \\ A_2(\Omega - \Omega_0)^q & \text{for } \Omega_1 \le \Omega. \end{cases} \quad (12)$$

In all aforementioned PSD roughness, A, A_1, A_2, p, q, k, w_1, w_2, C_{sp}, α β, S_0, and Ω_0 are the real and positive parameters estimated from field test.

Although various PSD functions have been proposed for fitting the measured PSD roughness, it has been found that most pavement profiles including road surface, runway surface, and rail-track surface exhibit very similar trends of PSD curves. Some researchers have even reduced the PSD curves to only two families: one for rigid (cement concrete pavements), and the other for flexible (asphalt concrete pavements) [25, 26].

2.1.2 Effect of periodic joints of rigid pavement

Thermal expansion joints are commonly set on rigid pavement surface to reduce thermal stress in pavement. Thermal joints of rigid pavement can be a significant source of excitation of vehicle vibration, especially when the joint sealing material gets lost. This could cause undesirable riding qualities at a high speed [43]. Figure 4 shows an idealized rigid pavement surface with periodic joints. An actual rigid pavement surface might be considered as a combination of this periodic profile and a random field.

In the spatial domain, pavement roughness $\xi(x)$ in Fig. 4 can be described by a periodic function within a period $[-d/2,d/2]$.

$$\xi(x) = \begin{cases} 0, & x \in [-d/2, -\Delta/2) \\ c(x), & x \in [-\Delta/2, \Delta/2]. \\ 0, & x \in (\Delta/2, d/2] \end{cases} \quad (13)$$

Here spatial period along the longitudinal direction, d, represents slab length of rigid pavement, Δ and h, respectively, represent maximum width and depth of joint, $c(x)$ is named shape function of joint, which is a symmetrical function and describes the shape of the joint. With this description, pavement roughness $\xi(x)$ becomes an even function with period d.

A period function can be legitimately expanded using Fourier series, which is sometimes called frequency spectrum analysis of periodic function [53]. As such, coefficients of the expanded Fourier series of $\xi(x)$ a_n and b_n $(n = 0,1,2,)\cdots$ are given by

$$a_n = \frac{1}{d/2}\int_{-d/2}^{d/2} \xi(x)\cos\frac{n\pi x}{d/2}\mathrm{d}x = \frac{2}{d}\int_{-\Delta/2}^{\Delta/2} c(x)\cos\frac{2n\pi x}{d}\mathrm{d}x, \quad (14a)$$

$$b_n = 0,\ n = 0, 1, 2, \ldots \text{ (for } \xi(x) \text{ is an even function).} \quad (14b)$$

Wavenumber spectrum of roughness is thus expressed as

$$S_{\xi\xi}(\Omega_n) = a_n^2 + b_n^2 = \frac{4}{d^2}\left[\int_{-\Delta/2}^{\Delta/2} c(x)\cos\frac{2n\pi x}{d}\mathrm{d}x\right]^2, \quad (15a)$$

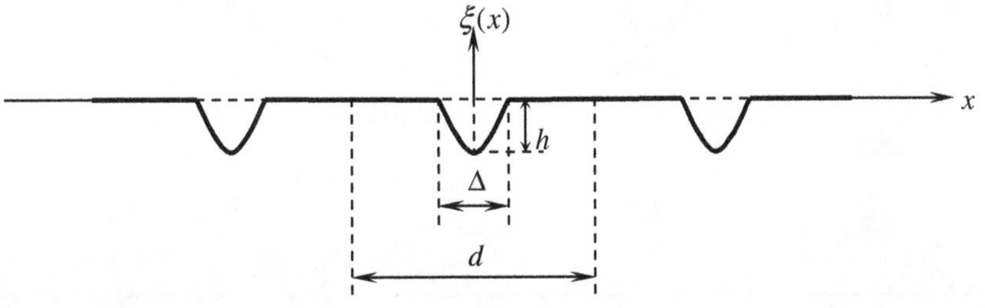

Fig. 4 Periodic joints of rigid pavement surface

$$\Omega_n = n/d \; (n = 0, 1, 2, \cdots\cdots). \qquad (15b)$$

In which Ω_n represents discrete wavenumber and $S_{\xi\xi}(\Omega_n)$ is discrete wavenumber spectrum. It is evident from (18) that different shapes of joints determine different discrete spatial spectrum. Four types of shape function, namely rectangular curve, parabolic curve, cosine curve, and triangular curve have been observed and investigated by Sun [54].

Table 1 gives the discrete spatial spectrums of these four types of shape functions computed from (15a). The comparison between these discrete spectrums for specified rigid pavement with slab length 5 m and joint width 0.03 m is plotted in Fig. 5 where $S = h\Delta/d$, $\Omega_n = 2\pi n/d$, $n = 0, 1, 2, \cdots$, and $\Delta/d = 0.006$. From this figure it can be seen that the spectrum of rectangular joint is the greatest in four types of joints, while the spectrum of triangular joints is the smallest. The ratio of the magnitude of four types of joints based on $S_{\xi\xi}(\Omega_0)$ approximates to 4:1.78:1.62:1. Because $S^2 = (h\Delta/d)^2$, the effect of joint of rigid pavement with long slab length d is much less than that of the pavement with short slab length.

2.2 Two-dimensional description

One-dimensional random field model of pavement surface is adequate for two-wheel vehicles, such as bicycles and motor-cycles but inadequate for cars and trucks having two or more wheel per axle. Actual highway and airfield pavement consists of a two-dimensional surface of finite width, a nominal camber and grader. The elevation of pavement surface exhibit random fluctuations about the nominal geometry and therefore should be more accurately treated as a two-dimensional random field, $\xi(x, y)$, with space coordinates x and y as the indexing parameters as shown in Fig. 6.

When ignoring those isolated large fluctuations such as potholes, fluctuations of pavement surface can be approximate by a homogenous, Gaussian random field with a zero mean [49, 55]. Probabilistic structure of a two-dimensional random field can be completely defined either by two-dimensional autocorrelation function

$$\begin{aligned} R_{AC}(X, Y) &= E[\xi(x_1, y_1)\xi(x_2, y_2)] \\ &= E[\xi(x_1, y_1)\xi(x_1 + X, y_1 + Y)] \end{aligned} \qquad (16)$$

or by two-dimensional PSD, $S_{\xi\xi}(\Omega, \mathrm{K})$, where $X = x_2 - x_1$, $Y = y_1 - y_2$, Ω and K represent wavenumber in x-axis and y-axis directions, respectively. Here $S_{\xi\xi}(\Omega, \mathrm{K})$ is defined as a double-sided Fourier transform of autocorrelation function (16).

$$S(\Omega, \mathrm{K}) = (2\pi)^{-2} \int_{-\infty}^{\infty}\int_{-\infty}^{\infty} R(X, Y)e^{-i(\Omega X + \mathrm{K}Y)} \mathrm{d}X\mathrm{d}Y. \qquad (17)$$

Table 1 Discrete wavenumber spectrum resulting from periodic joints of rigid pavement

Joints	Shape function	Diagram	Discrete spatial spectrum
Rectangular curve	$-h$	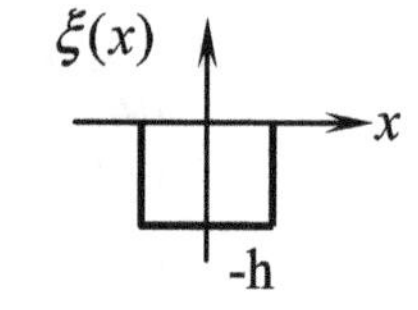	$S_{\xi\xi}(\Omega_0) = 4S^2$ $S_{\xi\xi}(\Omega_n) = [\frac{\sin(\Delta\Omega_n/2)}{\Delta\Omega_n/2}]^2 S_{\xi\xi}(\Omega_0).$
Parabolic curve	$\frac{4h}{\Delta^2}x^2 - h$	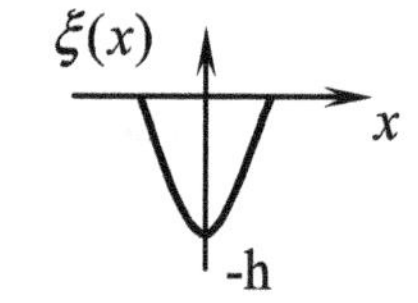	$S_{\xi\xi}(\Omega_0) = 16S^2/9$ $S_{\xi\xi}(\Omega_n) = \frac{9[\cos(\Delta\Omega_n/2) - \sin(\Delta\Omega_n/2)/(\Delta\Omega_n/2)]^2}{(\Delta\Omega_n/2)^4} S_{\xi\xi}(\Omega_0).$
Cosine curve	$-h\cos\frac{\pi}{\Delta}x$	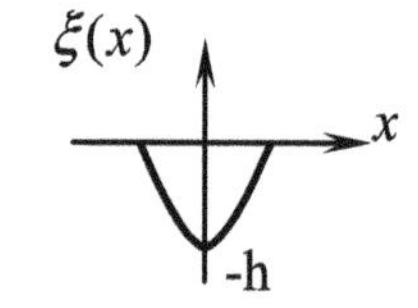	$S_{\xi\xi}(\Omega_0) = 16S^2/\pi^2$ $S_{\xi\xi}(\Omega_n) = [\frac{\cos(\Delta\Omega_n/2)}{1 - (\Delta\Omega_n/\pi)^2}]^2 S_{\xi\xi}(\Omega_0).$
Triangular curve	$\frac{2h}{\Delta}\lvert x\rvert - h$	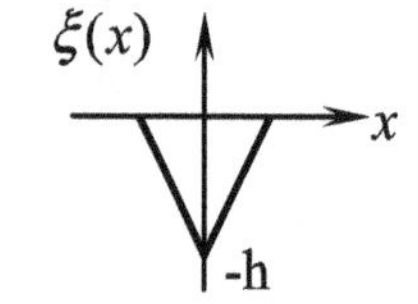	$S_{\xi\xi}(\Omega_0) = S^2$ $S_{\xi\xi}(\Omega_n) = \frac{4[1 - \cos(\Delta\Omega_n/2]^2}{(\Delta\Omega_n/2)^4} S_{\xi\xi}(\Omega_0).$

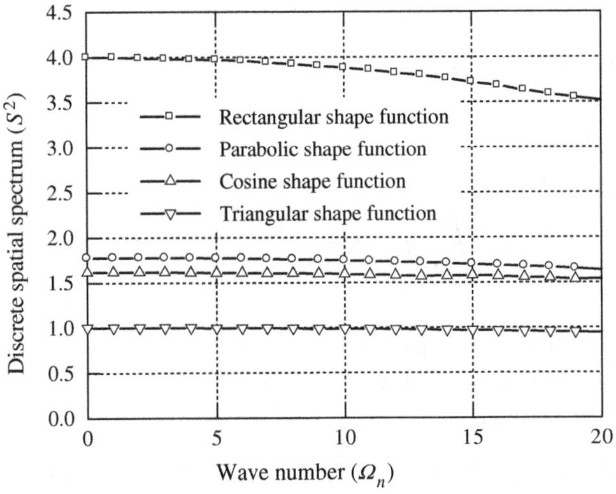

Fig. 5 Discrete wavenumber spectrum resulting from periodic joints of rigid pavement

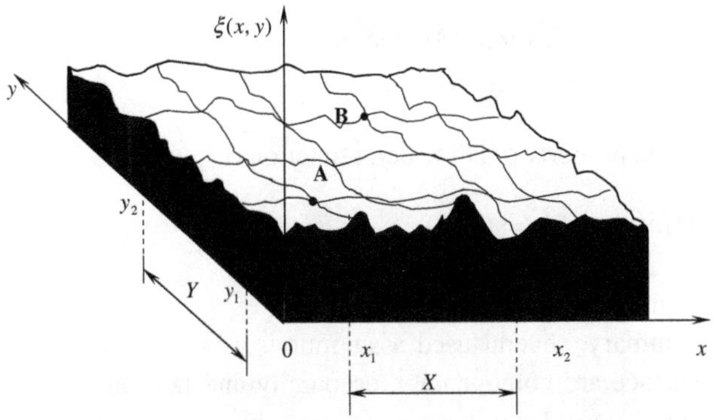

Fig. 6 Two-dimensional random field model of pavement surface

The determination of the two-dimensional autocorrelation function or two-dimensional PSD relies on the measured elevation of the entire pavement surface, which is a possible but formidable task in data acquisition, numerical computation and data storage for hundreds of thousands of miles of roadways. Efforts have, therefore, been made to simplify the two-dimensional random field such that it can be uniquely generated from the one-dimensional random field models by imposing certain hypothetic properties to pavement surface [49, 50, 55].

2.2.1 The hypothesis of isotropy

Consider a two-dimensional random field model of pavement surface as shown in Fig. 7. There are two parallel wheel paths separated by a constant distance Y along the transverse direction. Longitudinal profiles along each wheel path can be derived from the two-dimensional random field $\xi(x, y)$ as follows.

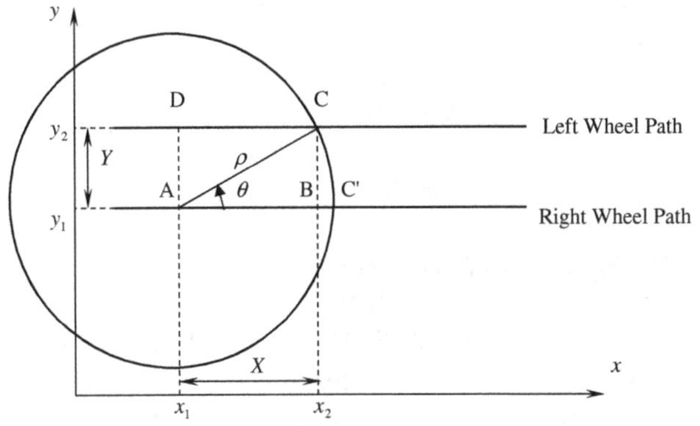

Fig. 7 A plane view of two-dimensional isotropic random field

From (16) the autocorrelation functions between A and B and between D and C are, respectively, given by

$$R_{AB}(X,0) = E[\xi(x_1,y_1)\xi(x_1+X,y_1)] \tag{18a}$$

and

$$R_{DC}(X,0) = E[\xi(x_1,y_2)\xi(x_1+X,y_2)]. \tag{18b}$$

Since $\xi(x,y)$ is homogenous,

$$R_{AB}(X,0) = R_{DC}(X,0) = R_{\xi\xi}(X). \tag{19}$$

Crosscorrelation functions, $R_{AC}(X,Y)$ and $R_{CA}(X,Y)$, are even functions of X and Y:

$$R_{AC}(X,Y) = R_{CA}(X,-Y) = R_{AC}(-X,Y). \tag{20}$$

Now, assume that $\xi(x, y)$ be an isotropic random field [49, 56]. The property of isotropy requires that for any profile making an angle θ with x-axis, the following condition holds:

$$R(\rho\cos\theta, \rho\sin\theta) = R(\rho,0). \tag{21}$$

From (16), (19), and (21), it follows that

$$\begin{aligned} R_{AC}(X,Y) &= R_{CA}(X,-Y) = R_{AC}(-X,Y) \\ &= R_{AC}(\sqrt{X^2+Y^2},0) = R_{\xi\xi}(\sqrt{X^2+Y^2}). \end{aligned} \tag{22}$$

Since autocorrelation function and PSD form a Fourier pair, cross-PSD is given by

$$S_{AC}(\Omega) = S_{CA}(\Omega) = (2\pi)^{-1}\int_{-\infty}^{\infty} R_{\xi\xi}(\sqrt{X^2+Y^2})e^{-i\Omega X}\mathrm{d}X. \tag{23}$$

Equations (19) and (23) are general results for spectral analysis under the hypothesis of isotropy. When using the definition of a normalized cross-PSD developed by Dodds and Robson [49], cross-PSD can also be written as

$$S_{AC}(\Omega) = S_{CA}(\Omega) = g(\Omega)S_{\xi\xi}(\Omega). \tag{24}$$

where $g(\Omega)$ is coherence function between direct-PSD and cross-PSD of two parallel wheel paths, which is always not greater than 1 [57]. Heath [58] further presented a closed-form $g(\Omega)$

$$g(\Omega) = 1 - \frac{Y}{S_{\xi\xi}(\Omega)} \int_0^\infty S_{\xi\xi}(\sqrt{\Omega^2 + \zeta^2}) J_1(Y\zeta) d\zeta, \quad (25)$$

where $J_1(\cdot)$ is the first-order Bessel function and $\Omega \geq 0$.

2.2.2 The hypothesis of uncorrelation

Parkhilovskii [55] proposed the hypothesis of uncorrelation. It assumes that the two profiles of two parallel wheel paths, $\xi(x, y_1)$ and $\xi(x, y_2)$ $\xi(x, y_2)$, can be derived from two uncorrelated random fields, that is

$$\xi(x, y_1) = \zeta(x) + Y\gamma(x), \quad (26a)$$

$$\xi(x, y_2) = \zeta(x) - Y\gamma(x). \quad (26b)$$

It then follows that

$$S_{AB}(\Omega) = S_{DC}(\Omega) = S_{\zeta\zeta}(\Omega) + Y^2 S_{\gamma\gamma}(\Omega), \quad (27a)$$

$$S_{AC}(\Omega) = S_{DB}(\Omega) = S_{\zeta\zeta}(\Omega) - Y^2 S_{\gamma\gamma}(\Omega). \quad (27b)$$

Thus, direct- and cross-PSD can be expressed in terms of direct PSDs of $\zeta(x)$ and $\gamma(x)$. Robson [59] has examined the physical and mathematical basis of Parkhilovskii model. He concluded that Parkhilovskii model can be made compatible with isotropic model for a profile-pair description of pavement (i.e., two parallel railway tracks), and may be used where isotropy assumption is not valid.

2.2.3 The Hypothesis of Shift

Sun and Su [60] proposed the hypothesis of shift to construct two-dimensional PSD using one-dimensional PSD. According to homogeneity in (19), autocorrelations of $\xi(x, y_1)$ and $\xi(x, y)$ equal to each other. Sun and Su [60] assumed $\xi(x_1 + X, y_1 + Y) = \xi(x_1 + X + s, y_1 + Y)$. In other words, the elevation of one wheel path is equal to the elevation of a parallel but shifted wheel path with a spatial lag s. With

$$\begin{aligned} R_{AC}(X, Y) &= E[\xi(x_1, y_1)\xi(x_1 + X, y_1 + Y)] \\ &= E[\xi(x_1, y_1)\xi(x_1 + X + s, y_1 + Y)] \\ &= R_{\xi\xi}(X + s), \end{aligned} \quad (28)$$

where spatial lag s is a parameter that can be estimated from field test of autocorrelation functions of parallel wheel path. Some properties about the spatial lag can be $s = 0$ as $Y = 0$ where $R_{AC}(X, 0) = R_{\xi\xi}(X)$ and $s \to \infty$ as $Y \to \infty$ where $R_{AC}(X, \infty) = 0$.

Under the hypothesis of shift, since PSD is the Fourier transform of correlation function, we have

$$S_{AC}(\Omega) = S_{CA}(\Omega) = (2\pi)^{-1} \int_{-\infty}^{\infty} R_{\xi\xi}(X + s) e^{-i\Omega X} dX. \quad (29)$$

Note that

$$R_{\xi\xi}(X + s) = \int_{-\infty}^{\infty} S_{\xi\xi}(K) e^{iK(X+s)} dK. \quad (30)$$

Replacing $R_{\xi\xi}(X + s)$ in (29) by (30) we obtain

$$\begin{aligned} S_{AC}(\Omega) &= S_{CA}(\Omega) \\ &= (2\pi)^{-1} \int_{-\infty}^{\infty} \int_{-\infty}^{\infty} S_{\xi\xi}(K) e^{iK(X+s)} e^{-i\Omega X} dX dK. \end{aligned} \quad (31)$$

Reversing the order of integration and using the convergence concept of a generalized function we have the inner integral

$$e^{iKs} \int_{-\infty}^{\infty} S_{\xi\xi}(K) e^{i(K-\Omega)X} dX = 2\pi\delta(K - \Omega) e^{iKs}. \quad (32)$$

Here the following integral is used [61]

$$\int_{-\infty}^{\infty} e^{iZX} dX = 2\pi\delta(Z). \quad (33)$$

Substituting (32) into (31) we get

$$S_{AC}(\Omega) = S_{CA}(\Omega) = S_{\xi\xi}(\Omega) e^{is\Omega}. \quad (34)$$

Here the property of Dirac delta function (Lighthill 1958)

$$\int_{-\infty}^{\infty} f(t)\delta(t - t_0) dt = f(t_0). \quad (35)$$

is used in the derivation of (34).

In summary, several used assumptions related to pavement surface are homogeneity, ergodicity and normal distribution. These have been supported by a number of measurements of pavement surface. However, there are still counterexamples demonstrating the existence of pavement surfaces that do not satisfy the homogeneous and ergodic properties [62]. In those situations, pavement surfaces are perceived by a vehicle as nonstationary stochastic processes [47, 63, 64]. It is worthwhile noting that nowadays with technology advancement, one can measure and obtain entire two-dimensional topology of pavement surface using remote sensing, line scanning laser, and synthetic aperture radar. For more accurate applications, it is no longer necessary to use hypothetical random field model of pavement surface.

3 Contact force on vehicle–pavement interface

Wheel loads in specifications of highway and airport pavement design are presently treated as a static load pressure with uniform distribution [65]

$$P(r) = P_0/\pi r_0^2 \text{ for } |r| \leq r_0, \quad (36)$$

where r_0 is the radium of circle distribution of wheel loads and P_0 represents the total loads applied. Such a simplification of actual vehicle/aircraft load in design can be adequate when the speed of travel is low and pavement surface is smooth. However, when the speed of travel is high and pavement surface is uneven due to deterioration, influence of dynamic load generated by vehicle–pavement interaction can become very significant. As a matter of fact, pavement damage is directly resulting from a long-term effect of dynamic traffic loading. Therefore, contact force on vehicle–pavement interface needs to be considered in pavement design to more closely reflect actual loading condition.

All pavement surfaces exhibit irregularities that cause vehicle vibration, which, in turn, results in moving stochastic contact forces on pavement [66]. At low speed of travel vehicle vibration is insignificant but at high speed of travel vehicle may vibrate significantly due to poor pavement surface conditions. It does not only reduce the ride quality but also generates additional damage to vehicle and pavement structures beyond static load.

The interaction between vehicles and pavements has been investigated extensively by the automobile industry for improving ride quality and reducing the mechanical fatigue of vehicle [5, 25, 50, 67–69]. Hedric [9], Abbo et al. [10], and Markow et al. [70] identified some critical factors that affect dynamic loads on pavement. These factors include vehicle and axle configuration, vehicle load, suspension characteristics (stiffness, damping), speed of travel, pavement roughness, faults, joint spacing and slab warping. From the perspective of pavement design and maintenance, however, little has been known about the effects of vehicle vibration on response, damage and performance of pavement structures [20, 54]. Although some efforts have been devoted to the measurement and prediction of dynamic wheel loads [4, 6, 11, 12, 71, 72], few of them provides a complete theoretical foundation for describing contact force induced by vehicle–pavement interaction. This can be a crucial element leading to a more precise dynamic analysis of pavement structures [54, 73, 74].

Of three approaches for studying vibration-generated contact forces, namely, analytic, experimental, and numerical simulation methods, only the last two have been used widely. Experimental method offers real results, it is however very costly and limited by safety requirements. Moreover, results from experimental method only suits in some degree for specific test conditions (e.g., vehicle and pavement types, speed of travel) [54]. In many circumstances, numerical simulation associated with a limited field tests serves as the most prevailing approach.

Numerical simulation has the advantage of capable of extrapolating experimental results over a range of test conditions where the experiment would be too dangerous or too expensive. To study the contact forces between vehicle and pavement using numerical simulation, a vehicle must be simplified to a vehicle model so as to simulate the real operation conditions of the vehicle. Then, based on the vehicle model associated with measured or simulated pavement surface roughness, dynamic contact forces can be generated using a validated vehicle simulation program.

3.1 Pavement surface as a source of excitation to vehicle

As far as the vehicle vibration is concerned, pavement surface serves as a source of excitation when the vehicle travels along the road. As such, spatial fluctuation of pavement surface gets converted to temporal random excitation of vehicle suspension system. In this article, we assume that the speed of travel is constant. Different stochastic process of excitation can be resulting from different hypotheses on two-dimensional random field of pavement surface, which are present in this section.

3.1.1 General two-dimensional random field

Let $\mathbf{x} = (x, y)$ be the fixed coordinates in which the spatial random field is defined. Let $\mathbf{x}' = (x', y')$ be the moving coordinates connected with the moving vehicle and travel at the constant speed $\mathbf{v}$ (vector quantity) with respect to the fixed coordinates $\mathbf{x}$. Denote $\mathbf{R}_\xi(\mathbf{x}_1, \mathbf{x}_2)$ the correlation function that describes the spatial random field $\xi(\mathbf{x})$. Denote $\mathbf{R}_\eta(\mathbf{x}'_1, \mathbf{x}'_2; t_1, t_2)$ the correlation function that describes the temporal random excitation $\eta(\mathbf{x}', t)$. Since

$$\eta(\mathbf{x}', t) = \xi(\mathbf{x}' + \mathbf{v}t), \tag{37}$$

we have

$$\mathbf{R}_\eta(\mathbf{x}'_1, \mathbf{x}'_2; t_1, t_2) = \mathbf{R}_\xi(\mathbf{x}'_1 + \mathbf{v}t_1, \mathbf{x}'_2 + \mathbf{v}t_2). \tag{38}$$

From the assumption mentioned above, $\xi(\mathbf{x})$ is a homogeneous random field. Accordingly, its correlation function $\mathbf{R}_\xi(\mathbf{x}_1, \mathbf{x}_2)$ becomes $\mathbf{R}_\xi(\mathbf{X})$ where$\mathbf{X} = \mathbf{x}_2 - \mathbf{x}_1$. Since vehicle velocity is a constant vector, the homogenous random field $\eta(\mathbf{x}', t)$ in spatial domain is converted to a stationary stochastic process in time domain. In other words, (38) can be replaced by

$$\mathbf{R}_\eta(\mathbf{X}', \tau) = \mathbf{R}_\xi(\mathbf{X}' + \mathbf{v}\tau), \tag{39}$$

where $\mathbf{X}' = \mathbf{x}'_2 - \mathbf{x}'_1$ and $\tau = t_2 - t_1$. Now we define $S_\xi(\mathbf{\Omega})$ as the spatial PSD of two-dimensional random field, i.e.,

$$\mathbf{S}_\xi(\mathbf{\Omega}) = (2\pi)^{-2} \int_{R^2} \mathbf{R}_\xi(\mathbf{X}) e^{-i\mathbf{\Omega}\mathbf{X}} d\mathbf{X}. \tag{40}$$

The PSD of random excitation $\eta(\mathbf{x}', t)$ can be expressed as

$$\mathbf{S}_\eta(\boldsymbol{\Omega}, \omega) = \mathbf{S}_\xi(\boldsymbol{\Omega})\delta(\omega - \boldsymbol{\Omega}\mathbf{v}). \tag{41}$$

Equation (41) can be proved as follows. Since

$$\mathbf{R}_\eta(\mathbf{X}', \tau) = \int_{-\infty}^{\infty}\int_{R^2} \mathbf{S}_\eta(\boldsymbol{\Omega}, \omega)e^{i(\boldsymbol{\Omega}\mathbf{X}'+\omega\tau)}\mathrm{d}\boldsymbol{\Omega}\mathrm{d}\omega. \tag{42}$$

Applying Eqs. (41) to (42) gives

$$\mathbf{R}_\eta(\mathbf{X}', \tau) = \int_{R^2} [\mathbf{S}_\eta(\boldsymbol{\Omega})e^{i\boldsymbol{\Omega}\mathbf{x}'}\int_{-\infty}^{\infty} \delta(\omega - \boldsymbol{\Omega}\mathbf{v})e^{i\omega\tau}\mathrm{d}\omega]\mathrm{d}\boldsymbol{\Omega}$$
$$= \mathbf{R}_\xi(\mathbf{X}' + \mathbf{v}\tau). \tag{43}$$

Figure 8 shows a plan view of a vehicle with N tires. We now consider the cross-PSD of between i th and j th tires. It is clear that if i th and j th tires are located in the same side of the vehicle, either in the left side or in the right side, then we should have $X' = X_{\mathrm{ij}}$ and $Y = 0$. If i th tire is located in right side and j th tire is located in left side, or vise verse, then we should have $X' = X_{\mathrm{ij}}$ and $Y = 0$.

Given Fig. 7 condition, (48) can be written in terms of spatial PSD roughness:

$$S_{\eta_i\eta_j}(X_{ij}, \omega) = \begin{cases} \frac{1}{v}[\int_{-\infty}^{\infty} \mathbf{S}_\xi(\frac{\omega}{v}, K)\mathrm{d}K]\ \mathrm{e}^{i\omega X_{ij}/v} & \text{if } i \text{ and } j \text{ are located in the same side} \\ \frac{1}{v}[\int_{-\infty}^{\infty} \mathbf{S}_\xi(\frac{\omega}{v}, K)\mathrm{e}^{iKY_0}\mathrm{d}K]\ \mathrm{e}^{i\omega X_{ij}/v} & \text{if } i \text{ and } j \text{ are located in the different side} \end{cases} \tag{50}$$

This is the same as Eq. (39). It is necessary to represent the PSD of random excitation $\eta(\mathbf{x}', t)$ in terms of angular frequency ω and spatial lag $\mathbf{X}'$, i.e.,

$$\mathbf{S}_\eta(\mathbf{X}', \omega) = \frac{1}{2\pi}\int_{-\infty}^{\infty} \mathbf{R}_\xi(\mathbf{X}' + \mathbf{v}\tau)\mathrm{e}^{-i\omega\tau}\mathrm{d}\tau. \tag{44}$$

If the direction of vehicle velocity vector is the same as that of the x-axis, then the following transformation applies:

$$X' + v\tau = X, Y' = Y \text{ and } v = |\mathbf{v}|. \tag{45}$$

Substituting (45) into (44), (9) is converted to

$$\mathbf{S}_\eta(\mathbf{X}', \omega) = \frac{1}{2\pi v}\int_{-\infty}^{\infty} \mathbf{R}_\xi(\mathbf{X})\mathrm{e}^{-i\omega X/v}\mathrm{d}X\mathrm{e}^{i\omega X'/v}, \tag{46}$$

where $\mathbf{X} = (X, Y)$ and $\mathbf{X}' = (X', Y')$. Corresponding to (40), spatial correlation function $\mathbf{R}_\xi(\mathbf{X})$ is the inverse Fourier transform of the spatial PSD $\mathbf{S}_\xi(\boldsymbol{\Omega})$, i.e.,

$$\mathbf{R}_\xi(\mathbf{X}) = \int_{R^2} \mathbf{S}_\xi(\boldsymbol{\Omega})\mathrm{e}^{i\boldsymbol{\Omega}\mathbf{X}}\mathrm{d}\boldsymbol{\Omega}, \tag{47}$$

where $\boldsymbol{\Omega} = (\Omega, K)$. Combining Eqs. (47) and (46), we obtain

$$\mathbf{S}_\eta(\mathbf{X}', \omega) = \frac{1}{v}\left[\int_{-\infty}^{\infty} \mathbf{S}_\xi\left(\frac{\omega}{v}, K\right)\mathrm{e}^{iKY}\mathrm{d}K\right]\mathrm{e}^{i\omega X'/v}. \tag{48}$$

In the derivation of (13), the following equation is used.

$$\int_{-\infty}^{\infty} \mathrm{e}^{-iX(\frac{\omega}{v}-\Omega)}\mathrm{d}X = 2\pi\delta\left(\frac{\omega}{v} - \Omega\right). \tag{49}$$

Equation (48) is a general result applicable to any form of two-dimensional random field. Under the restriction of the specified vehicle systems, we can represent (48) furthermore.

where $X_{ij} = x_j - x_i\,(i,j = 1, 2, \ldots, N)$. As for direct-PSD excitation, we have $X_{ij} = 0(i = j)$ and $Y_0 = 0$. From Eq. (50) we get

$$S_{\eta_i\eta_i}(X_{ii}, \omega) = \frac{1}{v}\left[\int_{-\infty}^{\infty} \mathbf{S}_\xi\left(\frac{\omega}{v}, K\right)\mathrm{d}K\right], \quad i = 1, 2, \ldots, N. \tag{51}$$

Since the correlation function $\mathbf{R}_\xi(\mathbf{X})$ and PSD function $\mathbf{S}_\xi(\boldsymbol{\Omega})$ of the two-dimensional random field are both known from (40) and (47), there is no obstacle to compute the excitation PSD of (40) and (41).

3.1.2 *Two-dimensional random field under hypothesis of isotropy*

When pavement surface is treated as a two-dimensional random field under hypothesis of isotropy [56, 58], its two-dimensional PSD can be represented in terms of one-dimensional PSD. In other words, in light of (26) and the definition of two-dimensional PSD (40), the two-dimensional PSD can be constructed from the following Fourier transform:

$$\mathbf{S}_\xi(\boldsymbol{\Omega}) = (2\pi)^{-2}\int_{R^2} R_{\xi\xi}(\sqrt{X^2 + Y^2})\mathrm{e}^{-i(\Omega X+KY)}\mathrm{d}X\mathrm{d}Y. \tag{52}$$

Notice that

$$\int_{-\infty}^{\infty} \mathbf{S}_\xi(\boldsymbol{\Omega})\mathrm{e}^{iKY_0}\mathrm{d}K$$
$$= (2\pi)^{-2}\int_{-\infty}^{\infty}\int_{R^2} R_{\xi\xi}(\sqrt{X^2 + Y^2})\mathrm{e}^{-i(\Omega X+KY)}\mathrm{e}^{iKY_0}\mathrm{d}X\mathrm{d}Y\mathrm{d}K$$
$$= (2\pi)^{-1}\int_{-\infty}^{\infty} R_{\xi\xi}(\sqrt{X^2 + Y_0^2})\mathrm{e}^{-i\Omega X}\mathrm{d}X \tag{53}$$

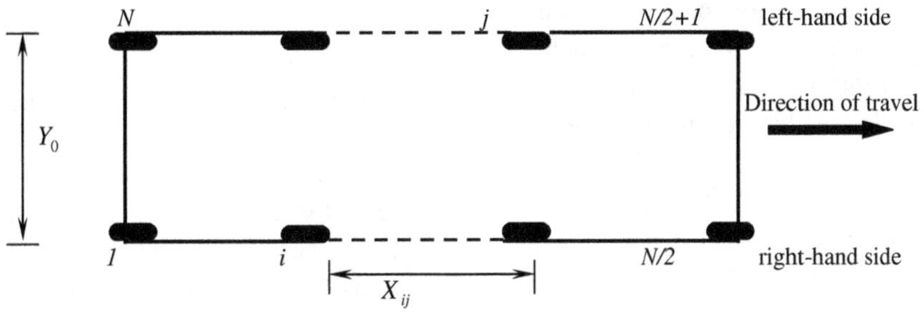

Fig. 8 A plane view of an N-tire vehicle model

Here, the general integration [61]

$$\int_{-\infty}^{\infty} \mathrm{e}^{-iK(Y-Y_0)}\mathrm{d}K = 2\pi\delta(Y - Y_0) \tag{54}$$

and the property of Dirac delta function [75]

$$\int_{-\infty}^{\infty} f(x)\delta(x - x_0)\mathrm{d}x = f(x_0) \tag{55}$$

are used in the derivation of Eq. (53).

Comparing (53) with (27), it is clear that the left side of (53) is just the PSD given by $S_{AC}(\Omega)$. Since it has been proved that $S_{AC}(\Omega) = g(\Omega)S_{\xi\xi}(\Omega)$ where $g(\Omega)$ is an ordinary coherence function and $S_{\xi\xi}(\Omega)$ is one-dimensional PSD roughness. Hence, (40) and (41) can be further expressed as

$$S_{\eta_i\eta_j}(X_{ij}, \omega) = \begin{cases} \frac{1}{v}S_{\xi\xi}(\frac{\omega}{v})\ \mathrm{e}^{i\omega X_{ij}/v} & \text{if } i \text{ and } j \text{ are located in the same side} \\ \frac{1}{v}g(\frac{\omega}{v})S_{\xi\xi}(\frac{\omega}{v})\mathrm{e}^{i\omega X_{ij}/v} & \text{if } i \text{ and } j \text{ are located in the different side} \end{cases} \tag{56}$$

and

$$S_{\eta_i\eta_i}(X_{ii}, \omega) = \frac{1}{v}S_{\xi\xi}\left(\frac{\omega}{v}\right), \quad i = 1, 2, \ldots, N. \tag{57}$$

3.1.3 *Two-dimensional random field under hypothesis of uncorrelation*

When pavement surface is treated as a two-dimensional random field under hypothesis of uncorrelation [55], the integration of the left side of (53) becomes either $S_{AB}(\Omega)$ if i and j are located on the same side of the vehicle, or $S_{AC}(\Omega)$ if i and j are located on the different side of the vehicle, which are further given by (30a) and (30b), respectively. Without any difficulty, it is straightforward to see that (40) and (41) can be further written as

$$S_{\eta_i\eta_j}(X_{ij}, \omega) = \begin{cases} \frac{1}{v}[S_{\zeta\zeta}(\frac{\omega}{v}) + Y_0^2 S_{\gamma\gamma}(\frac{\omega}{v})]\ \mathrm{e}^{i\omega X_{ij}/v} & \text{if } i \text{ and } j \text{ are located in the same side} \\ \frac{1}{v}[S_{\zeta\zeta}(\frac{\omega}{v}) - Y_0^2 S_{\gamma\gamma}(\frac{\omega}{v})]\mathrm{e}^{i\omega X_{ij}/v} & \text{if } i \text{ and } j \text{ are located in the different side} \end{cases} \tag{58}$$

and

$$S_{\eta_i\eta_i}(X_{ii}, \omega) = \begin{cases} \frac{1}{v}[S_{\zeta\zeta}(\frac{\omega}{v}) + Y_0^2 S_{\gamma\gamma}(\frac{\omega}{v})] & \text{if } i \text{ is located in the right side} \\ \frac{1}{v}[S_{\zeta\zeta}(\frac{\omega}{v}) - Y_0^2 S_{\gamma\gamma}(\frac{\omega}{v})] & \text{if } i \text{ is located in the left side} \end{cases} \quad i = 1, 2, \ldots, N. \tag{59}$$

3.1.4 *Two-dimensional random field under hypothesis of shift*

When pavement surface is constructed from a two-dimensional random field under hypothesis of shift, (40) and (41) can be written as

$$S_{\eta_i\eta_j}(X_{ij}, \omega) = \begin{cases} \frac{1}{v}S_{\xi\xi}(\frac{\omega}{v})\ \mathrm{e}^{i\omega X_{ij}/v} & \text{if } i \text{ and } j \text{ are located in the same side} \\ \frac{1}{v}S_{\xi\xi}(\frac{\omega}{v})\mathrm{e}^{i\omega(X_{ij}+s)/v} & \text{if } i \text{ and } j \text{ are located in the different side} \end{cases} \tag{60a}$$

and

$$S_{\eta_i\eta_i}(X_{ii}, \omega) = \begin{cases} \frac{1}{v}S_{\xi\xi}(\frac{\omega}{v}) & \text{if } i \text{ is located in the right side} \\ \frac{1}{v}S_{\xi\xi}(\frac{\omega}{v})\mathrm{e}^{i\omega s/v} & \text{if } i \text{ is located in the left side} \end{cases} \quad i = 1, 2, \ldots, N. \tag{60b}$$

3.2 Vehicle models

Here, we consider vehicle models for simulating dynamic contact forces. Three kinds of vehicle models are commonly used nowadays, i.e., quarter-vehicle, half-vehicle, and full-vehicle models. Making a reasonable choice between a complex vehicle model and a simple vehicle model relies on the nature of the problem. The advantage of the complex vehicle model is that it can accurately predict the acceleration response at different locations of the vehicle. The complex vehicle model is thus suitable for studying influences of vehicle vibrations on human body and the fatigue life of vehicles caused by vehicle vibrations. However, the complex vehicle model is highly sophisticated and requires detailed input and long execution times even for simple problems. Because vehicle sizes and loads vary greatly, it is more difficult to select representative parameter values for the complex vehicle model than for the simple vehicle model, which increases the difficulty to compare computer simulation results conducted by different research groups. This should be always kept in mind.

In Sect. 3.1, pavement surface is converted into a stationary stochastic process as input to excite vehicle vibration when the vehicle travels at a constant speed. To proceed further, we also assume that vehicle's suspension system is linear, under which the equations of motion of vehicle suspension systems for small oscillations can be derived using the Lagrange equations.

Let **Y** be a set of M independent generalized coordinates (e.g., the absolute displacement of components of vehicle systems) that completely specify the configuration of the system measured from the equilibrium position. Then, the kinetic energy T, potential energy U, and dissipative energy D, can be expressed as

$$T = \frac{1}{2}\dot{\mathbf{Y}}^T \mathbf{M}^{abs}\dot{\mathbf{Y}}, \tag{61a}$$

$$U = \frac{1}{2}\dot{\mathbf{Y}}^T \mathbf{K}^{abs}\dot{\mathbf{Y}}, \tag{61b}$$

$$D = \frac{1}{2}\dot{\mathbf{Y}}^T \mathbf{C}^{abs}\dot{\mathbf{Y}}, \tag{61c}$$

where $\mathbf{M}^{abs}$, $\mathbf{K}^{abs}$ and $\mathbf{C}^{abs}$ are mass, stiffness, and viscous damping matrices with respect to absolute displacement vector $Y = \{Y_1, Y_2, \ldots, Y_M\}^T$, respectively. The Lagarange equations of motion are

$$\frac{d}{dt}\left(\frac{\partial T}{\partial \dot{\mathbf{Y}}_j}\right) + \frac{\partial D}{\partial \dot{\mathbf{Y}}_j} + \frac{\partial U}{\partial \mathbf{Y}_j} = 0v = |\mathbf{v}| \quad j = 1, 2, \ldots, M. \tag{62}$$

Under the assumption that all components of the vehicle system be linear, (61) and (62) yield a set of simultaneous second-order differential equations with constant coefficients:

$$\mathrm{M}^{abs}\left\{\ddot{\mathrm{Y}}\right\} + \mathrm{C}^{abs}\left\{\dot{\mathrm{Y}}\right\} + \mathrm{K}^{abs}\left\{\mathrm{Y}\right\} = \mathrm{C}_f^{abs}\{\dot{\eta}\} + \mathrm{K}_f^{abs}\{\eta\} = \left\{\mathrm{F}^{abs}(t)\right\}, \tag{63}$$

where $\boldsymbol{\eta}$ and $\dot{\eta}$ are, respectively, the pavement surface elevation vector and its derivative process. Clearly, here we assume that the contact between tires and pavement surface is in the form of point distribution.

Without loss of generality, let components of the vehicle system directly contacting with ground in the moving coordinates $\mathbf{x}'$ be numbered from 1 to N (see Fig. 7), where N represents the total number of tires. Let

$$\begin{cases} Z_i = Y_i - \eta_i & \text{for } i = 1, 2, \ldots, N \\ Z_i = Y_i & \text{for } i = N+1, N+2, \ldots, M \end{cases} \tag{64}$$

Equation (63) can be rewrote as

$$\mathbf{M}\{\mathbf{Z}\} + \mathbf{C}\{\mathbf{Z}\} + \mathbf{K}\{\mathbf{Z}\} = \mathbf{C}_f\{\dot{\eta}\} + \mathbf{K}_f\{\eta\} = \{\mathbf{F}(t)\}, \tag{65}$$

where **M**, **K** and **C** are mass, stiffness, and viscous damping matrices with respect to relative displacement $Z = \{Z_1, Z_2, \ldots, Z_M\}^T$, respectively. Since the total number of degree of freedom is assumed to be M, and η is N-dimensional temporal excitation vector representing pavement surface-induced displacements in coordinates $\mathbf{x}'$, **M**, **K**, and **C** are M by M matrices, $\mathbf{C}_f$, and $\mathbf{K}_f$ are M by N matrices, and force vector $\mathbf{F}(t)$is an M by 1 matrix. Equations (63) and (64) represent a linear mathematical model of vehicle systems. Taking Fourier transform to both sides of Eq. (65), we have

$$\begin{aligned}&[-\omega^2\mathbf{M} + i\omega\mathbf{C} + \mathbf{K}]_{M\times M}\{\tilde{\mathbf{Z}}\}_{M\times 1} \\ &\quad = [i\omega\mathbf{C}_f + \mathbf{K}_f]_{M\times N}\{\tilde{\eta}\}_{N\times 1},\end{aligned} \tag{66}$$

in which $\tilde{\mathbf{Z}}$ and $\tilde{\eta}$ are, respectively, the Fourier transform of **Z** and η, i.e.,

$$\tilde{\mathbf{Z}}(\omega) = \frac{1}{2\pi}\int_{-\infty}^{\infty} \mathbf{Z}(t)\mathrm{e}^{-i\omega t}\mathrm{d}t, \tag{67a}$$

$$\tilde{\eta}(\omega) = \frac{1}{2\pi}\int_{-\infty}^{\infty} \boldsymbol{\eta}(t)\mathrm{e}^{-i\omega t}\mathrm{d}t. \tag{67b}$$

Multiplying the inverse matrix of $[-\omega^2\mathbf{M} + i\omega\mathbf{C} + \mathbf{K}]$, it is straightforward to see

$$\{\tilde{\mathbf{Z}}\} = [-\omega^2\mathbf{M} + i\omega\mathbf{C} + \mathbf{K}]^{-1}[i\omega\mathbf{C}_f + \mathbf{K}_f]\{\tilde{\eta}\}. \tag{68}$$

The primary aim of establishing vehicle models is to find the scalar response of $\mathbf{Z}(t)$ as well as $\tilde{\mathbf{Z}}(\omega)$. This can be accomplished by using stochastic process theory of linear systems. The dynamic characteristics of the vehicle at angular frequency ω are defined by $M \times N$ frequency response function (FRF) matrix, $H(\omega)$, of the vehicle system, where $H(\omega) = [H(\omega)_{ij}]$, $i = 1, 2, \ldots, M$ for the M outputs and $j = 1,2,\ldots,N$ for the N inputs of pavement excitation to tires. These functions are defined as follows. If a vertical displacement $\eta_j(t) = \mathrm{e}^{i\omega t}$ is applied to the vehicle at jth tire, with all the other inputs being zero, then the response of the vehicle is given by $Z(t)_{ij} = H(\omega)_{ij} e^{i\omega t}$. Based on Eq. (68), it is straightforward to see that FRF matrix can be given by

$$\mathbf{H}(\omega) = [-\omega^2\mathbf{M} + i\omega\mathbf{C} + \mathbf{K}]^{-1}_{M\times M}[i\omega\mathbf{C}_f + \mathbf{K}_f]_{M\times N}. \quad (69)$$

Let

$$\mathbf{A} = \mathbf{AR} + i\mathbf{AI} = [-\omega^2\mathbf{M} + i\omega\mathbf{C} + \mathbf{K}]^{-1} \quad (70a)$$

and

$$\mathbf{Q} = \mathbf{QR} + i\mathbf{QI} = [i\omega\mathbf{C}_f + \mathbf{K}_f], \quad (70b)$$

where **AR** and **AI**, respectively, represent real and imaginary parts of matrix **A**, **QR** and **QI**, respectively, represent real and imaginary parts of matrix **Q**. It is convenient to express FRF matrix in terms of real part **HR** and imaginary part **HI**

$$\mathbf{H}(\omega) = \mathbf{HR} + i\mathbf{HI}, \quad (71)$$

where

$$\mathbf{HR} = \mathbf{AR}\cdot\mathbf{QR} - \mathbf{AI}\cdot\mathbf{QI},$$

$$\mathbf{HI} = \mathbf{AI}\cdot\mathbf{QR} + \mathbf{AR}\cdot\mathbf{QI}.$$

From the stochastic process theory, PSD response of a linear system and PSD excitation satisfy the following relationship [76]:

$$\underset{M\times M}{\mathbf{S_Z}(X_{ij},\omega)} = \underset{M\times N}{\mathbf{H}^*(\omega)}\underset{N\times N}{\mathbf{S}_\eta(X_{ij},\omega)}\underset{N\times M.}{\mathbf{H}^T(\omega)} \quad (72a)$$

or equivalently,

$$S_{Z_iZ_j}(X_{ij},\omega) = \sum_{l=1}^{N}\sum_{k=1}^{N} H^*_{ik}(\omega)H_{jl}(\omega)S_{\eta_k\eta_l}(X_{kl},\omega), \quad (72b)$$

where $\mathbf{S_Z}(X_{ij},\omega) = [S_{Z_iZ_j}(X_{ij},\omega)]_{M\times M}$, $\mathbf{S}_\eta(X_{ij},\omega) = [S_{\eta_i\eta_j}(X_{ij},\ \omega)]_{N\times N}$, $\mathbf{H}^*(\omega)$ and $\mathbf{H}^T(\omega)$ are, respectively, the conjugate matrix and the transposed matrix of FRF matrix $\mathbf{H}(\omega)$. It may be noticed that $\mathbf{H}^*(\omega)$ and $\mathbf{H}^T(\omega)$ can be, respectively, expressed in the form of real and imaginary parts by means of Eqs. (72a) and (72b)

$$\mathbf{H}^*(\omega) = \mathbf{HR} - i\mathbf{HI}, \quad (73a)$$

$$\mathbf{H}^T(\omega) = \mathbf{HR}^T + i\mathbf{HI}^T. \quad (73b)$$

Similarly, we may write $\mathbf{S_Z}(X_{ij},\omega)$ as

$$\mathbf{S_Z}(X_{ij},\omega) = \mathbf{SR_Z}(X_{ij},\omega) + i\mathbf{SI_Z}(X_{ij},\omega) \quad (74)$$

in which $\mathbf{SR_Z}(X_{ij},\omega)$ and $\mathbf{SI_Z}(X_{ij},\omega)$ are, respectively, the real and imaginary parts of response spectral density $\mathbf{S_Z}(\omega)$. If the pavement excitation spectral density $\mathbf{S}_\eta(\omega)$ is a real spectrum matrix, then by expanding (72b) using (73a) and (73b), we have

$$\mathbf{SR_Z}(X_{ij},\omega) = \mathbf{HR}\cdot\mathbf{S}_\eta(X_{ij},\omega)\cdot\mathbf{HR}^T + \mathbf{HI}\cdot\mathbf{S}_\eta(X_{ij},\omega)\cdot\mathbf{HI}^T. \quad (75a)$$

and

$$\mathbf{SI_Z}(X_{ij},\omega) = \mathbf{HR}\cdot\mathbf{S}_\eta(X_{ij},\omega)\cdot\mathbf{HI}^T - \mathbf{HI}\cdot\mathbf{S}_\eta(X_{ij},\omega)\cdot\mathbf{HR}^T. \quad (75b)$$

If the pavement excitation spectral density $\mathbf{S}_\eta(\omega)$ is a complex spectrum matrix, say

$$\mathbf{S}_\eta(X_{ij},\omega) = \mathbf{SR}_\eta(X_{ij},\omega) + i\mathbf{SI}_\eta(X_{ij},\omega) \quad (76)$$

then by expanding Eqs. (72b) using (73a), (73b), and (76), we have

$$\begin{aligned}\mathbf{SR_Z}(X_{ij},\omega) = {} & [\mathbf{HR}\cdot\mathbf{SR}_\eta(X_{ij},\omega) + \mathbf{HI}\cdot\mathbf{SI}_\eta(X_{ij},\omega)]\cdot\mathbf{HR}^T, \\ & - [\mathbf{HR}\cdot\mathbf{SI}_\eta(X_{ij},\omega) - \mathbf{HI}\cdot\mathbf{SR}_\eta(X_{ij},\omega)]\cdot\mathbf{HI}^T,\end{aligned} \quad (77a)$$

and

$$\begin{aligned}\mathbf{SI_Z}(X_{ij},\omega) = {} & [\mathbf{HR}\cdot\mathbf{SI}_\eta(X_{ij},\omega) - \mathbf{HI}\cdot\mathbf{SR}_\eta(X_{ij},\omega)]\cdot\mathbf{HR}^T \\ & + [\mathbf{HR}\cdot\mathbf{SI}_\eta(X_{ij},\omega) - \mathbf{HI}\cdot\mathbf{SR}_\eta(X_{ij},\omega)]\cdot\mathbf{HI}^T.\end{aligned} \quad (77b)$$

3.3 Dynamic load

3.3.1 Time-domain analysis

Figure 9 shows a sketch of the *i*th tire contacted with rough pavement surface. From Newton's second law of motion it is direct to know that the contact force between *i*th tire and pavement, $P_i(t)$, can be given by

$$P_i(t) = k_iZ_i(t) + c_i\dot{Z}_i(t), \quad i = 1,2,..,N, \quad (78)$$

where k_i and c_i, respectively, the *i*th tire spring stiffness and viscous damping, both being constant. According to stochastic process theory, if the input of a linear time-invariable system is a stationary random process, then the output of the system is also a stationary random process [76]. As was noted, since temporal random excitation of pavement surface Gaussian ergodic random process with zero mean, the response $Z_i(t)$ and its derivative process $\dot{Z}_t(t)$ are both zero mean Gaussian stationary stochastic processes. Taking expectation to both sides of (78), we get the mean function of dynamic contact forces, m_P

$$m_{P_i}(t) = E[P_i(t)] = k_iE[Z_i(t)] + c_iE[\dot{Z}_i(t)] = 0. \quad (79)$$

It should be pointed out that the mean function $m_P(t)$ here, not containing the static load of vehicle, only represents the statistical average of dynamic effect. If the static load is considered, then the complete mean function of dynamic contact forces, $\bar{P}_i$, is given by

$$\bar{P}_i = m_{P_i}(t) + P_{0i} = m_i g, \tag{80}$$

where g is the acceleration due to gravity, P_{0i} and m_i are, respectively, the effective weight and effective mass distributed by the vehicle on ith tire. Without special statements, the derivation of formulation in the context is based on (79).

It is of interest to know the correlation function of dynamic contact forces between vehicle and pavement because this information is critical for analyzing the dynamic response of pavement structure under vehicle loads. According to the definition of correlation function and using Eq. (78), we have

$$\begin{aligned} R_{P_iP_j}(X_{ij},\tau) &= E[P_i(t)P_j(t+\tau)] \\ &= k_ik_jR_{Z_iZ_j}(X_{ij},\tau) + c_ic_jR_{\dot{Z}_i\dot{Z}_j}(X_{ij},\tau) \\ &\quad + k_ic_jR_{Z_i\dot{Z}_j}(X_{ij},\tau) + c_ik_jR_{\dot{Z}_iZ_j}(X_{ij},\tau). \end{aligned} \tag{81}$$

It is known from stochastic differential principles that there exists exchangeability between expectation and square differential [31]. In other words, if a random process $Z(t)$ is differentiable for any order, it is proved that

$$\frac{\partial^{n+m}}{\partial t^n \partial s^m} R_{ZZ}(t,s) = R_{Z^{(n)}Z^{(m)}}(t,s), \tag{82}$$

where $Z^{(n)} = \mathrm{d}^n Z(t)/\mathrm{d}t^n$. In addition, if $Z(t)$ is a stationary process, then (73a) becomes

$$(-1)^n \frac{\partial^{n+m}}{\partial \tau^{n+m}} R_{ZZ}(\tau) = R_{Z^{(n)}Z^{(m)}}(\tau). \tag{83}$$

Applying Eq. (83) into Eq. (81) gives

$$\begin{aligned} R_{P_iP_j}(X_{ij},\tau) &= k_ik_jR_{Z_iZ_j}(X_{ij},\tau) - c_ic_j\frac{\partial^2}{\partial \tau^2}R_{Z_iZ_j}(X_{ij},\tau) \\ &\quad + k_ic_j\frac{\partial}{\partial \tau}R_{Z_iZ_j}(X_{ij},\tau) - c_ik_j\frac{\partial}{\partial \tau}R_{Z_iZ_j}(X_{ij},\tau). \end{aligned} \tag{84}$$

It is clearly seen that autocorrelation function is obtained as $i = j$

$$R_{P_iP_i}(X_{ij},\tau) = k_i^2R_{Z_iZ_i}(X_{ij},\tau) - c_i^2\frac{\partial^2}{\partial \tau^2}R_{Z_iZ_i}(X_{ij},\tau), \tag{85}$$

and crosscorrelation function is obtained as $i \neq j$.

3.3.2 Frequency-domain analysis

Having obtained correlation function, it is not difficult to represent the power spectral density (PSD), which is defined as the Fourier transform of correlation function. From Eq. (81) we have

$$\begin{aligned} S_{P_iP_j}(X_{ij},\omega) &= k_ik_jS_{Z_iZ_j}(X_{ij},\omega) + c_ic_jS_{\dot{Z}_i\dot{Z}_j}(X_{ij},\omega) \\ &\quad + k_ic_jS_{Z_i\dot{Z}_j}(X_{ij},\omega) + c_ik_jS_{\dot{Z}_iZ_j}(X_{ij},\omega), \end{aligned} \tag{86}$$

where

$$S_{P_iP_j}(X_{ij},\omega) = \frac{1}{2\pi}\int_{-\infty}^{\infty} R_{P_iP_j}(X_{ij},\tau)e^{-i\omega t}\mathrm{d}\tau, \tag{86a}$$

$$S_{Z_iZ_j}(X_{ij},\omega) = \frac{1}{2\pi}\int_{-\infty}^{\infty} R_{Z_iZ_j}(X_{ij},\tau)\mathrm{e}^{-i\omega t}\mathrm{d}\tau, \tag{86b}$$

$$S_{\dot{Z}_i\dot{Z}_j}(X_{ij},\omega) = \frac{1}{2\pi}\int_{-\infty}^{\infty} R_{\dot{Z}_i\dot{Z}_j}(X_{ij},\tau)\mathrm{e}^{-i\omega t}\mathrm{d}\tau, \tag{86c}$$

$$S_{Z_i\dot{Z}_j}(X_{ij},\omega) = \frac{1}{2\pi}\int_{-\infty}^{\infty} R_{Z_i\dot{Z}_j}(X_{ij},\tau)\mathrm{e}^{-i\omega t}\mathrm{d}\tau, \tag{86d}$$

$$S_{\dot{Z}_iZ_j}(X_{ij},\omega) = \frac{1}{2\pi}\int_{-\infty}^{\infty} R_{\dot{Z}_iZ_j}(X_{ij},\tau)\mathrm{e}^{-i\omega t}\mathrm{d}\tau. \tag{86e}$$

Correlation function $R_{P_iP_j}(\tau)$ is the Fourier inverse transform of PSD function $S_{Z_iZ_j}(\omega)$:

$$R_{Z_iZ_j}(X_{ij},\omega) = \int_{-\infty}^{\infty} S_{Z_iZ_j}(X_{ij},\tau)\mathrm{e}^{i\omega t}\mathrm{d}\omega. \tag{87}$$

Taking derivative to both sides of Eq. (87) gives [31, 182]

$$\frac{\mathrm{d}^n}{\mathrm{d}\tau^n}R_{Z_iZ_j}(X_{ij},\omega) = i^n\int_{-\infty}^{\infty} \omega^n S_{Z_iZ_j}(X_{ij},\tau)\mathrm{e}^{i\omega t}\mathrm{d}\omega. \tag{88}$$

By combining Eqs. (83) and (88), we find that

$$S_{\dot{Z}_i\dot{Z}_j}(X_{ij},\omega) = \omega^2 S_{Z_iZ_j}(X_{ij},\omega), \tag{88a}$$

$$S_{Z_i\dot{Z}_j}(X_{ij},\omega) = i\omega S_{Z_iZ_j}(X_{ij},\omega), \tag{88b}$$

$$S_{\dot{Z}_iZ_j}(X_{ij},\omega) = -i\omega S_{Z_iZ_j}(X_{ij},\omega). \tag{88c}$$

Substituting (88) into (84), we eventually get the PSD forces in terms of PSD of relative displacement response:

$$S_{P_iP_j}(X_{ij},\omega) = [k_ik_j + c_ic_j\omega^2 + i\omega(k_ic_j - c_ik_j)]S_{Z_iZ_j}(X_{ij},\omega). \tag{89}$$

Equation (89) applies to both direct-spectrum and cross-spectrum. In addition, as far as direct-spectrum is concerned, we can rewrite PSD forces in more concise form:

$$S_{P_iP_i}(\omega) = (k_i^2 + c_i^2\omega^2)S_{Z_iZ_i}(\omega). \tag{90}$$

Standard deviation of contact forces can be expressed in terms of PSD forces. Applying Eq. (90) to the definition of standard deviation directly shows that

$$\sigma_{P_i}^2 = \int_{-\infty}^{\infty} S_{P_iP_i}(\omega)\mathrm{d}\omega = \int_{-\infty}^{\infty} (k_i^2 + c_i^2\omega^2)S_{Z_iZ_i}(\omega)\mathrm{d}\omega. \tag{91}$$

Furthermore, if the standard deviations of displace response and velocity response, σ_{Z_i} and $\sigma_{\dot{Z}_i}$, are deployed, Eq. (91) can be rewritten as

$$\sigma_{P_i}^2 = k_i^2\sigma_{Z_i}^2 + c_i^2\sigma_{\dot{Z}_i}^2, \tag{92}$$

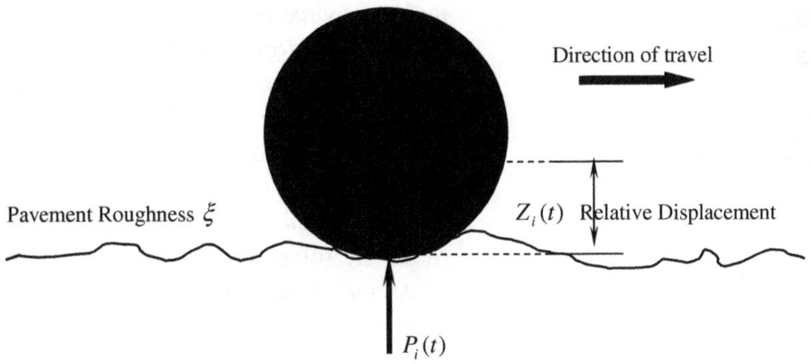

Fig. 9 A sketch of the *i*th tire on rough pavement surface

in which

$$\sigma_{Z_i}^2 = \int_{-\infty}^{\infty} S_{Z_iZ_i}(\omega)\mathrm{d}\omega \text{ and } \sigma_{\dot{Z}_i}^2 = \int_{-\infty}^{\infty} \omega^2 S_{Z_iZ_i}(\omega)\mathrm{d}\omega. \tag{93}$$

3.3.3 A Case Study of Walking-Beam Suspension System

As a case study, a walking-beam suspension system is shown in Fig. 10. Governing equations of this vehicle's suspension system are also given by (65) in matrix form, where parameters and their values related to this walking-beam system are listed in Table 2 with $m_1 = 1{,}100$ kg, $m_2 = 3{,}900$ kg, $\mathrm{I} = 465$ kgm^2, $s = 1.30$ m, $a_1 = a_2 = 0.5$, $k_1 = k_2 = 1.75$ (MN/s), $k_3 = 1.0$ (MN/s), $c_3 = 15.0$ (kNs/m), and $c_1 = c_2 = 2.0$ (kNs/m).

$$\mathbf{M} = \begin{bmatrix} 0 & 0 & m_2 \\ m_1a_2 & m_1a_1 & 0 \\ -I/b & I/b & 0 \end{bmatrix},$$

$$\mathbf{C} = \begin{bmatrix} -c_3a_2 & -c_3a_1 & c_3 \\ c_1 + c_3a_2 & c_2 + c_3a_1 & -c_3 \\ c_1a_2b & -c_2a_1b & 0 \end{bmatrix},$$

$$\mathbf{K} = \begin{bmatrix} -k_3a_2 & -k_3a_1 & k_3 \\ k_1 + k_3a_2 & k_2 + k_3a_1 & -k_3 \\ k_1a_2b & -k_2a_1b & 0 \end{bmatrix},$$

$$\{\mathbf{F}(t)\} = \begin{bmatrix} 0 & c_3a_2 & k_3a_2 & 0 & c_3a_1 & k_3a_1 \\ -m_1a_2 & -c_3a_2 & -k_3a_2 & -m_1a_1 & -c_3a_1 & -k_3a_a \\ I/b & 0 & 0 & -I/b & 0 & 0 \end{bmatrix} \times \begin{Bmatrix} \ddot{\eta}_1 \\ \dot{\eta}_1 \\ \eta_1 \\ \ddot{\eta}_2 \\ \dot{\eta}_2 \\ \eta_2 \end{Bmatrix}.$$

A two-dimensional isotropic random field is used for numerical simulation. The one-dimensional PSD roughness proposed by the International Organization for Standardization (ISO) in the power form [49, 52]

$$S_{\xi\xi}(\Omega) = S_0\Omega^{-\gamma}, \tag{94}$$

with parameters $S_0 = 3.37 \times 10^{-6}$ m^3/cycle and $\gamma = 2.0$ is adopted for numerical computation. Figure 11 shows the PSD contact force of the right tire at the speed of 20 m/s.

In aforementioned study, the contact area between the tire and the pavement surface is assumed as a point contact. There is no difficulty to extend this point contact to a distributed contact by considering the footprint as a weighted integration of contacting points [77–79]. It should be noted that the effect of nonlinearity in vehicle suspension and variable speed of travel (e.g., acceleration, deceleration, etc.), and inhomogeneity in pavement surface on contact force between vehicle and pavement have not been addressed here, which have been considered in various studies [80].

4 Pavement response under moving stochastic loads

4.1 Background

A large class of time-dependent sources such as vehicle, submarines, aircraft, and explosion-induced waves belongs to moving sources. The study of response of media (pavement, runway, rail-track, bridge, air, and sea) to moving sources is named moving source problem (MSP), which is of particular interest to structural design, noise assessment, target detection, etc. [81]. A number of studies have been addressed to MSP in various fields of physics. For instance, the response of an ice plate of finite thickness caused by moving loads was discussed by Strathdee et al. [82]. Wells and Han [83] analyzed the noise generated by a moving source in a moving medium. As for the aspect of

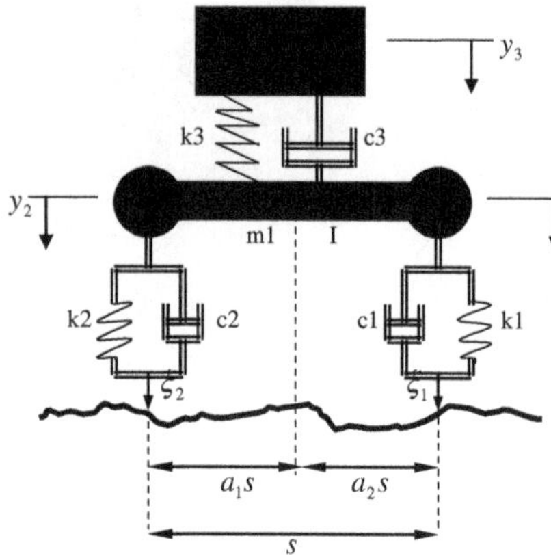

Fig. 10 A walking-beam suspension system

Table 2 Parameters used in the case study of walking-beam suspension system

Parameter	Description
m_1	Unsprung mass
m_2	Sprung mass
I	Moment inertia of unsprung mass
k_1	Spring stiffness of right side tire
k_2	Spring stiffness of left side tire
k_3	Spring stiffness of suspension
c_1	Damping of right side tire
c_2	Damping of left side tire
c_3	Damping of suspension
s	Effective width of the vehicle
y	Absolute vertical displacement of center of unsprung mass
y_1	Absolute vertical displacement of right side of unsprung mass
y_2	Absolute vertical displacement of left side of unsprung mass
y_3	Absolute vertical displacement of sprung mass
ξ_1	Absolute height of pavement profile along right wheel path
$\xi_2 nn$	Absolute height of pavement profile along left wheel path

elastodynamics, even more investigations are being done [84–94]. One may refer to Sun [95] for detailed review of MSP.

Methods for solving MSP primarily include integral transformation, characteristic curve, and modal analysis [96–100]. A common characteristic of previous studies is to utilize Galilean transform. The advantage of using Galilean transform is that the governing field equations, usually partial differential equations, can be reconstructed in a moving coordination so that the effect of source velocity may be reflected in parametric ordinary differential equations. However, since Galilean transform requires that the source is steadily moving, the methods based on Galilean transform only apply to the steady-state solution. As for transient solution, a feasible way is to directly apply high-order integral transformation to the field equations. Since velocity parameter is included in boundary or initial condition, there is intractable obstacle when integrating the field equations. This can be the reason why most of the MSPs are solved only for steady-state solution.

As far as pavement response is concerned, some pioneers have dedicated their research to the subject of effects of moving loads on pavement structure. To perform the dynamic analysis of pavement structures, one of the indispensable considerations is the dynamic traffic loading, which is the excitation source of pavement structures. Since pavement loads are caused by vehicle vibration, it is necessary to include the vehicle in the investigations of pavement loads. Employing MIT heavy truck simulation program Hedric and his associates, Abbo and Markow [9–11], examined the influence of joint spacing, step faulting, vehicle suspension characteristics, and vehicle velocity on pavement damage as defined by fatigue cracking in concrete slab. The response of pavement under their consideration is achieved using PMARP, a static finite element program for pavement analysis developed by the Purdue. The conclusions obtain provide a valuable approximation of pavement response.While PMARP is a static finite element program, the properties of inertia and damping of the pavement structure were not included in consideration.

An advanced pavement model called MOVE program was developed by Chen [101] and Monismith et al. [73] at UC-Berkeley. This model can take the motion of load into account by means of finite element methods and is an important dedication to pavement structural analysis. The vehicle load is assumed to be an infinite line load to simplify the analysis from a three-dimensional problem to a two-dimensional problem. In addition, the model is based on the deterministic elastodynamics; therefore, neither pavement surface roughness nor vehicle suspension can be considered in the presented framework.

Since dynamic effects have been increasingly important in the prediction of pavement response, damage, and performance [14, 16, 18], it has become necessary to develop better mathematical models to account the effects of motion and fluctuation of contact forces caused by various types of vehicle suspensions [73]. The author has carried out a number of studies on deterministic and stochastic MSPs over the last two decades by providing a general approach that can include surface roughness and vehicle suspensions in the response of continuum media [54, 74].

The complexity in MSP is the variable load position, which not only makes the representation of the field equation in time-invariant fixed coordinates difficult but also increases the difficulty of integration due to the appearance of speed parameter. In most of the previous MSP studies, some simple cases with deterministic conditions, for example, a constant load with uniform speed, are considered. However, the research on dynamic response of continuum media to moving stochastic vehicle loads has not yet been studied in the literature.

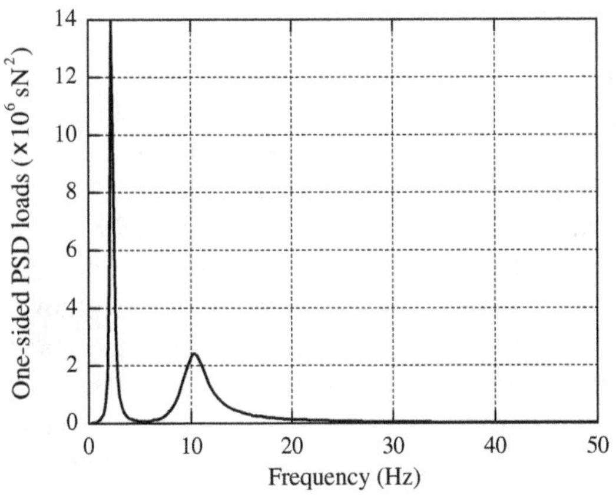

Fig. 11 PSD of contact forces against frequency (speed of travel = 20 m/s)

4.2 Problem statement of MSP

4.2.1 Pavement as a continuum medium

Pavement structures are traditionally classified into two categories: rigid pavement and flexible pavement. The former often refers to Portland cement concrete (PCC) pavement, the latter often refers to asphalt concrete (AC) pavement. This classification is not very strict because PCC pavements are not "rigid," or are AC pavements "flexible" in many situations. From the standpoint of structural nature PCC pavements behave more like a slab, while AC pavement acts more like a multi-layered system. Therefore, it is more reasonable to classify pavement on the base of mechanic properties. There have been developed many models to simulate the behavior of pavement structures. For instance, Hardy and Cebon [16, 18] simplified AC pavement into an Euler–Bernoulli beam. From the theoretical perspective, most of the proposed pavement models belong to continuum media [102, 183–187].

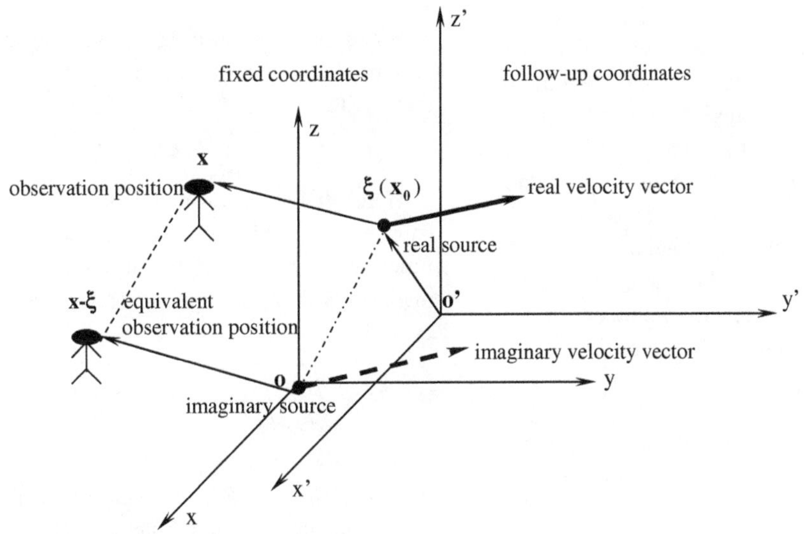

Fig. 12 Schematic sketch of a stationary coordinate and a moving coordinate

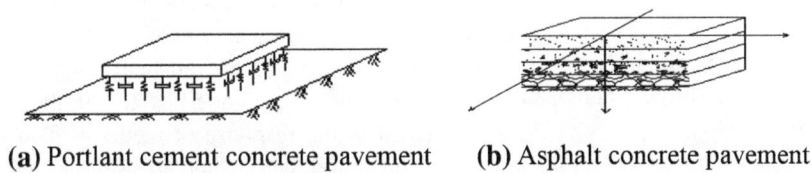

Fig. 13 Physical models of two general types of pavement structures

Consider a linear medium with the region R or the boundary B being at rest initially. A load is applied when the medium is at rest and moves according to a given law of motion (Fig. 12). The MSP is to solve the response of the medium to the moving load. The governing equation of such a system thus belongs to linear partial differential equation. According to the theory of linear equation, the solution of the equation can be constructed by the integration of the fundamental solution of the equation or the so-called Green's function. It is usually described as the superposition principle or, equivalently, the reciprocal principle [103].

Besides the above-mentioned description of the continuum media, the following assumptions are made here. One assumption is that, comparing to the mass of vehicle, the mass of pavement structures (including surface layer, base, subgrade, and soil foundation) is large enough such that pavement vibration is much smaller than vehicle vibration. Another assumption is that wave velocity excited by a dynamic load in pavement structure is much faster than vehicle's speed of travel. With these two assumptions, the couple effect of vehicle–pavement interaction can be negligible.

4.2.2 Moving stochastic vehicle loads

Contact forces applied on pavements by vehicles are moving stochastic loads. The statistical characteristics of the dynamic contact forces have been discussed in detail previously. Dynamic contact forces follow with two aspects of meanings. One is that the location of the force is changing continuously with traveling of vehicle, and another is that the amplitude of the force is varying due to vehicle vibrations.

Without any loss of generality, a vehicle is assumed to be a linear system traveling along the x-axis at a constant speed. The contact forces between the vehicle and pavement can be given by a concentrated moving load:

$$F(\mathbf{x},t) = P_{sto}(t)\delta(x - vt)\delta(y) \tag{95}$$

where t and v are time variable and vehicle velocity, respectively, δ() is the Dirac-delta function, which is defined by

$$\delta(x - x_0) = \begin{cases} 0, & \text{for } x \neq x_0 \\ \infty, & \text{for } x = x_0 \end{cases} \text{ and } \int_{-\infty}^{\infty} f(x)\delta(x - x_0)\mathrm{d}x = f(x_0). \tag{96}$$

In addition, $P_{sto}(t)$, a function of time, represents amplitudes of stochastic contact forces and can be expressed as two terms below:

$$P_{sto}(t) = P_0 + P(t). \tag{97}$$

Here P_0, representing the static load applied at a tire when the vehicle is at rest, is a constant quantity, and $P(t)$ represents the dynamic portion of the stationary stochastic contact forces with a zero mean. Its correlation function, power spectral density and standard deviation are, respectively, denoted as $R_{PP}(\tau)$, $S_{PP}(\omega)$, and σ_P.

4.3 Representation theory of MSP

It is convenient to assume a three-dimensional configuration with observation variable x = (x,y,z), source variable $\xi = (\xi, \eta, \zeta)$, and time $t \geq 0$. Suppose a linear differential operator **O** describes the dynamic property of a physical system and appropriate interface and boundary conditions relate the field quantities of specified problems. Obviously, for a concrete elastodynamic problem, the linear differential operator **O** is given by the well-known Navier–Stoke's field equations. The Green's function is then defined as the fundamental solution of the system. In other words, for the problem discussed here, the Green's function corresponds to the solution of the governing equations as the point source takes the form of a Dirac delta function in both spatial and temporal domains.

4.3.1 General Formulation

Without loss of generality, vanishing initial conditions are assumed here. According to the causality of a physical system, the Green's function $\mathbf{G}(\mathbf{x}, t) = 0$ for $t < 0$. We may then write

$$\mathbf{O}[\mathbf{G}(\mathbf{x} - \xi, t - \tau)] = \delta(\mathbf{x} - \xi)\delta(t - \tau). \tag{98}$$

Here, the initiation of the source is delayed by τ. The causality of a physical system requires that for Green's function $t \leq \tau$. Since the initial condition of the linear medium is zero, the dynamic response could be expressed as

$$\mathbf{u}(\mathbf{x}, t) = \int_{-\infty}^{t} \int_{S} F(\xi, \tau)\mathbf{G}(\mathbf{x}, t - \tau; \xi)d\xi d\tau, \tag{99}$$

where $\mathbf{x} = (x, y, z)$, $\xi = (\xi, \eta, \zeta)$, $\mathbf{v} = (v_x, v_y, v_z)$, $d\xi = d\xi d\eta d\zeta$, and S is the region R or the boundary B. It is know that the Green's function $\mathbf{G}(\mathbf{x}, t; \xi)$ corresponds to the solution of the equation when a unit point impulse is applied at position ξ. Assume that the medium is infinite in those dimensions of interest. It is known that the governing equations are linear. According to the reciprocal principle, the response of media at field point $\mathbf{x}$ in the fixed coordinates when the source lies at ξ in the fixed coordinates is equal to the response of media at field point $\mathbf{x} - \xi$ in the fixed coordinates when the source lies at $\mathbf{0}$ in the same coordinate.

Define the impulse response function (IRF) $\mathbf{h}(\mathbf{x}, t)$ as the solution of

$$\mathbf{O}[\mathbf{h}(\mathbf{x}, t-\tau)] = \delta(\mathbf{x})\delta(t-\tau). \qquad (100)$$

According to the above mentioned analysis, we have

$$\mathbf{h}(\mathbf{x}-\boldsymbol{\xi}, t) = \mathbf{G}(\mathbf{x}, t; \boldsymbol{\xi}) = \mathbf{G}(\mathbf{x}-\boldsymbol{\xi}, t; \mathbf{0}). \qquad (101)$$

Substituting (101) into (99), we get

$$\mathbf{u}(\mathbf{x}, t) = \int_{-\infty}^{t} \int_{S} F(\boldsymbol{\xi}, \tau)\mathbf{h}(\mathbf{x}-\boldsymbol{\xi}, t-\tau)\mathrm{d}\boldsymbol{\xi}\mathrm{d}\tau. \qquad (102)$$

Furthermore, if the concrete form $F(\boldsymbol{\xi}, \tau)$ in (103) is considered, we may rewrite (102) as

$$\mathbf{U}(\mathbf{x}, t) = \int_{-\infty}^{t} P_{sto}(\tau)\mathbf{h}(\mathbf{x}-\mathbf{v}\tau, t-\tau)\mathrm{d}\tau. \qquad (103)$$

in which $\mathbf{v}\tau = (v_x\tau, v_y\tau, v_z\tau)$. Also, the property of Dirac delta function (102) is used here. If the transformation $\theta = t-\tau$ is used, then (103) can be expressed as

$$\mathbf{u}(\mathbf{x}, t) = \int_{0}^{\infty} P_{sto}(t-\theta)\mathbf{h}(\mathbf{x}-\mathbf{v}t+\mathbf{v}\theta, \theta)\mathrm{d}\theta. \qquad (104)$$

Equations (103) and (104) are general results for MSP and are named generalized Duhamel's integral (GDI). From the point of view of time history, we might regard a moving load as a series of impact on continuum media during a number of tiny time intervals. The integration of the response of the medium excited by each impulse is thus equal to the cumulative effect of the moving load. Although solving for the Green's function is still a nontrivial task, convolution (103) does provide a sound theoretical representation, which can be very powerful when combining with numerical computation such as finite element method.

4.3.2 Deterministic analysis for a moving constant load

The solution of the problem described here can be constituted using the solutions of two individual problems because of the superposition principle of the solution of linear equations. One problem deals with the deterministic response of the medium under the moving constant load $P_0\delta(x-vt)\delta(y)$. The other problem deals with the random response of the medium under the moving stochastic load $P(t)\delta(x-vt)\delta(y)$. In this section, the first problem is analyzed. In the next section, the second problem is analyzed. The summary of the solution is provided in the sections followed. If the load is a constant with amplitude P_0, the response are given by

$$\mathbf{u}(\mathbf{x}, t) = P_0 \int_{0}^{\infty} \mathbf{h}(x-vt+v\theta, y, z, \theta)\mathrm{d}\theta. \qquad (105)$$

It is obvious that the response is no longer a constant independent on the time t. If the source is a fixed load, according to our example, the solution should be a static quantity. Actually, this idea is straightforward demonstrated by putting the velocity variable $v = 0$ into (105). The result gives that

$$\mathbf{u}(\mathbf{x}, t) = P_0 \int_{0}^{\infty} \mathbf{h}(\mathbf{x}, \theta)\mathrm{d}\theta. \qquad (106)$$

Clearly, the response is a constant without depending on the time variable.

4.3.3 Stochastic analysis for a moving stochastic vehicle load

In this section, a stochastic moving source is analyzed. In the derivation of the GDI in (103), we require no special assumptions on $P_{sto}(t)$. Therefore, if $P_{sto}(t)$ is a stochastic process, and (103) becomes an integral in the sense of Stieltjes integration. Meanwhile, the response $\mathbf{u}(\mathbf{x}, t)$ becomes a stochastic process. We now consider the response of media to $P(t)\delta(x-vt)\delta(y)$.

As mentioned before, $P(t)$ is a zero mean stationary process with autocorrelation function $R_{PP}(\tau)$, PSD $S_{PP}(\omega)$ and standard deviation σ_P.Taking the expectation of both sides of (103) and using the exchangeability of expectation and integration, we obtain the mean function of the response, i.e.,

$$\mathbf{u}(\mathbf{x}, t) = \int_{0}^{\infty} E[P(t-\theta)\mathbf{h}(x-vt+v\theta, y, z, \theta)]\mathrm{d}\theta. \qquad (107)$$

It is not difficult to obtain the spatial-time correlation functions for the response, i.e.,

$$\begin{aligned} R_{\mathbf{u}}(\mathbf{x}_1, \mathbf{x}_2; t_1, t_2) &= \int_{-\infty}^{t_2} \int_{-\infty}^{t_1} E[P(\tau_1)P(\tau_2)\mathbf{h}(x_1-v\tau_1, y_1, z_1, t_1-\tau_1) \\ &\quad \times \mathbf{h}(x_2-v\tau_2, y_2, z_2, t_2-\tau_2)]\mathrm{d}\tau_1\mathrm{d}\tau_2, \qquad (108) \end{aligned}$$

where $R_{\mathbf{u}}(\mathbf{x}_1, \mathbf{x}_2; t_1, t_2)$ is the correlation function of response $\mathbf{u}(\mathbf{x}, t)$. Let $\mathbf{x}_1 = \mathbf{x}_2 = \mathbf{x}$, then we obtain the time autocorrelation function

$$\begin{aligned} R_{\mathbf{u}}(\mathbf{x}; t_1, t_2) = \int_{0}^{\infty} \int_{0}^{\infty} & E[P(t_1-\theta_1)P(t_2-\theta_2) \\ & \times \mathbf{h}(x-vt_1+v\theta_1, y, z, \theta_1) \\ & \times \mathbf{h}(x-vt_2+v\theta_2, y, z, \theta_2)]\mathrm{d}\theta_1\mathrm{d}\theta_2. \qquad (109) \end{aligned}$$

where $\theta_j = t_j - \tau_j (j = 1, 2)$. By substituting $t_1 = t_2 = t$ into (108) and (109), it is straightforward to find second moment functions, i.e., the mean square functions of the random response.

It has been known that for a linear system with a stationary stochastic excitation at a fixed position, the response of that system is still a stationary stochastic process [76]. However, this conclusion only applies to the fixed source problem. For a linear system with a moving stochastic source, the random response of that system is a nonstationary stochastic process even if the random excitation $P(t)$ is a stationary stochastic process [54, 74]. To show this, we inspect (107) and (109), which are apparently not stationary because time variable t is contained in the kernel function of the solution. In other words, in a situation where a source is moving with respect to a receiver a nonstationary signal will be recorded at the observer position, even when the source produces a stationary output. This effect is known as the Doppler shift. It is also useful to realize that although the system is a linear system, it essentially becomes a time varying system when a moving source is applied.

In many circumstances, there may be a demand to push the analysis further into the frequency domain. For instance, response information of amplitude distributions and frequency components of media to moving vehicle loads is needed for vehicle optimum control and pavement performance prediction. To fulfill this purpose, it is necessary to use spectral analysis techniques to obtain the required information. As such, one may encounter difficulties when performing Fourier spectral analysis technique because this technique has been devised primarily for stationary signals. Although some variations of Fourier spectral analysis technique have been introduced to dealing with nonstationary stochastic processes, Fourier spectral analysis are not ideal tools for nonstationary signals induced by MSP. This shortcoming has led Sun [104]to develop the so-called follow-up spectral analysis, by which the commonly used spectral analysis technique is still applicable with sound theoretical foundation.

Let coordinates $oxyz$ and $o'x'y'z'$ be, respectively, fixed coordinates and follow-up coordinates moving with the moving source. The relationship between the two coordinates is

$$\mathbf{x}' = \mathbf{x} - \mathbf{v}t. \tag{110}$$

Therefore, a moving source $\mathbf{x_0} + \mathbf{v}t$ in the fixed coordinate $oxyz$ becomes a fixed source $\mathbf{x}' = \mathbf{x_0}$ in the follow-up coordinate $o'x'y'z'$. Here $\mathbf{x_0}$ is a constant vector. Now consider the response of the medium at a moving field point $\mathbf{x} + \mathbf{v}t + \mathbf{x_0}$ in the fixed coordinates. Utilizing (104), we have

$$\mathbf{u}(\mathbf{x} + \mathbf{v}t + \mathbf{x_0}, t) = \int_0^\infty P_{sto}(t - \theta)\mathbf{h}(\mathbf{x} + \mathbf{x_0} + \mathbf{v}\theta, \theta)\mathrm{d}\theta. \tag{111}$$

The mean function and autocorrelation function of the response described by (111) are, respectively, given by

$$\begin{aligned} E[\mathbf{u}(\mathbf{x} + \mathbf{v}t + \mathbf{x_0}, t)] &= \int_0^\infty E[P_{sto}(t - \theta)\mathbf{h}(\mathbf{x} + \mathbf{x_0} + \mathbf{v}\theta, \theta)]\mathrm{d}\theta \\ &= E[P_{sto}(t - \theta)] \int_0^\infty \mathbf{h}(\mathbf{x} + \mathbf{x_0} + \mathbf{v}\theta, \theta)d\theta \\ &= P_0 \int_0^\infty \mathbf{h}(\mathbf{x} + \mathbf{x_0} + \mathbf{v}\theta, \theta)\mathrm{d}\theta. \end{aligned} \tag{112}$$

and

$$\begin{aligned} R_{\mathbf{u}}(\mathbf{x} + \mathbf{v}t + \mathbf{x_0}, \tau) &= \int_0^\infty \int_0^\infty R_{PP}(\tau + \theta_1 - \theta_2)\mathbf{h}(\mathbf{x} + \mathbf{x_0} + \mathbf{v}\theta_1, \theta_1) \\ &\quad \times \mathbf{h}(\mathbf{x} + \mathbf{x_0} + \mathbf{v}\theta_2, \theta_2)\mathrm{d}\theta_1\mathrm{d}\theta_2. \end{aligned} \tag{113}$$

According to the definition of a stationary stochastic process [31], (112), and (113) indicate that the response at a moving field point $\mathbf{x} + \mathbf{v}t + \mathbf{x_0}$ in the fixed coordinates becomes a stationary process. It is evident that the moving field point $\mathbf{x} + \mathbf{v}t + \mathbf{x_0}$ in the fixed coordinates becomes a fixed field point $\mathbf{x} + \mathbf{x_0}$ in the follow-up coordinates as illustrated in Fig. 12. To show this, we replace x in (110) by $\mathbf{x} + \mathbf{v}t + \mathbf{x_0}$

$$\mathbf{x}' = \mathbf{x} + \mathbf{x_0}. \tag{114}$$

In other words, the response of a fixed position $\mathrm{x} + \mathrm{x}_0$ in the follow-up coordinates to a moving stationary stochastic load possesses the stationary property. Thus, Fourier spectral analysis technique is still applicable here. It should be pointed out that the explanation of the stationary process $\mathbf{u}(\mathbf{x} + \mathbf{v}t + \mathbf{x_0}, t)$ is essentially different from the commonly described stationary process. In general, the mean function of a commonly described stationary process refers to the time average of the random process, while the mean in (112) is indeed interpreted as a spatial average of the random response. The same explanation applies to the autocorrelation function shown in (113).

In light of the aforementioned explanation, there is no difficulty to define PSD in the follow-up coordinates. The following spectral analysis is performed in the follow-up coordinates. We may rewrite the autocorrelation function in the follow-up coordinates as

$$\begin{aligned} R_{\mathbf{u}}(\mathbf{x}', \tau) = \int_0^\infty \int_0^\infty & R_{PP}(\tau + \theta_1 - \theta_2)\mathbf{h}(\mathbf{x} + \mathbf{x_0} + \mathbf{v}\theta_1, \theta_1) \\ & \times \mathbf{h}(\mathbf{x} + \mathbf{x_0} + \mathbf{v}\theta_2, \theta_2)\mathrm{d}\theta_1\mathrm{d}\theta_2, \end{aligned} \tag{115}$$

where $\mathbf{x}'$ is expressed by (114). Let $\tau = 0$ and put it into Eq. (115), we obtain the mean square function of the random response in the follow-up coordinate

$$\begin{aligned} \psi_{\mathbf{u}}^2(\mathbf{x}', t) = \int_0^\infty \int_0^\infty & \psi_P^2(\theta_2 - \theta_1)\mathbf{h}(\mathbf{x} + \mathbf{x_0} + \mathbf{v}\theta_1, \theta_1) \\ & \times \mathbf{h}(\mathbf{x} + \mathbf{x_0} + \mathbf{v}\theta_2, \theta_2)\mathrm{d}\theta_1\mathrm{d}\theta_2. \end{aligned} \tag{116}$$

In addition, since $P(t)$ is a zero mean stationary process, we have

$$\sigma_P^2 = Var[P(t)] = \psi_P^2 = R_{PP}(0) = const, \tag{117}$$

where σ_p and $Var[P(t)]$ are, respectively, the standard deviation and variance of $P(t)$. So we can rewrite (117) as

$$\sigma_{\mathbf{u}}^2(\mathbf{x}',t) = \psi_{\mathbf{u}}^2(\mathbf{x}',t) = \left[\int_0^\infty \mathbf{h}(\mathbf{x}+\mathbf{x_0}+\mathbf{v}\theta,\theta)\mathrm{d}\theta\right]^2 R_{PP}(0), \tag{118}$$

where $\sigma_{\mathbf{u}}$ and $\psi_{\mathbf{u}}^2$ are the standard deviation and variance of the response field, respectively. Define the relationship between the frequency response function and the impulse unit response function in the follow-up coordinates as

$$\mathbf{H}(\mathbf{x}';\omega) = \int_0^\infty \mathbf{h}(\mathbf{x}+\mathbf{x_0}+\mathbf{v}\theta,\theta)e^{-i\omega\theta}\mathrm{d}\theta. \tag{119}$$

According to Wiener-Khintchine theory, the PSD and the autocorrelation function form a Fourier transform pair. Taking Fourier transform to both sides of (115) and noticing the (119), we obtain the expression of PSD in follow–up coordinates, i.e.,

$$S_{\mathbf{u}}(\mathbf{x}';\omega;v) = |\mathbf{H}(\mathbf{x}';\omega)|^2 S_{PP}(\omega). \tag{120}$$

Similarly, an expression for the time autocorrelation function can be obtained by taking the Fourier inverse transform of (120)

$$R_{\mathbf{u}}(\mathbf{x}';\tau) = (2\pi)^{-1}\int_{-\infty}^{\infty} |\mathbf{H}(\mathbf{x}';\omega)|^2 S_{PP}(\omega)e^{i\omega\tau}\mathrm{d}\omega. \tag{121}$$

Hence, the mean square function is also given by

$$\psi_{\mathbf{u}}^2(\mathbf{x}') = R_{\mathbf{u}}(\mathbf{x}';\tau) = (2\pi)^{-1}\int_{-\infty}^{\infty} |\mathbf{H}(\mathbf{x}';\omega)|^2 S_{PP}(\omega)\mathrm{d}\omega. \tag{122}$$

4.4 Pavement Models

There are mainly two types of pavement structures: Portlant cement concrete pavement and asphalt concrete pavement (Fig. 13). These pavement structures can be modeled by a beam, a slab, a layered medium on a half-space or rigid bedrock. A number of studies have been carried out lately by Sun and his associates using analytic method and analytic–numerical method [105–163]. Beskou and Theodorakopoulos [164] provided a recent review on numerical methods for studying dynamic effects of moving loads on road pavements. Sun and Greenberg [74], Sun and Luo [125–128], Sun et al. [153], Luo et al. [165] and Sun et al. [123] provide concrete examples of pavement models subject to moving loads.

The response of pavement systems under dynamic loads may be expressed in partial-differential equations. A generic description of the governing equations is:

$$\varphi[\mathbf{u}(\mathbf{x},t;\theta)] = \mathbf{P}(\mathbf{x},t), \tag{123}$$

where $\varphi[\cdot]$ is a partial-differential operator, $\mathbf{x} = (x,y,z)$ a spatial vector in Cartesian coordinates, t is the time variable, $\mathbf{P}(\mathbf{x},t)$ is the applied dynamic load (i.e., the input), which can be recorded by data acquisition system during laboratory experiments and field tests, $\mathbf{u}(\mathbf{x},t;\boldsymbol{\theta})$ the pavement response vector (i.e., the output in the form of displacements, stresses, and strains), and $\theta = (\theta_1,\theta_2,\ldots,\theta_n)$ is the parameter vector to be identified.

A pavement structure usually consists of a surface course, base courses and subgrade. Within each course, there may be several sub-layers made up of different materials. A two-dimensional Kirchhoff thin slab resting on a Winkler foundation is the common model for PCC pavements. The operator $\varphi[\cdot]$ for a Kirchhoff thin slab is

$$\varphi[\cdot] = D\nabla^2\nabla^2 + K + C\partial/\partial t + \rho h\partial^2/\partial t^2, \tag{124}$$

where $D = Eh^3/[12(1-\mu^2)]$, and ρ, μ and h are the density, Poisson's ratio and thickness of the slab, E is the Young's elastic modulus, K is the modulus of subgrade reaction, C is the radiation damping coefficient, and $\nabla^2 = \partial^2/\partial x^2 + \partial^2/\partial y^2$ is the Laplace operator. The parameter vector is $\theta = (E,\ h,\ \mu, K,\ C,\ \rho)$. For a multilayered flexible pavement system, the governing equation is controlled by a three-dimensional Navier–Stokes's equation for each layer

$$G^*\nabla^2\mathbf{u} + (\lambda^* + G^*)\nabla\nabla\cdot\mathbf{u} + \rho\mathbf{f} = \rho\partial^2\mathbf{u}/\partial t^2, \tag{125}$$

where $\nabla = \partial/\partial x + \partial/\partial y + \partial/\partial z$, $\mathbf{f}$ the body force vector, the Laplace operator $\nabla^2 = \partial^2/\partial x^2 + \partial^2/\partial y^2 + \partial^2/\partial z^2$, $G^* = G(1+i\eta_d)$, $\lambda^* = \lambda(1+i\eta_d)$ in which $i = \sqrt{-1}$, η_d is the hysteretic damping coefficient, Lamb constants λ and G the bulk modulus and shear modulus, respectively, λ^* and G^* the complex counterparts of λ and G, respectively. The subgrade may be artificially divided into a number of thin layers. Within each layer the soil is characterized to be isotropic, homogenous and have the same structural and material properties, while these properties vary for different layers. Furthermore, physical nonlinearity may possibly be presented in asphalt surface layer and soil subgrade using nonlinear constitutive models involving viscoelasticity-viscoplasticity. Eq. (125) adopts the simplest model to account for viscoelasticity. A more generic model is the generalized viscoelastic model, which includes the Burgers model, Maxwell model and Kelvin model as its special cases. Fig. 14 presents a schematic plot of a list of viscoelastic models.

To solve a viscoelastic problem, elastic solutions is sought first and then the correspondence principle is applied to convert the elastic solution into a viscoelastic solution. Two elastic/viscoelastic subgrade models will be studied: a half-space and a layer resting on bedrock. Clearly, the parameter vector $\theta = (E,\ h,\ \lambda, G,\ \rho)$ varies from layer to layer. When viscoelasticity is considered, more parameters will appear in the parameter vector. In principle, the adoption of a generalized viscoelastic model in forward dynamic analysis introduces no significant

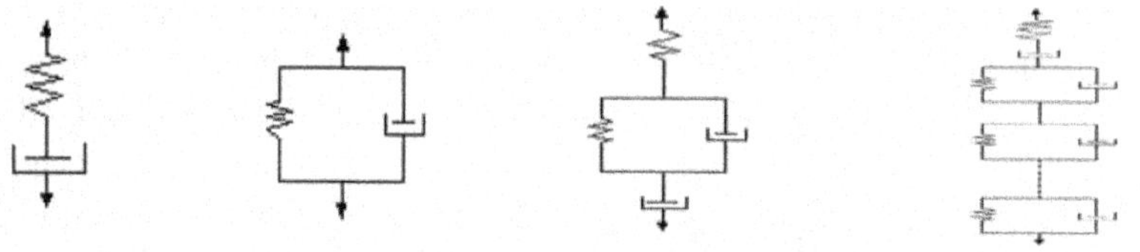

(a) Kelvin model **(b)** Maxwell model **(c)** Burgers model **(d)** Generalized model

Fig. 14 A schematic plot of different viscoelasic models

difficulty. The number of Kelvin components in this model can be estimated through a thorough investigation. With properly specified initial and boundary conditions, Eqs. (123)–(125) constitute a complete mathematical description of the forward dynamic problem. Forward analysis aims to solve for the response $\mathbf{u}(\mathbf{x}, t; \boldsymbol{\theta})$ provided that the excitation $\mathbf{P}(\mathbf{x}, t)$ and the parameter vector θ are known. These mathematical, physical models describe the behavior of different types of pavement systems and they will be studied in great depth.

Equation (123) belongs to a wave equation from a mathematical physics point of view. Its solution can be obtained in the form of a Lebesgue–Stieltjes integral using proper integral transformation, depending upon the nature of the problem (e.g., steady-state vs. transient) and upon how the problem is formulated (e.g., in Cartesian or cylindrical coordinates). Let the Green's function (the fundamental solution) of Eq. (1) be $\mathbf{G}(\mathbf{x}, t; \theta) = \varphi^{-1}[\boldsymbol{\delta}(\mathbf{x}, t)]$, in which $\boldsymbol{\delta}(\cdot)$ is the Dirac-delta function and $\varphi^{-1}[\cdot]$ denotes the inverse operator of $\varphi[\cdot]$. Let pavements be at rest prior to the NDE test, leading to a vanishing initial condition. The solution of Eq. (123) under loading condition $\mathbf{P}(\mathbf{x}, t)$ can be constructed as

$$\mathbf{u}(\mathbf{x}, t; \theta) = \varphi^{-1}[\mathbf{P}(\mathbf{x}, t)] = \int_S \mathbf{G}(\mathbf{x}, t; \theta) \mathrm{d}S, \tag{126}$$

where S is the region where $\mathbf{P}(\mathbf{x}, t)$ is defined. Equation (126) can also be equivalently represented in the transformed domain

$$\tilde{\mathbf{u}}(\boldsymbol{\xi}, \omega; \theta) = T\{[\mathbf{u}(\mathbf{x}, t; \theta)]\} = \int_{\tilde{S}} \tilde{\mathbf{G}}(\boldsymbol{\xi}, \omega; \theta) \mathrm{d}\tilde{S}, \tag{127}$$

where $T\{\cdot\}$ is a transformation operator, $\tilde{\mathbf{u}}(\boldsymbol{\xi}, \omega; \theta)$ and $\tilde{\mathbf{G}}(\boldsymbol{\xi}, \omega; \theta)$ are the response and the Green's function in the transformed domain, respectively, $\tilde{S}$ is the region in the transformed domain where the transformed dynamic load $\tilde{\mathbf{P}}(\boldsymbol{\xi}, \omega)$ is defined; $\boldsymbol{\xi} = (\xi, \eta, \zeta)$ and ω are the counterpart of spatial vector $\mathbf{x} = (x, y, z)$ and time variable t.

The Thomas-Haskell method relates a transformed response at the bottom of a layer, in the form of a transfer matrix, to a corresponding quantity at the top of a lower layer. For the last half century, this method has served as the cornerstone for numerous studies in multilayered elastic analysis. Another benchmark proposed by Kausel and Roesset contributes an alternative, in which the dynamic stiffness matrix is expanded in terms of wave number and approximated by taking terms only up to the second order of the wave number. The Thomas-Haskell method is more accurate than the Kausel-Roesset method but demands more computational effort. Forward dynamic analysis for multilayered viscoelastic media will not add significant difficulty because both methods are still applicable, though it is a demanding task. Since each method has its own strengths and weaknesses, both methods will be adopted in the project to tackle wave propagation through multilayered viscoelastic media. Equations (126) and (127) are amenable to numerical evaluation of the dynamic response in the time–space domain and in the transformed domain, respectively, provided that $\mathbf{P}(\mathbf{x}, t)$ and θ are known. Without doubt, the computation here involves intensive numerical evaluation of multifold integration of complex functions with unstable characteristics in time and space.

5 Discussion and Future Research

In this section, discussion and future research of the unified theory of dynamics of vehicle–pavement interaction under moving and stochastic load is carried out from the following aspects: (a) nature of the problem, (b) modeling, (c) methodology, and (d) further extension and engineering application.

(a) Nature of the problem. The setting of vehicle–pavement interaction can be categorized either in a deterministic framework or a stochastic framework. Stochasticity may be presented in speed, magnitude and position of the loading, structural models of vehicle–pavement system and constitutive models. A stochastic framework provides a more realistic setting but exhibit more complexity. The setting of vehicle–pavement interaction can also be categorized either as a steady-state problem or as a transient problem. Except very few studies [104, 145], almost all existing literatures belong to a steady-state problem described in a deterministic framework.

(b) Modeling. The modeling of vehicle–pavement interaction involves the load model, the vehicle model and the pavement model. The load model addresses spatial distribution of the load (e.g., concentrated load, distributed load, multiple load, etc.), speed of the load (e.g., constant speed, varying speed, etc.), trajectory of the load (e.g., straight line, curve, etc.), and magnitude of the load (e.g., constant load, impact load, sinusoidal load, varying load, and random load). The vehicle model addresses vehicle suspension system, which involves mass distribution, dimension, and configuration of the vehicle. Quarter-vehicle model, half-vehicle model, and whole vehicle model have all been established with increasing complexity involving different number of spring and dashpot elements. Regardless of how many spring and dashpot elements are being used in the built vehicle model, almost all the literature studies tend to use linear suspension system due to its ease of computation when integrated with a pavement model. The pavement model addresses the simplification of structural system of pavement using beam model, slab model, and layered medium on a half space or bedrock. In terms of constitutive models of the material, almost all studies only considered linear elastic and linear viscoelastic materials when tackling dynamics of vehicle–pavement interaction because of the complexity of the problem. Asphalt concrete pavement actually exhibits complex viscoelastic–viscoplastic-damage properties [166–172], which should be integrated into the subject. Also, it is rare to consider the coupling effect of vehicle–pavement interaction, while this is not the case in train–railway interaction because of the mechanism of the interaction and the relative mass difference between vehicles (e.g., car, truck, train, and airplane) and transportation infrastructure (e.g., highway, bridge, and railway). None of the studies has addressed vehicle–pavement interaction using deteriorated pavement models as well as long-term pavement damage and failure due to vehicle–pavement interaction.

(c) Methodology. The methods for solving vehicle–pavement interaction problem can be classified into analytic approach and numerical application, though the implementation of analytic approach still requires numerical computation. For vehicle dynamics, equation of motion is first established as a set of linear differential equation system and solved in the frequency domain using frequency response function or in the time domain using numerical sequential integration. For pavement dynamics, analytic approach makes use of integral transform to treat wave propagation in continuum media, while numerical approach such as finite difference, finite element and boundary element methods makes use of discretization in time and in space. The advantage of analytic approach is that it provides insights for revealing physics of the wave propagation in continuum media and can be highly efficient in terms of computation implementation, particularly when the spatial scale of the problem involves hundreds of kilometers (e.g., seismology) or the speed of the load is very high and close to various critical speeds of waves in media. The advantage of numerical approach is that they can deal with pavement having complex geometric structure as well as nonlinear constitutive model of the material.

(d) Further extension and engineering application. The study of dynamics of vehicle–pavement interaction provides a deep understanding for improving vehicle design (e.g., road-friendly vehicle suspension system), road transportation safety [173–177], long-lasting pavement structures design [178–181], ride quality and infrastructure asset management. An improved quantitative understanding on effects and mechanisms of various factors on dynamics of vehicle–pavement interaction is the fundamentals for increased application and accuracy and reliability of structural health monitoring, nondestructive testing and evaluation, environmental vibration mitigation and weight-in-motion. It will also benefit the nation's transportation economy by reducing operation and maintenance costs of vehicles and transportation infrastructure as well as increasing transportation productivities.

6 Conclusions

Irregularities of pavement surface, from the small-scale unevenness of material on pavement surface to the large-scale undulating of vertical curve of a highway or an airport, all belong to spatial fluctuation of pavement surface at different scale. Extensive study has been accomplished both domestically and internationally, toward measuring the physical aspects of pavement roughness, analyzing the resulting data, and evaluating the riding performance of pavements. Instrumentation and analysis technique including the spectral analysis approaches have been summarized in the article. A number of the PSD functions including the effect of thermal joints on PSD roughness have been presented here. These PSD functions are similar in their shapes and only different in their mathematical descriptions. Under the condition of constant speed of travel and linear vehicle suspension, it is proven that dynamic contact force between vehicle and pavement is a stationary stochastic process. Its mean function is given by (79) without the consideration of static loads or by (80) with the consideration of static loads. The correlation function, PSD forces, and standard deviation are, respectively, given by (84), (89) and (92). The concept and the methodology present in the article are not restricted to specific surface roughness and/or vehicle models. They are generally applicable to all kinds of linear vehicle models and measured pavement surface conditions. The response of linear continuum media to moving stochastic vehicle loads is analyzed herein. We show that there exists predicable relation among surface roughness, vehicle suspensions and speed, and the response of continuum media. The theory developed here is widely applicable to moving vehicle loads.

Acknowledgments This study is sponsored in part by the National Science Foundation, by National Natural Science Foundation of China, by Ministry of Communication of China, by Jiangsu Natural Science Foundation to which the author is very grateful. The author is also thankful to Professor Wanming Zhai, Editor Mr. Yao Zhou and Editor-in-Chief Yong Zhao for their invitations.

References

1. Pavement Roughness and Rideability, Project 1–23 (1981) National Cooperative Highway Research Program, Project Statement
2. Abbo E, Hedric JK, Markow M, Brademeyer B (1987) Analysis of moving loads on rigid pavements. International symposium on heavy truck suspensions, Canberra, Australia
3. Hudson WR et al. (1992) Impact of truck characteristics on pavements: truck load equivalent factors. Report No. FHWA-RD-91-064, Federal Highway Administration, Washington, DC
4. Gillepie TD et al. (1993) Effects of heavy-vehicle characteristics on pavement response and performance, Report 353, National Cooperative Highway Research, National Academy Press, Washington, DC
5. Barrodal I, Erickson RE (1980) Algorithms for least-square linear prediction and maximum entropy spectral analysis. Geophysics 45:420–446
6. Bendat JS, Piersol AG (1971) Random data: analysis and measurement procedure. Wiely-Interscience, New York
7. Beskou ND, Theodorakopoulos DD (2011) Dynamic effects of moving loads on road pavements: a review. Soil Dyn Earthq Eng 31(4):547–567
8. BSI proposals for generalized road inputs to vehicles (1972) *ISO/TC, 108/WG9*, document no. 5, International Organization for Standardization
9. Captain KE, Boghani AB, Wormley DN (1979) Analytical tire models for dynamic vehicle simulation. Veh Syst Dyn 8:1–32
10. Cebon D (1989) Vehicle-generated road damage: a review. Veh Syst Dyn 18(1–3):107–150
11. Cebon D (1993) Interaction between heavy vehicles and roads. *SAE* Technical Paper No. 93001, Society of Automotive Engineers
12. Chen, S.S. (1987). The response of multi-layered systems to dynamic surface loads. Ph.D Dissertation, Department of Civil Engineering, University of California, Berkeley, California
13. Cole DJ (1990) Measurement and analysis of dynamic tire forces generated by lorries. Ph.D Dissertation, University of Cambridge, Cambridge United Kingdom
14. Cole J, Huth J (1958) Stresses produced in a half-plane by moving loads. J Appl Mech, ASME 25:433–436
15. De Barro FCP, Luco JE (1994) Response of a layered visco-elastic half-space to a moving point load. Wave Motion 19(2):189–210
16. Deng X and Sun L (1996) Dynamic vertical loads generated by vehicle–pavement interaction. CSME 12th proc. symp. on advances in transportation system, Canadian Society for Mechanical Engineering, Hamilton
17. Deng X, Sun L (1996) The Euclid norm weight model for and its application in pavement evaluation. China J Highw Transp 9(1):21–29
18. Dodds CJ (1974) The laboratory simulation of vehicle service stress. J Eng Ind Trans, ASME 96(3):391–398
19. Dodds CJ, Robson JD (1973) The description of road surface roughness. J. Sound Vib 31:175–183
20. Eason G (1965) The stresses produced in a semi-infinite solid by a moving surface force. Int J Eng Sci 2:581–609
21. Elattary MA (1991) Moving loads on an infinite plate strip of constant thickness. J Phys D 24(4):541–546
22. Eringen AC, Suhubi ES (1975) Elastodynamics, vol I. Academic Press, New York
23. Felszegpy SF (1996) The Timoshenko beam on an elastic foundation and subject to a moving step loads, Part I: steady-state response. J Vib Acous, ASME 118:227–284
24. Felszegpy SF (1996) The Timoshenko beam on an elastic foundation and subject to a moving step loads, Part II: transient response. J Vib Acous, ASME 118:285–291
25. Fryba L (1972) Vibration of solids and structures under moving loads. Noordhoff International Publishing, Gronigen
26. Fryba L, Nakagiri S And, Yoshikawa N (1993) Stochastic finite elements for a beam on a random foundation with uncertain damping under a moving force. J Sound Vib 163(1):31–45
27. Gakenheimer DC, Miklowitz J (1969) Transient excitation of an elastic half space by a point load traveling on the surface. J Appl Mech, ASME 37:505–522
28. Galaitsis AG, Bender EK (1976) Wheel/rail noise, part V: measurement of wheel and rail roughness. J Sound Vib 46:437–451
29. Gelfand JM, Shilov GE (1964) Generalized functions. Volume I: properties of operations. Academic Press, New York
30. Gillespie TD (1985) Heavy truck ride: *SP-607*. Society of Automotive Engineers, Warrendale
31. Gillespie TD (1986) Developments in road roughness measurement and calibration procedures. University of Michigan Transportation Research Institute, Ann Arbor
32. Hall AW, Hunter PA, Morris GJ (1971) Status of research on runway roughness. Report SP-270, National Aeronautics and Space Administration, p 127–142
33. Hardy MSA, Cebon D (1993) Response of continues pavement to moving dynamic loads. J Eng Mech, ASCE 119(9):1762–1780
34. Hardy MSA, Cebon D (1994) Importance of speed and frequency in flexible pavement response. J Eng Mech, ASCE 120(3):463–482
35. Harrison RF, Hammond KJ (1986) Approximate, time domain, non-stationary analysis of stochastically excited, non-linear systems with particular reference to the motion of vehicles on rough ground. J Sound Vib 105(3):361–371
36. Hass R, Hudson WR, Zaniewski J (1994) Modern pavement management. Kriger, Malabar
37. Heath AN (1987) Application of the isotropic road roughness assumption. J Sound Vib 115(1):131–144
38. Heath AN, Good MG (1985) Heavy vehicle design parameters and dynamic pavement loading. Australian Road Research, 15(4)
39. Hedric JK and Markow M (1985) Predictive models for evaluating load impact factors of heavy trucks on current pavement conditions, Interim Report to USDOT Office of University Research under Contract DTRS5684-C-0001
40. Honda H, Kajikawa Y, Kobori T (1982) Spectra of surface roughness on bridge. J Struct Eng, ASCE, 108(ST9), 1956–1966

41. Houbolt JC (1962) Runway roughness studies in aeronautical fields. Paper No. 3364, ASCE, New York
42. Hsueh TM, Panzien J (1974) Dynamic response of airplanes in ground operation. J Transp Eng ASCE 100:743–756
43. Huang YH (1993) Pavement analysis and design. Englewood Cliffs, New Jersey
44. Hudson WR and Senvner FH (1962) AASHO road test principal relationships—performance versus stress, Rigid Pavement, Special Report, Highway Research Board
45. Hwang ES, Nowak AS (1991) Simulation of dynamic load for bridges. J Struct Eng ASCE 117(5):1413–1434
46. Iyengar RN, Jaiswal OR (1995) Random field modeling of railway track irregularities. J Transp Eng, ASCE, 121(4): 303–308
47. Kamesh KM, Robson JD (1978) The application of the isotropic road roughness assumption. J Sound Vib 57:80–100
48. Kenney JT (1954) Steady-state vibration of beam on elastic foundation for moving load. J Appl Mech, ASME 21(4):359–364
49. La Barre RP, Forkes RT (1969) The measurement and analysis of road surface roughness. MIRA Report No. 1970/5
50. Lee HP (1994) Dynamic response of a beam with intermediate point constrains subject to a moving load. J Sound Vib 171(3):361–368
51. Lee HR, Scheffel JL (1968) Runway roughness effects on new aircraft types. J Aerosp Eng, ASCE 94(AT1):1–17
52. Lighthill MJ (1958) Introduction to Fourier analysis and generalized functions. Cambridge University Press, London
53. Luo F, Sun L, Gu W (2011) Elastodynamic inversion of multilayered media via surface deflection—Part II: implementation and numerical verification. J Appl Mech 78(5):169–206
54. Macvean DB (1980) Response of vehicles accelerating on a random profile. Ingenieur-Archiv 49:375–380
55. Marcondes J, Burgess GJ, Harichandran R, Snyder MB (1991) Spectral analysis of highway pavement roughness. J Transp Eng, ASCE 117(5):540–549
56. Marcondes JA, Snyder MB, Singh SP (1992) Predicting vertical acceleration in vehicles through road roughness. J Transp Eng ASCE 118(1):33–49
57. Markow M, Hdric JK, Bradmeyer BD, Abbo E (1988) Analyzing the interactions between vehicle loads and highway pavements, paper presented at the 67th Annual meeting, Transp Res Board, Washington, DC
58. Markow M, Hedric JK, Brademeyer B, Abbo E (1988) Analyzing the interaction between vehicle loads and highway pavements. Paper Presented at 67th Annual Meeting of the Transportation Research Board, Washington, DC
59. McCullough BF, Steitle DC (1975) Criteria development to evaluation runway roughness. J Transp Eng ASCE 101(TE2): 345–363
60. Monismith CL, Mclean DB (1971) Design considerations for asphalt pavements. Report No. TE 71-8, Institute of Transportation and Traffic Engineering, University of California, Berkeley
61. Monismith CL, Lysmer J, And Sousa J, Hedric JK (1988) Truck pavement interactions—requisite research. Transc SAE SP-765 881849:87–96
62. Monismith CL, Sousa J, Lysmer J (1988) Modern pavement design technology including dynamic load conditions, SAE conference on vehicle/pavement interaction, SAE Trans, Society of Automotive Engineers. SP-765, No. 881856, 33–52
63. Morse PM, Keshbach H (1953) Methods of theoretic physics. Part I and II. McGraw-Hill Book Company Inc, New York
64. Nasirpour F, Kapoor SG, Wu SM (1978) Runway roughness characterization by DDS approach. J Transp Eng ASCE 104(TE2):213–226
65. Newland DE (1986a) General theory of vehicle response to random road roughness, random vibration—status and recent developments, Elishakoff I and Lyon RH (eds.), Amsterdam: Elsevier
66. Newland DE (1986) The effect of a footprint on perceived surface roughness. Proc Roy Soc (London) A 405:303–327
67. Newland DE (1993) An introduction to random vibration, spectral and wavelet analysis, 3rd edn. Longman, New York
68. Nigam NC, Narayanan S (1994) Applications of random vibrations. Narosa Publishing House, New Delhi
69. Orr LW (1988) Truck pavement factors—the truck manufacturer's viewpoint. Transc. SAE SP-765 881842:1–4
70. Pan G, Atluri SN (1995) Dynamic response of finite sized elastic runways subjected to moving loads: a coupled bem/fem approach. Int J Numer Methods Eng 38(18):3143–3166
71. Pan T, Sun L (2012) Sub-microscopic phenomena of metallic corrosion studied by a combined photoelectron spectroscopy in air (PESA) and scanning Kelvin probe force microscopy (SKPFM) approach. Int J Electrochem Sci 7:9325–9344
72. Pan T, Sun L, Yu Q (2012) An atomistic-based chemophysical environment for valuating asphalt oxidation and antioxidants. J Mol Model 18:5113–5126.
73. Parkhilovskii IG (1968) Investigation of the probability characteristics of the surface of distributed types of roads. Avtom Prom 8:18–22
74. Payton RG (1964) An application of the dynamic Betti-Rayleigh reciprocal theorem to moving-point loads in elastic media. Quart Appl Math 21:299–313
75. Rauhut JB, Kennedy TW (1982) Characterizing fatigue life for asphalt concrete pavements, Transportation Research Record No. 888, Transportation Research Board
76. Road surface profile—reporting measured data (1984) Draft proposal ISO/DP 8608, mechanical vibration, international organization for standardization
77. Robson JD (1978) The role of Parkhilovskii model in road description. Veh Systm Dyn 7:153–162
78. Robson JD, Dodds CJ (1970) The response of vehicle components to random road surface indications. Proc. IUTAM Symp., Frankfurt. (ORDER): Academic
79. Sayer MW, Gillespie TD and Queiroz CAV (1986) The international road roughness experiment—establishing correlation and calibration standard for measurements. World Bank Tech. Paper No. 45, World Bank, Washington, DC, 25–29
80. Sayers MW (1985) Characteristic power spectral density functions for vertical and roll components of road roughness. Proc. symp. on simulation control of ground vehicles and transp. Systems, ASME, New York
81. Sayles RS, Thomas TR (1978) Surface topography as a nonstationary random process. Nature 271:431–434
82. Snyder JE, Wormley DN (1977) Dynamic interactions between vehicles and elevated, flexible randomly irregular guideways. Transactions of the American Society of Mechanical Engineers. J Dyn Syst Meas Control 99:23–33
83. Sobczyk K and Macvean DB, (1976). Non-stationary random vibration of road vehicles with variable velocity. Symposium on stochastic problems in dynamics, University of Southampton, England, (editor Clarkson BL)
84. Spiegel MB (1971) Theory and problems of advanced mathematics for engineers and scientists. McGraw-Hill Book Company, New York
85. Steele CR (1967) The finite beam with a moving load. J Appl Mech, ASME 34(2):111–118
86. Strathdee J, Robinson WH, Haines EM (1991) Moving loads on ice plates of finite thickness. J Fluid Mech 226:37–61
87. Sun L, Deng X, (1995) Dynamic loads caused by vehicle–pavement interactions, J Southeast Univ, 26(5): 142–145; also in Proc. China Forum 95 on Transportation, Chongqing University, (Chongqing, Sichuan, China), 862–868 (in Chinese)

88. Sun L (1998) Theoretical investigation on dynamics of vehicle–pavement interaction. Final Technical Report prepared for the National Science Foundation, Southeast University, Nanjing, China
89. Sun L (2001) A closed-form solution of Bernoulli–Euler beam on viscoelastic foundation under harmonic line loads. J Sound Vib 242(4):619–627
90. Sun L (2001) Closed-form representation of beam response to moving line loads. J Appl Mech, ASME 68(2):348–350
91. Sun L (2001) Computer simulation and field measurement of dynamic pavement loading. Math Comput Simul 56(3):297–313
92. Sun L (2001) Developing spectrum-based models for international roughness index and present serviceability index. J Transp Eng ASCE 127(6):463–470
93. Sun L (2001) Dynamic response of beam-type structures to moving line loads. Int J Solids Struct 38(48–49):8869–8878
94. Sun L (2001) On human perception and evaluation to road surfaces. J Sound Vib 247(3):547–560
95. Sun L (2001) Time-harmonic elastodynamic Green's function of plates for line loads. J Sound Vib 246(2):337–348
96. Sun L (2002) A closed-form solution of beam on viscoelastic subgrade subjected to moving loads. Comput Struct 80(1):1–8
97. Sun L (2002) Optimum design of road-friendly vehicle suspension systems subject to rough road surface. Appl Math Model 26(5):635–652
98. Sun L (2002) Time-frequency analysis of thin slabs subjected to dynamic ring loads. Acta Mechanica Nos 153(3–4):207–216
99. Sun L (2003) An explicit representation of steady state response of a beam resting on an elastic foundation to moving harmonic line loads. Int J Numer Anal Methods Geomech 27:69–84
100. Sun L (2003) Dynamic response of Kirchhoff plate on a viscoelastic foundation to harmonic circular loads. J Appl Mech, ASME 70(4):595–600
101. Sun L (2003) Simulation of pavement roughness and IRI based on power spectral density. Math Comput Simul 61:77–88
102. Sun L (2005) Dynamics of plate generated by moving harmonic loads. J Appl Mech, ASME 72(5):772–777
103. Sun L (2006) Analytical dynamic displacement response of rigid pavements to moving concentrated and line loads. Int J Solids Struct 43:4370–4383
104. Sun L (2007) Steady-state dynamic response of a Kirchhoff's slab on viscoelastic Kelvin's foundations to moving harmonic loads. J Appl Mech, ASME 74(6):1212–1224
105. Sun L and Deng X (1996). Roughness and spectral analysis of runway profile. Eastern China Highway, No. 2 (in Chinese)
106. Sun L, Deng X (1996) Analysis of pavement wave-number spectral density and random dynamic pressure generated by vehicle–pavement interactions. J Xi'an Highw Univ 16(2):17–21 (in Chinese)
107. Sun L, Deng X (1996) General theory for steady dynamic problem of infinite plate on an elastic foundation. Acta Mech Sin 28(6):756–760 (in Chinese)
108. Sun L, Deng X (1996) Mathematical model and experiment design of pavement dynamic loads. J Xi'an Highw Univ 16(4):50–53
109. Sun L, Deng X (1996) Mathematical model of weight in evaluation systems. J Southeast Univ 12(2):111–118 (in Chinese)
110. Sun L, Deng X (1996) Random pressure generated by airplane-airfield interactions. J Chongqing Jiaotong Inst 15(4):14–20 (in Chinese)
111. Sun L, Deng X (1996) Spectral analysis about surface evenness of airport pavement. East China Highw 2:35–38 (in Chinese)
112. Sun L, Deng X (1997) Transient response for infinite plate on Winkler foundation by a moving distributed load. Chin J Appl Mech 14(2):72–78 (in Chinese)
113. Sun L, Deng X (1997) Steady response of infinite plate on viscoelastic Kelvin foundation to moving loads. Chin J Geotech Eng 19(2):14–22 (in Chinese)
114. Sun L, Deng X (1997) Random response of beam under a moving random load in the line source form. Acta Mech Sin 29(3):365–368
115. Sun L, Deng X (1997) Influence of velocity, frequency and characteristics of vehicle on dynamic pavement loads. China Civ Eng J 30(6):137–147 (in Chinese)
116. Sun L, Deng X (1997) Random response of beam under a moving random load in the line source form. Acta Mech Sin 29(3):365–368 (in Chinese)
117. Sun L, Deng X (1997) Transient response for infinite plate on Winkler foundation by a moving distributed load. Chin J Appl Mech 14(2):72–78 (in Chinese)
118. Sun L, Deng X (1997) Transient response of bridge to traveling random vehicle loads. J Vib Shock 16(1):62–68 (in Chinese)
119. Sun L, Deng X (1997) Weight analysis in evaluation system. J Syst Sci Syst Eng (English Edition) 6(2):137–147
120. Sun L, Deng X (1997) Steady response of infinite plate on viscoelastic Kelvin foundation to moving loads. Chin J Geotech Eng 19(2):14–22 (in Chinese)
121. Sun L, Deng X (1998) Dynamic analysis of infinite beam under the excitation of moving line loads. Appl Math Mech 19(4): 367–373 (in Chinese)
122. Sun L, Deng X (1998) Dynamic analysis of infinite beam under the excitation of moving line loads. Appl Math Mech 19(4):367–373 (in Chinese)
123. Sun L, Deng X (1998) Predicting vertical dynamic loads caused by vehicle–pavement interaction. J Transp Eng, ASCE 126(5): 470–478
124. Sun L, Deng X (1998) Predicting vertical dynamic loads caused by vehicle–pavement interaction. J Transp Eng, ASCE 124(5): 470–478
125. Sun L, Deng X (1998) Recent advances on spectral analysis of pavement surface. J Southeast Univ 14(4):223–234 (in Chinese)
126. Sun L, Duan Y (2013) Dynamic response of top-down cracked asphalt concrete pavement under a half-sinusoidal impact load. Acta Mechanica.
127. Sun L, Greenberg B (1999) Dynamic response of linear systems to moving stochastic sources. J Sound Vib 229(4):957–972
128. Sun L, Greenberg BS (2000) Dynamic response of linear systems to moving stochastic sources. J Sound Vib 229(4):957–972
129. Sun L, Gu W (2011) Pavement condition assessment using fuzzy logic theory and analytic hierarchy process. J Transp Eng ASCE 137(9):648–655
130. Sun L, Kennedy TW (2002) Spectral analysis and parametric study of stochastic pavement loads. J Eng Mech, ASCE 128(3):318–327
131. Sun L, Luo F (2007) Arrays of dynamic circular loads moving on an infinite plate. Int J Numer Meth Eng 71(6):652–677
132. Sun L, Luo F (2007) Nonstationary dynamic pavement loads generated by vehicles traveling at varying speed. J Transp Eng, ASCE 133(4):252–263
133. Sun L, Luo F (2008) Steady-state dynamic response of a Bernoulli–Euler beam on a viscoelastic foundation subject to a platoon of moving dynamic loads. J Vib Acous, ASME 130(5): 051002.1–051002.19
134. Sun L, Luo F (2008) Transient wave propagation in multilayered viscoelastic media—theory, numerical computation and validation. J Appl Mech, ASME 75(3):031007.1–031007.15
135. Sun L, Ronald Hudson W (2005) Probabilistic approaches for pavement fatigue cracking prediction based on cumulative damage using Miner's law. J Eng Mech, ASCE 131(5):546–549
136. Sun L, Su J (2001) Modeling random fields of road surface irregularities. Int J Road Mater Pavement Des 2(1):49–70

137. Sun L, Wang Dengzhong (2012) Nondestructive testing and evaluation of subgrade compaction by portable falling deflectometer. J Highw Transp Res Dev 29(2):41–47 (in Chinese)
138. Sun L, Xin Xiantao GuW (2012) Comprehensive comparison of nano-materials for asphalt modification. J Transp Eng Inf 10(2):1–11 (in Chinese)
139. Sun L and Zheng KA (1998) Unified theory for solving moving source problem: general principle of the Duhamel integral, Academic Periodical Abstracts of China, Vol. 4, No. 4 (in Chinese)
140. Sun L, Zheng K (1998) Distress stress analysis of flexible pavement structure. J Chongqing Jiaotong Inst 17(4):1–7 (in Chinese)
141. Sun L, Zhu YT (2013) A two-stage serial viscoelastic-viscoplastic constitutive model for characterizing nonlinear time-dependent deformation behavior of asphalt mixtures. Constr Build Mater 40:584–595
142. Sun L, Deng X, Zhang J (1996) An optimal method for interchange design. East China Highw 3:27–31 (in Chinese)
143. Sun L, Deng X, Gu W (1997) Identification of typical asphalt pavement structures for category design of pavements. China Civ Eng J 30(3):55–62 (in Chinese)
144. Sun L, Deng X, Gu W (1997) Recent advancement of pavement roughness and dynamic loads—I. Pavement spectral analysis and its application, II. Theory and experiment of pavement dynamic loads. Acad Period Abstr China 3(12):1442–1443 (in Chinese)
145. Sun L, Deng X, Gu W (1998) Random response of Kirchhoff plate and Bernoulli–Euler beam under moving random loads. Acad Period Abstr China 4(1):127 (in Chinese)
146. Sun L, Deng X, Gu W (1998) Steady and transient response of infinite plate on Kelvin foundation generated by a moving distributed load. Acad Period Abstr China 4(1):126 (in Chinese)
147. Sun L, Zhang Z, Ruth J (2001) Modeling indirect statistics of surface roughness. J Transp Eng, ASCE 127(2):105–111
148. Sun L, Hudson WR, Zhang Z (2003) Empirical-mechanistic method based stochastic modeling of fatigue damage to predict flexible pavement fatigue cracking for transportation infrastructure management. J Transp Eng ASCE 129(2):109–117
149. Sun L, Luo F, Chen TH (2005) Transient response of a beam on viscoelastic foundation under impact loads during nondestructive testing. J Earthq Eng Eng Vib 4(2):325–333
150. Sun L, Cai X, Yang J (2006) Genetic algorithm-based optimum vehicle suspension design using minimum dynamic pavement load as a design criterion. J Sound Vib 301(1–2):18–27
151. Sun L, Kenis W, Wang W (2006) Stochastic spatial excitation induced by a distributed contact with homogenous Gaussian random fields. J Eng Mech, ASCE 132(7):714–722
152. Sun L, Gu W, Luo F (2009) Steady state response of multilayered viscoelastic media under a moving dynamic distributed load. J Appl Mech, ASME 75(4):0410011–04100115
153. Sun L, Gu W, Mahmassani H (2011) Estimation of expected travel time using moment approximation. Can J Civ Eng 38:154–165
154. Sun L, Li A, Zhang Y, Xin X, Shao J (2011) Study on gradation of cement-fly ash stabilized coal gangue. J Transp Eng Inf (in Chinese) 9(4):7–10
155. Sun L, Luo F, Gu W (2011) Elastodynamic inversion of multilayered media via surface deflection—Part I: methodologies. J Appl Mech 78(5):2323–2333
156. Sun L, Wang D, Zhang H (2012) Predictive models of subgrade deflection using data from portable falling weight deflectometer. J Southeast Univ 42(5):970–975 (in Chinese)
157. Sun L, Xin X, Wang H, Gu W (2012) Microscopic mechanism of modified asphalt by multi-dimensional and multi-scale nanomaterial. J Chin Ceram Soc 40(10):1437–1447 (in Chinese)
158. Sun L, Xin X, Wang H, Gu W (2012) Performance of nano-material modified asphalt as paving materials. J Chin Ceram Soc 40(8):1095–1101
159. Sun L, Pan Y, and Gu W (2013) High order thin layer method for viscoelastic wave propagation in stratified media, Comput Methods Appl Mech Eng, 256, 65–76, http://www.sciencedirect.com/science/article/pii/S0045782513000157. Accessed 17 Sep 2013
160. Sun L, Chen L and Gu W (2013) Stress and deflection parametric study of high-speed railways CRTS-II Blastless Track Slab on bridge foundation. ASCE J Transp Eng (in press)
161. Sun L, Gu W and Xu B (2013) Characterizing uncertainty in pavement performance prediction using Monte Carlo simulation. J Southeast Univ, (English Edition) (in press)
162. Sun L, You K, Wang D, Gu W (2013) A study on the influence of road conditions on vehicle rollover. J Southeast Univ (in press) (in Chinese)
163. Sun L, Zhu H, Wang H, Gu W (2013) Preparation of nano-modified asphalt and its road performance evaluation. China J Highw Transp 26(1):1–8 (in Chinese)
164. Sun L, Zhu H, Zhu Y (2013) A two-stage viscoelastic-viscoplastic damage constitutive model of asphalt mixtures. J Mater Civ Eng ASCE.
165. Sussman NE (1974) Statistical ground excitation model for high speed vehicle dynamic analysis. High Speed Ground Transp J 8:145–154
166. Sweatman PF (1983) A study of dynamic wheel forces in axle group suspensions of heavy vehicles, Special Report No. 27, Australia Road Research Board
167. Van Deusen BD and McCarron GE (1967) A new technique for classifying random surface roughness: Trans. SAE 670032, Society of Automotive Engineers, New York
168. Wang SY, Sun L, Tang CA, Yang TH (2009) Numerical study of hydraulic fracture initiation and propagation around injection cavity in stiff soil. Constr Build Mater 23(6):2196–2206
169. Wang SY, Sun L, Zhu WC, Tang CA (2013) Numerical study on static and dynamic fracture evolution around rock cavities. J Rock Mech Geotech Eng 5(3):970–975
170. Wells VL, Han Y (1995) Acoustics of a moving source in a moving medium with application to propeller noise. J Sound Vib 184(4):651–663
171. Woodrooffe and Leblanc PA (1986) The influence of suspension variations on dynamic wheel loads of heavy vehicles. SAE Technical Paper No. 861973, Society Automotive Engineers
172. Yadav D, Nigam NC (1978) Ground induced non-stationary response of vehicles. J Sound Vib 106(2):217–225
173. You K, Sun L, Gu W (2010) Quantitative assessment of roadside safety on mountain highway. J Transp Eng Inf 8(3):49–55
174. You K, Wu J, Sun L, Zhang H (2011) Influence of road geometry on vehicle handling stability. J Highw Transp Res Dev 28(10):109–117 (in Chinese)
175. You K, Sun L, Gu W (2012) Risk analysis-based identification of road hazard locations using vehicle dynamic simulation. J Southeast Univ 42(1):150–155 (in Chinese)
176. You K, Sun L, Gu W (2012) Reliability-based risk analysis of roadway horizontal curve. J Transp Eng, ASCE 138(8):1071–1081.
177. You K, Sun L and Gu W (2013) Reliability design of highway horizontal curve based on vehicle stability analysis. J Traffic Transp Eng, (in press) Chinese)
178. Zhu WQ (1992) Random vibration. Academic Press, Beijing (in Chinese)
179. Zhu H, Sun L (2013) A viscoelastic-viscoplastic damage constitutive model for asphalt mixture based on thermodynamics. Int J Plast 40:81–100

180. Zhu HR and Sun L (2013) A mechanistic predictive model of rutting based on a two-stage serial viscoelastic-viscoplastic damage constitutive model for asphalt mixtures. J Eng Mech, ASCE, (in press)
181. Zhu Y, Wang Y, Sun L, You K (2009) Thermodynamic formulations of different coupling conditions between damage and plasticity. J Southeast Univ 39(5):39–65 (in Chinese)
182. Zhu H, Sun L, Yang J, Chen Z (2011) Developing master curves and predicting dynamic modulus of polymer modified asphalt mixtures. J Mater Civ Eng, ASCE 23(131):9–19
183. Zhu Y, Sun L, Xu HL (2011) L-curve based Tikhonov's regularization method for determining relaxation modulus from creep test. J Appl Mech, ASME 78(2):031002
184. Zhu Y, Sun L, Zhu H, Yu Q (2013) Time-dependent constitutive model of asphalt mixtures based on constant strain rate tests. J Wuhan Univ Technol (in Chinese) 36(5):922–926
185. Zhu H, Sun L and Zhu Y (2013) A viscoelastic-viscoplastic damage constitutive model based on thermodynamics theory for asphalt mixtures. Chin J Highw Eng, 26(1) (in Chinese)
186. Zhu Y, Sun L, Zhu H, Xiang W (2013) A constitutive model of viscoelastic-viscoplastic solids based on thermodynamics theory. Chin Quart J Mech, 31(4) (in Chinese)
187. Zu JWZ, Han RPS (1994) Dynamic rsponse of a spining Timoshenko beam with general boundary conditions and subjected to a moving load. J Appl Mech, ASME 61(1):152–160

17

Analysis of steering performance of differential coupling wheelset

**Xingwen Wu · Maoru Chi · Jing Zeng ·
Weihua Zhang · Minhao Zhu**

Abstract In order to improve the curving performance of the conventional wheelset in sharp curves and resolve the steering ability problem of the independently rotating wheel in large radius curves and tangent lines, a differential coupling wheelset (DCW) was developed in this work. The DCW was composed of two independently rotating wheels (IRWs) coupled by a clutch-type limited slip differential. The differential contains a static pre-stress clutch, which could lock both sides of IRWs of the DCW to ensure a good steering performance in curves with large radius and tangent track. In contrast, the clutch could unlock the two IRWs of the DCW in a sharp curve to endue it with the characteristic of an IRW, so that the vehicles can go through the tight curve smoothly. To study the dynamic performance of the DCW, a multi-body dynamic model of single bogie with DCWs was established. The self-centering capability, hunting stability, and self-steering performance on a curved track were analyzed and then compared with those of the conventional wheelset and IRW. Finally, the effect of coupling parameters of the DCW on the dynamic performance was investigated.

Keywords Differential coupling wheelset · Independently rotating wheel · Conventional wheelset · Steering performance

1 Introduction

With the development of urban railway transportation, the metro and lower floor light rail vehicles have been widely used in many cities. Whereas, compared with the main line railway vehicles, the urban railway vehicles meet more challenges because of the limitation of circumstance [1–3], which means that the urban railway vehicles may encounter a large number of curved tracks in daily operations, especially tight curves. Therefore, urban railway vehicles require a good steering capability to negotiate the curve with small radius. However, according to the previous operation experience of urban railway vehicles, the conventional wheelset cannot provide sufficient self-steering capability to negotiate the sharp curve, which may leads to severe wheel/rail wear and noise [2–4].

It is well known that the self-steering capability of conventional wheelset mainly depends on the longitudinal creep forces of wheel and rail [1–9]. When the wheelset deviates from the central position of track, the longitudinal creep forces are generated at the wheel/rail contact point due to the conical profile of the wheel tread. With the help of longitudinal creep forces and gravitational restoring forces, the wheelset has the ability to steer itself and return to the central position of track. Thus, the longitudinal creep forces make the conventional wheelset have the self-steering capability in the tangent track and curves [6–8]. To the author's knowledge, conventional wheelsets have enough steering capability in tangent lines and curves with large radii. However, they cannot provide enough steering capability to pass through sharp curves smoothly. The reason is that the difference of rolling radius at the contact point is insufficient to compensate the longer path the outer wheel needs; therefore, the outer wheel begins to skid and continuously contact with flange [5, 8]. In addition, the longitudinal creep forces are also the cause of hunting motion for the conventional wheelset. Once the forward speed of the vehicle exceeds the critical speed, the vehicle

X. Wu (✉) · M. Chi · J. Zeng · W. Zhang · M. Zhu
State Key Laboratory of Traction Power, Southwest Jiaotong University, Chengdu, China
e-mail: xingwen_wu@163.com

would experience the hunting motion, extremely threatening the running safety of vehicles.

In order to resolve the problems of conventional wheelset, many efforts have been made. For example, the semi-active and active actuation systems have been adopted to enhance the dynamic performance of railway vehicles [10, 11]. The independently rotating wheel (IRW) that decouples the wheelset is proposed to eliminate the hunting motion of the conventional wheelset and reduce the wheel/rail wear in sharp curves. The IRWs for railway vehicles have been investigated for many years. However, the use of IRW would also eliminate the guidance capability of the railway vehicles in large radius curves and tangent lines. Consequently, a compromise should be achieved between the curving performance in sharp curves and that in large radius curves and tangent lines by use of active controls like yaw control, creep control of damping, and stiff control. IRWs with profiled tread, with partial coupling, and with a superimposition gearbox have been proposed by Kaplan et al. [12], Dukkipati [13], and Jaschinski et al. [14], respectively. Gretashel and Bose [15] investigated the separate drive motors with precise torque control to provide guidance and curving capability. Goodall and coworkers [16–18] studied the active steering and optimized control strategy for IRWs.

This paper presents a differential coupling wheelset (DCW) to solve the problems of poor curving performance for the conventional wheelset in the sharp curve and bad steering capability for the IRW in the large radius curve and tangent line. In the DCW, both sides of IRWs are coupled by a clutch-type limited slip differential. In the tangent track, the clutch locks the differential, which does not permit a difference in rotation motion of the two wheels, and thus the DCW's dynamic behavior is similar to that of a traditional wheelset. In curves with small radius, the clutch will unlock the differential, and the DCW's dynamic performance is similar to that of a IRW; which can dramatically eliminate the sliding friction between wheel and rail, and reduce the wheel/rail wear and noise in sharp curves. Furthermore, due to the differential, the total rotation speed of two wheels keeps constant. Once the rotation speed of one wheel increases, another wheel decreases at the same time. This difference of rotation speed between two wheels generates a yaw motion for the DCW to negotiate the curves in the radial position to improve the curving performance of urban railway vehicles.

2 Differential coupling wheelset

To investigate the DCW's dynamic performance, two types of DCWs are discussed in this paper: one for a trailer bogie (Fig. 1) and another for a motor bogie (Fig. 2). It can be seen that the DCW consists of two wheels, a solid axle, a hollow axle, and a clutch-type limited slip differential. One wheel is mounted on the left side of a solid axle rigidly, and another wheel is connected to the right side of the solid axle through a bearing. Consequently, two wheels can rotate independently, which means that the DCW has the characteristics of IRWs. However, the guidance capability of an IRW only depends on the gravitational restoring force, which cannot provide enough steering capability. Thus, the clutch-type limited slip differential is used to couple the two IRWs to improve the steering capability of the bogie in large radius curves and tangent lines. The differential has two output gears: one is fixed on the solid axle, and another is connected to the IRW's web through a hollow axle. Since the differential is equipped with a clutch-type limited slip device, it applies a clutch torque to resist the relative motion between the output shafts.

In the multi-body dynamic model, the clutch-type limited slip device is modeled as a torque element combining a spring-damper element with a friction element as shown in Fig. 3. In Fig. 3, K and d, respectively, represent the coupling spring stiffness and coupling damping of the clutch-type limited slip device; $M_{\text{stick(max)}}$ and M_{slip} denote the maximum adhesion torque and the friction torque in the case of slipping.

The characteristics of the DCW can be described as follows:

$$\text{Differential coupling wheelset} = \begin{cases} \text{Traditional wheelset} & M_{\text{w}} < M_{\text{stick (max)}}, \\ \text{Independently rotating wheel} & M_{\text{w}} \geq M_{\text{stick (max)}}, \end{cases}$$

where M_{w} denotes the torque differences of two wheels. When M_{w} exceeds the $M_{\text{stick(max)}}$, the DCW expresses features of an IRW. In contrast, when M_{w} is less than $M_{\text{stick(max)}}$, the DCW has characteristics of a traditional wheelset.

In order to compare the steering performance of DCWs with other types of wheelsets, three types of single bogies, i.e., the bogies with the DCW, IRW, and conventional wheelset, are modeled in this paper, and their steering capabilities are compared in terms of wheel/rail lateral force, friction power, position of contact point on the wheel tread, and so on. In addition, the influence of clutch torque on wheelsets is analyzed.

3 Dynamic performance of bogies with DCWs

3.1 Dynamic model of bogies with DCWs

The trailer bogie and motor bogie with DCWs are modeled as shown in Fig. 4. The trailer bogie consists of two DCWs and a bogie frame (Fig. 4a), whereas the motor bogie is

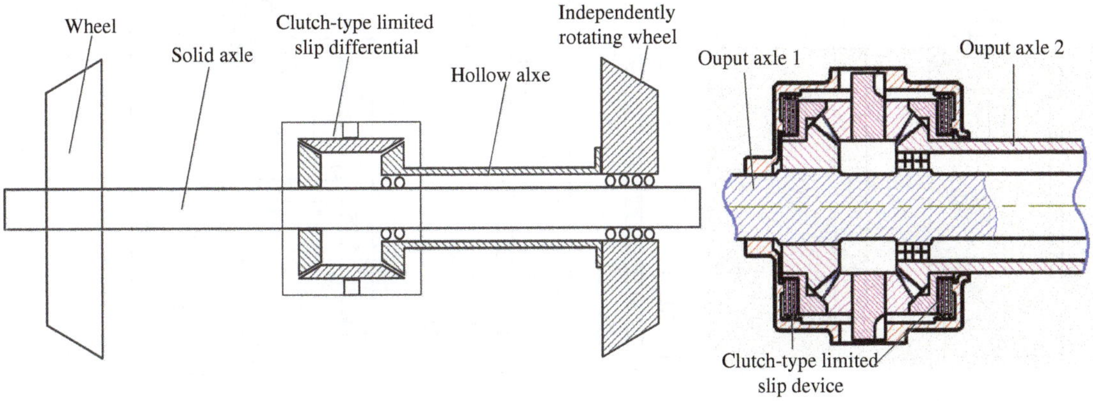

Fig. 1 DCW for trailer bogie

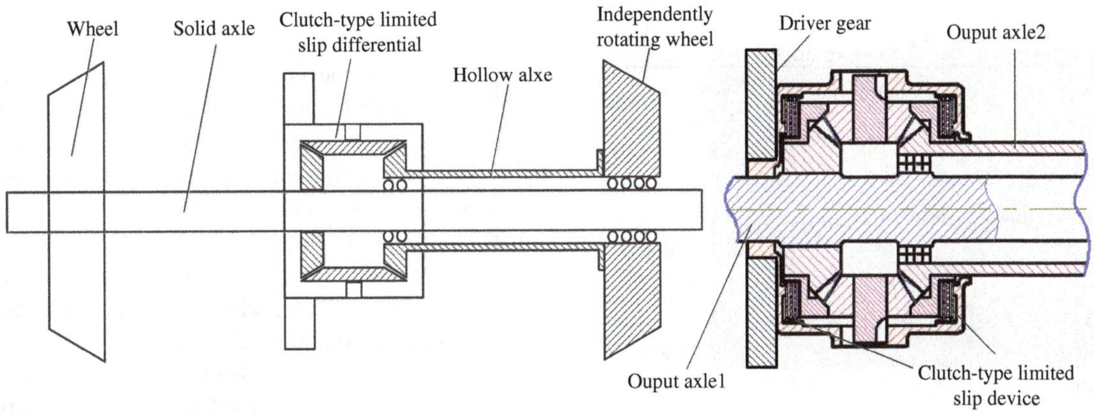

Fig. 2 DCW for motor bogie

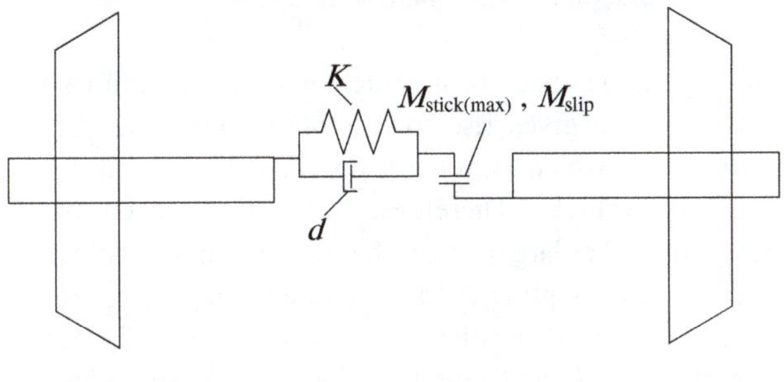

Fig. 3 Torque element of the DCW

composed of two DCWs, two motors, and a bogie frame (Fig. 4b). The DCWs and bogie frame are connected through primary suspensions. We built the dynamic models of the bogies using SIMPACK software. The motors are rigidly fixed on the bogie frame, which has only a pitch motion with respect to the bogie frame. The traction torque is transmitted from motors to the DCW. The gear constraint element is adopted to represent the meshing relationship between the differential and motor. The differential is modeled as a constraint element provided by SIMPACK. The clutch-type limited slip device is represented by a stick–slip rotational torque element. The FSATSIM algorithm is used for the calculation of wheel/rail contact forces. The parameters used in the dynamic models are listed in Table 1, and the degrees of freedom of bogies are shown in Table 2. Figure 5 indicates the wheel/rail contact point and conicity of S1002 wheel tread and 60 rail used in this work.

3.2 Self-centering capability of bogies with DCWs on the tangent line

Self-centering capability is a critical dynamic performance for the wheelset, which indicates the ability of returning to the central position of the track. Figure 6 illustrates a comparative analysis of the lateral displacement for five cases with an initial lateral displacement at the speed of 20 km/h on the tangent line. According to the results, the lateral displacement of the conventional wheelset and the DCW with limited slip device gradually converge to the central position of track. In contrast, the IRW and the DCW without the limited slip device travel to one side of rail from the beginning, and cannot return to the center of track,

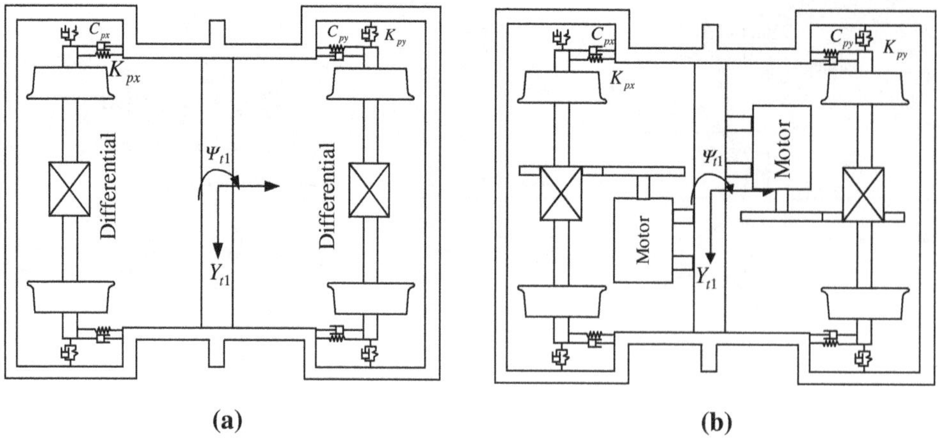

Fig. 4 Bogies with DCWs: **a** Trailer bogie; **b** Motor bogie

Table 1 Parameters used in the model

Bogie mass	3,200 kg
Wheelset mass	1,200 kg
Lateral and longitudinal stiffness of primary suspension	4 MN/m
Vertical stiffness of primary suspension	0.8 MN/m
Radius of wheel	0.325 m
Rail gage	1.435 m
Coefficient of friction	0.4
Coupling stiffness	60 kNm/rad
Coupling damping	60 kNms/rad
Max adhesion torque	500 Nm

which causes continuous flange contact, and severe wheel/rail wear and noise. The comparison analysis results indicate that the limited slip device plays a vital role in the dynamic performance of the DCW. The DCW could express the features of the IRW without the limited slip device. On the contrary, with the help of limited slip device, DCW could have a good self-centering capability of the conventional wheelset.

In order to acquire enough steering capability, the clutch-type limited slip device is applied into the differential for coupling two wheels. Figure 7 indicates the influence of coupling stiffness and damping on the lateral displacement of the DCW. As the coupling stiffness K and damping d increase, the lateral displacement of wheelset gradually converges to the central position of the track. This reflects that the increased coupling stiffness and damping is good for the improvement of steering performance. However, if the coupling stiffness and damping do not match reasonably, the DCW may show a "hunting motion." This motion is not a definite hunting motion but just a quasi-hunting motion, which is mainly induced by the self-excited oscillation of coupling stiffness and damping. Therefore, it is necessary to optimize the coupling parameters to ensure a good guidance capability of the DCW.

3.3 Stability analysis of the bogie with DCW

Once the operation speed of a vehicle exceeds the critical speed, the vehicle gives rise to a hunting motion in the lateral direction, which extremely threatens the operation safety of the vehicle. Therefore, the critical speed of vehicles should be larger than the maximum operation speed. Since low coupling stiffness and coupling damping cause the self-excited oscillation as shown in Fig. 7, the coupling stiffness K and coupling damping d are set to 100 kNm/rad and 100 kNms/rad, respectively, for stability analysis of the bogie. Figure 8 illustrates the bifurcation diagram of the bogie with DCW. It can be seen that the

Table 2 Degrees of freedom

Vehicle model	Type of motion					
	Longitudinal	Lateral	Vertical	Roll	Yaw	Pitch
Bogie frame	V	V	V	V	V	V
Differential coupling wheelset	V	V	V	V	V	V
Axle box	–	–	–	–	–	V
Motor	–	–	–	–	–	V

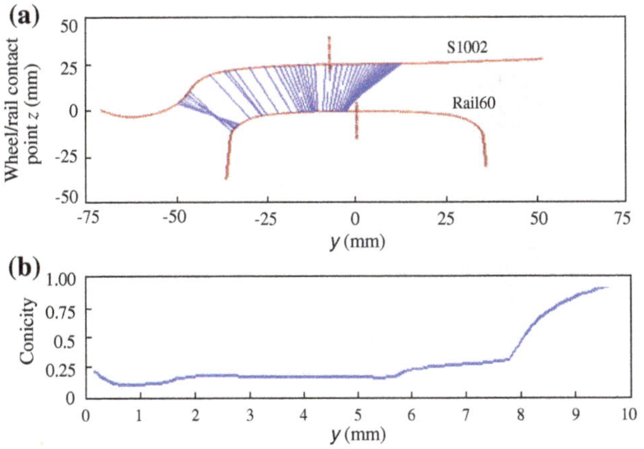

Fig. 5 Wheel/rail contact point (**a**); conicity (**b**) of S1002 and Rail 60

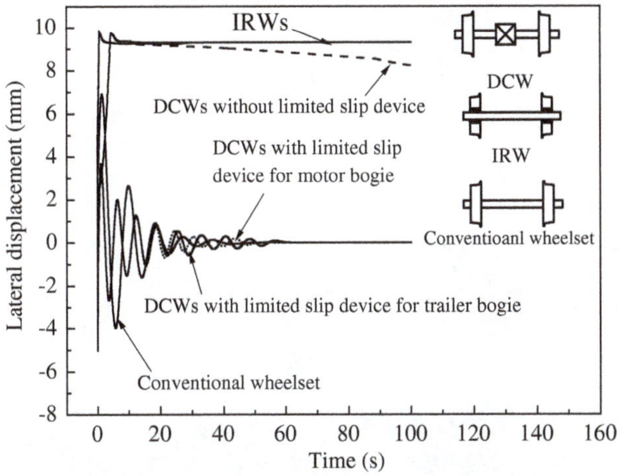

Fig. 6 Lateral displacement of wheelset

type of bifurcation is a typical supercritical Hopf bifurcation. In Fig. 8, point A represents the linear critical speed of bogie, and $V_A = 115$ km/h; point B denotes the nonlinear critical speed of bogie, and $V_B = 85$ km/h; the dash line indicates the unstable limited cycle; and the solid line indicates the stable limited cycle. When the vehicle speed V is less than V_B, the motion of the vehicle is always stable. When the vehicle speed is between V_B and V_A, the motion of the vehicle largely depends on the initial conditions. Figure 9 indicates the influence of coupling stiffness and coupling damping on the critical speed of the bogie with DCW. With increasing the coupling damping, the critical speed of the bogie increases sharply when the coupling damping is less than 50 kNms/rad. However, when the coupling damping exceeds 50 kNms/rad, the critical speed tends to be stable. In addition, the coupling stiffness has little influence on the critical speed.

3.4 Self-steering ability of the trailer bogie with DCW

To analyze the self-steering ability of the DCW, the curving performances of three types of bogies are compared in terms of wheel/rail lateral force, friction power, and position of contact point on the wheel tread. Figure 10 indicates the layout of curved track. The parameters of simulation track are listed in Table 3.

Generally, the bogie is guided in the curve section primarily by the lateral forces on the front wheelset. Thereby the lateral forces on the front wheelsets of the three types of bogies are analyzed, as shown in Fig. 11. It can be seen that the lateral forces on the outer IRW are smaller than the other two types of wheelsets. The reason is that the bogie with the conventional wheelset or DCW cannot adjust radially to full extent while the IRW can adapt better to the radial position of the curved track. Compared with the conventional wheelset, the DCW is much easier to negotiate the curve in radius position with the help of clutch-

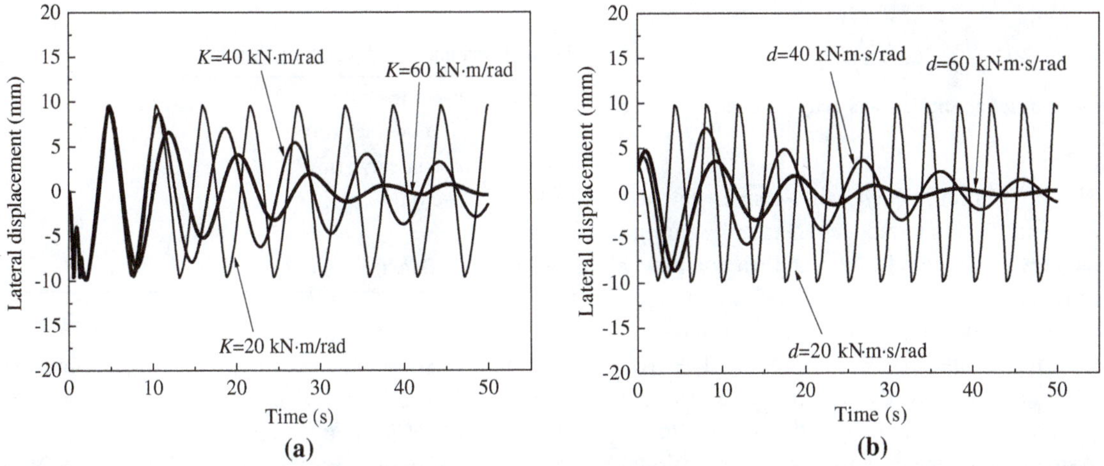

Fig. 7 Influence of coupling parameters on the lateral displacement of the DCW for different coupling stiffness *K* (**a**); different coupling damping *d* (**b**)

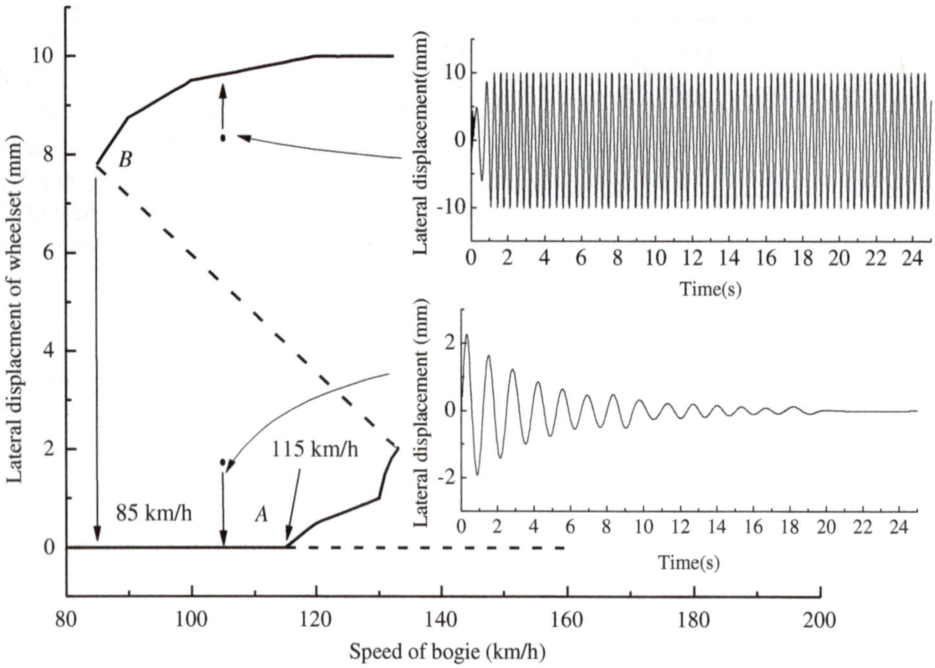

Fig. 8 Bifurcation diagram of the bogie with DCW

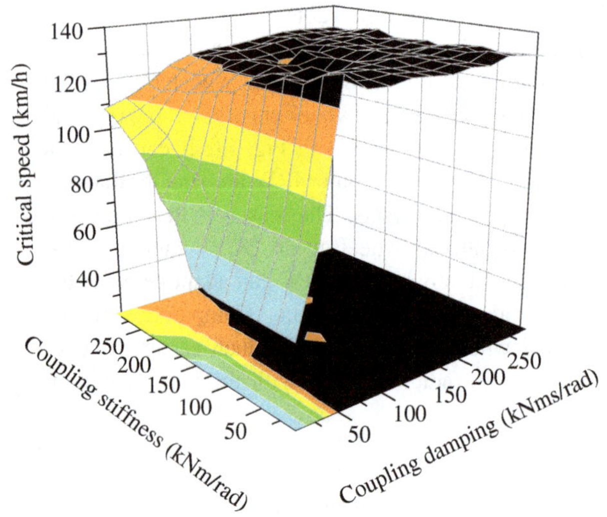

Fig. 9 Influences of coupling stiffness and coupling damping on the critical speed

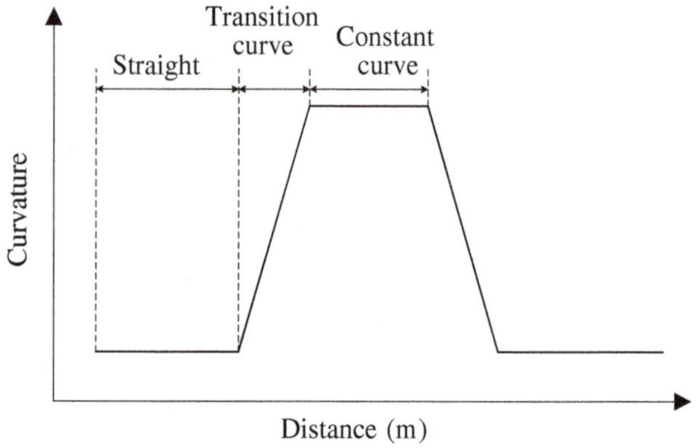

Fig. 10 Layout of simulation track

Table 3 Parameters of simulation track

Length of tangent track (m)	150
Length of transition track (m)	20
Length of constant curve (m)	50
Radius of curve (m)	30
Cant (m)	0
Running speed (km/h)	20

type limited slip differential, which could convert the slip friction to the rolling friction to reduce wheel/rail wear and noise, and generates small lateral forces and friction power to the solid wheelsets.

In addition, the frictional power as a wear index is investigated, and the result is shown in Fig. 12. The frictional power is calculated by the creep forces and the corresponding creep velocities within the local contact coordinate system. Compared with the conventional wheelset, the DCW has a better wear index because of its IRW characteristics.

Figure 13 shows the position of contact points on the wheel tread. The lateral displacement of contact points on the wheel of DCW is apparently smaller than that on the traditional wheelset. Furthermore, after the DCW goes through the curve section, the wheelset gradually returns to the central position of track. However, the IRW goes to one

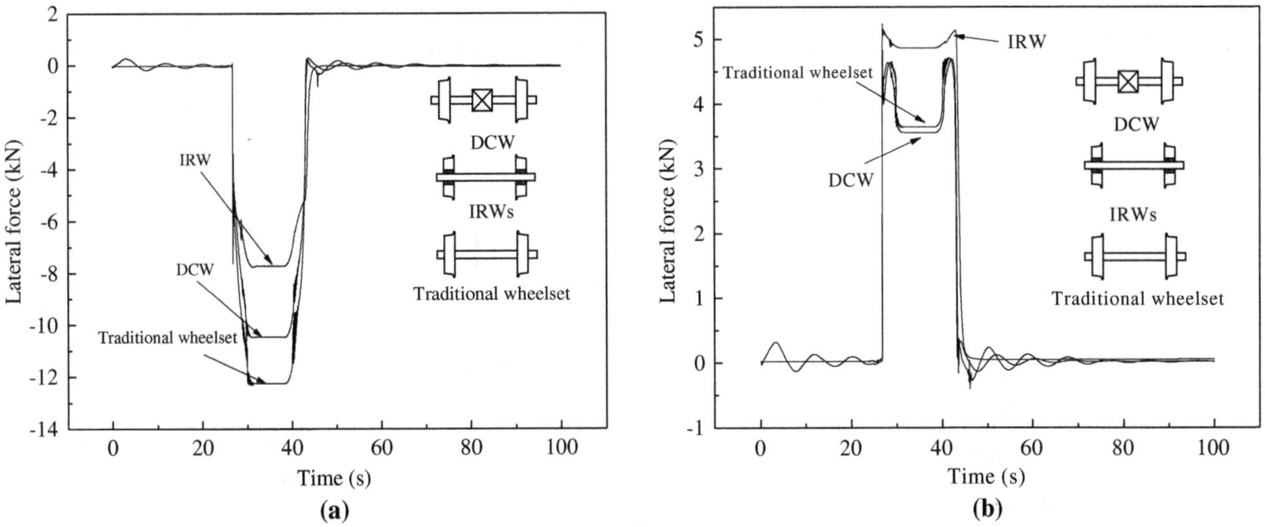

Fig. 11 Wheel/rail lateral forces: **a** Outer wheel; **b** Inner wheel

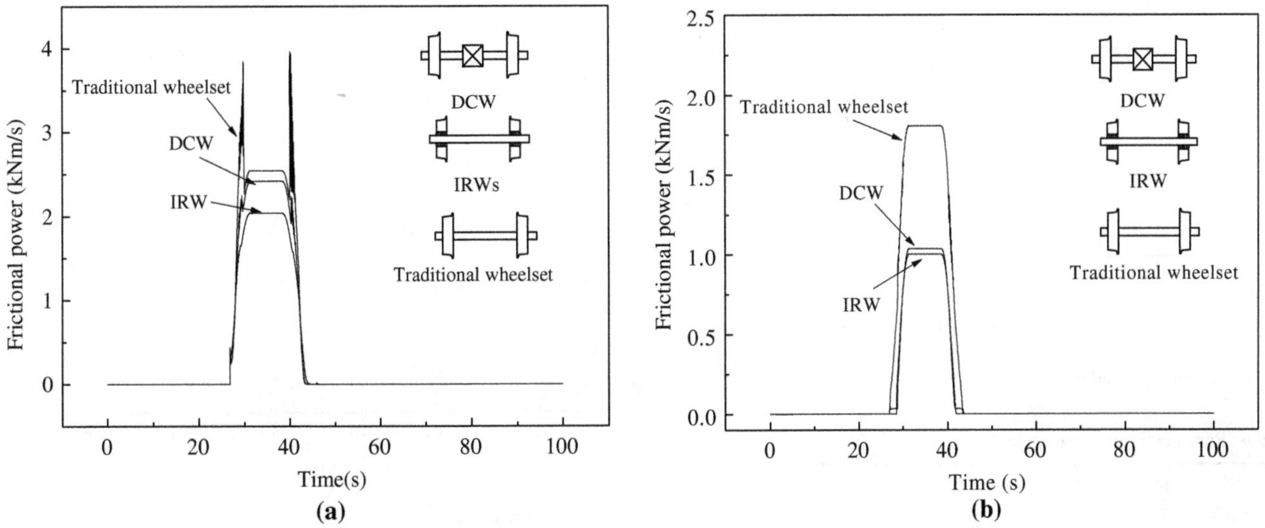

Fig. 12 Friction power: **a** Front wheelset; **b** Rear wheelset

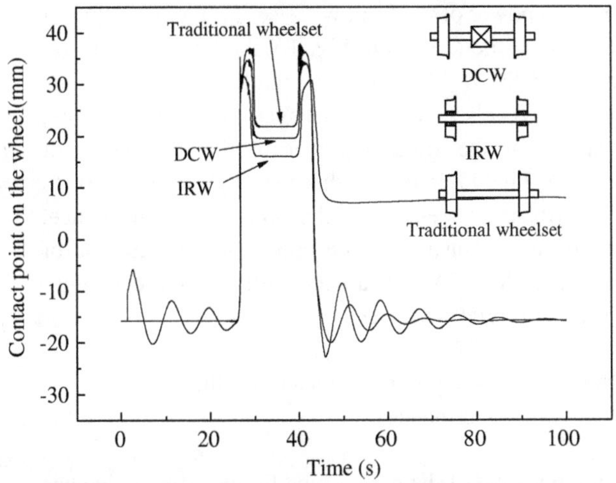

Fig. 13 Lateral displacement of contact points

side of rail and cannot return to the central position of track, resulting in eccentric wear of wheel and rail.

Figure 14 illustrates the rotation speed difference that occurs in the curve section due to the differential. As the rotation speed of the outer wheel increases, the inner wheel decreases. This endues the DCW with good self-steering performance and curving performance. When the wheelset gets out from the curve section, the clutch-type limited slip device locks the wheels at both sides so that the two wheels have the same rotation speed. In contrast, the IRW cannot return to the center of track, which makes the speeds of two wheels different.

From the above comparison, we can come to a conclusion that the DCW has better curving performance than the conventional wheelset. Due to the torque of the clutch-type limited slip device, the DCW can also express better self-

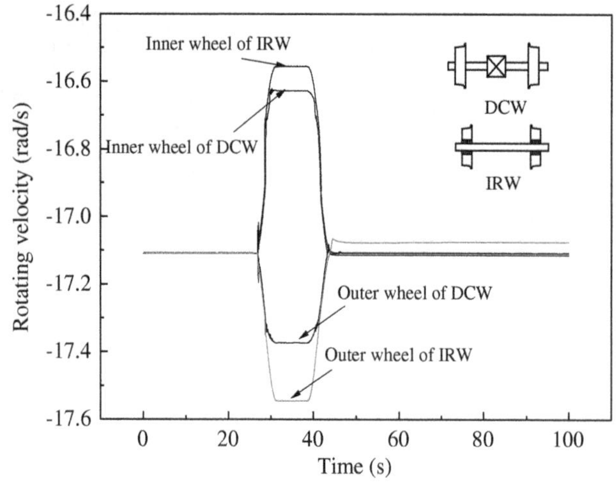

Fig. 14 Rotation speed of differential wheelsets

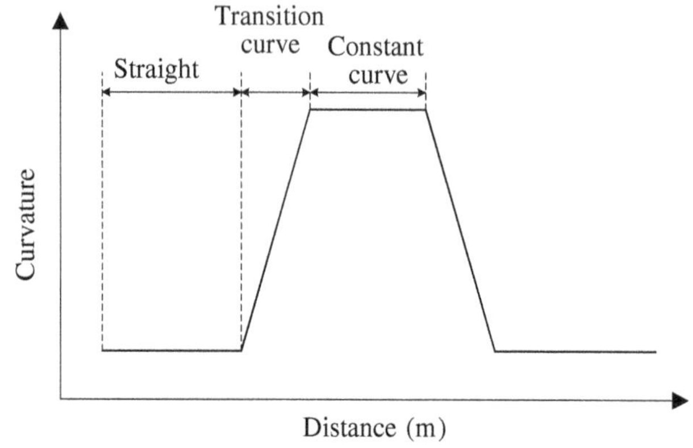

Fig. 15 Layout of simulation track

Table 4 Parameters of simulation track

Length of tangent track (m)	100
Length of transition track (m)	20
Length of constant curve (m)	50
Radius of curve (m)	50
Cant (m)	0

steering performance than the IRW. Therefore, the DCW processes the good curving performance of an IRW and the self-steering capability of the conventional wheelset.

3.5 Self-steering ability of motor bogie with the DCW

When the DCW is applied to a motor bogie, the differential is used to transmit the traction torque. It also allows both the wheels to rotate at different speeds, which differentiates it from the conventional wheelset. In the following, single motor bogies with DCW and traditional wheelset are analyzed and compared when the bogie goes through a curved track at a constant speed with the action of traction motor. The curved track is shown in Fig. 15, and the parameters are listed in Table 4.

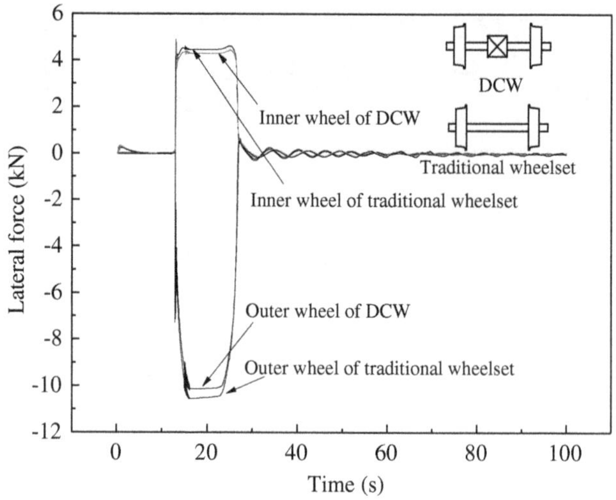

Fig. 16 Wheel/rail lateral force

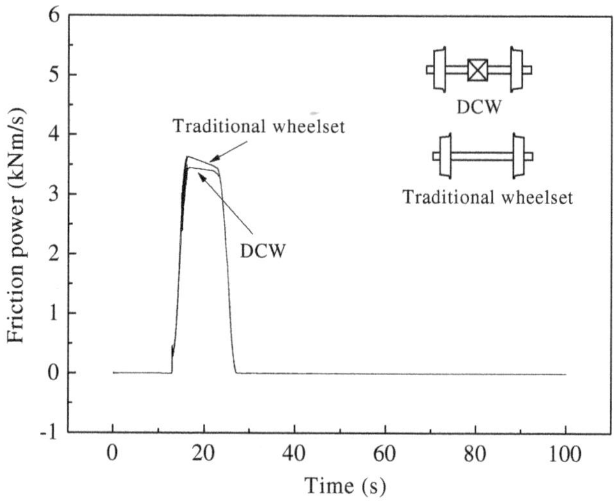

Fig. 17 Friction power

Figures 16 and 17 indicate the wheel/rail lateral force and friction power of the front wheelset for the two types of bogies. As can be seen from Fig. 16, the wheel/rail lateral force of the DCW is apparently smaller than that of the traditional wheelset in the curve section. Furthermore, comparison of the friction power of the two kinds of wheelset in Fig. 17 indicates that the DCW is superior to the traditional wheelset in the curving performance. Therefore, a conclusion can be drawn that in the case of motor bogie, the DCW has a better self-steering capability than the traditional wheelset.

3.6 Influence of coupling parameters on the DCW's dynamic performance

The clutch torque of the clutch-type limited slip device has a critical effect on the dynamic behavior of the DCW, and

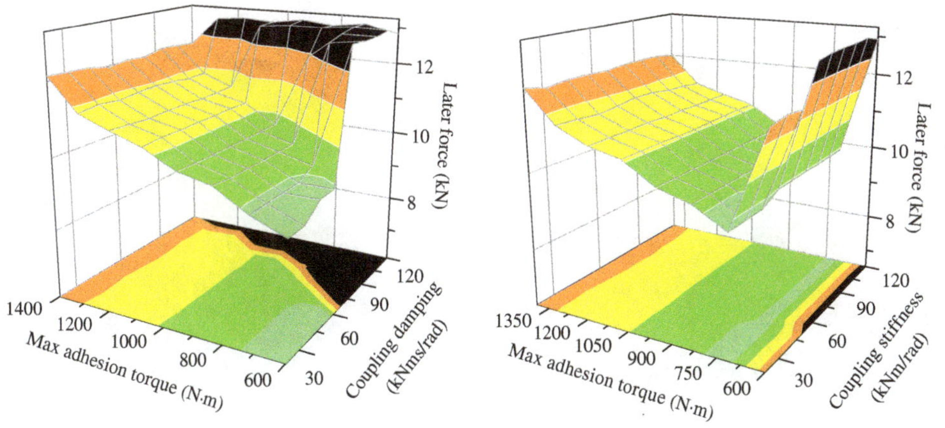

(a) (b)

Fig. 18 Influence of coupling parameters on the lateral force of DCW: **a** Maximum adhesion torque $M_{\text{stick(max)}}$ versus coupling damping with coupling stiffness $K = 60$ kNm/rad; **b** Maximum adhesion torque $M_{\text{stick(max)}}$ versus coupling stiffness with coupling damping $d = 60$ kNms/rad

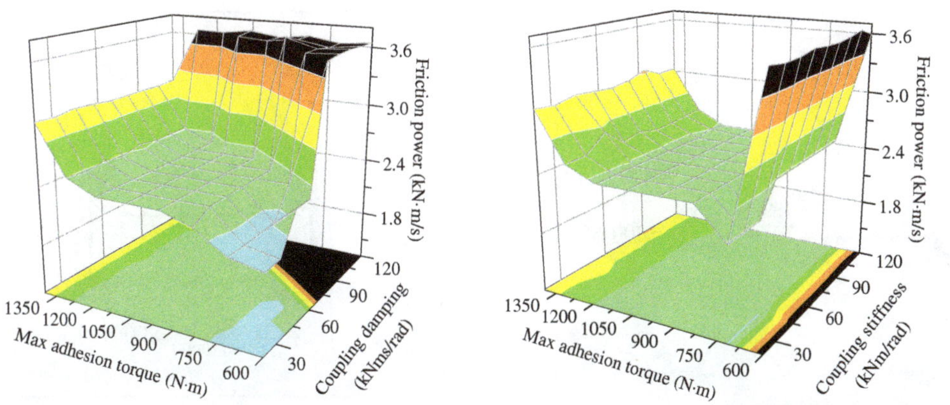

(a) (b)

Fig. 19 Influence of coupling parameters on the friction power of DCW: **a** Maximum adhesion torque $M_{\text{stick(max)}}$ versus coupling damping with coupling stiffness $K = 60$ kNm/rad; **b** Maximum adhesion torque $M_{\text{stick(max)}}$ versus coupling stiffness with coupling damping $d = 60$ kNms/rad

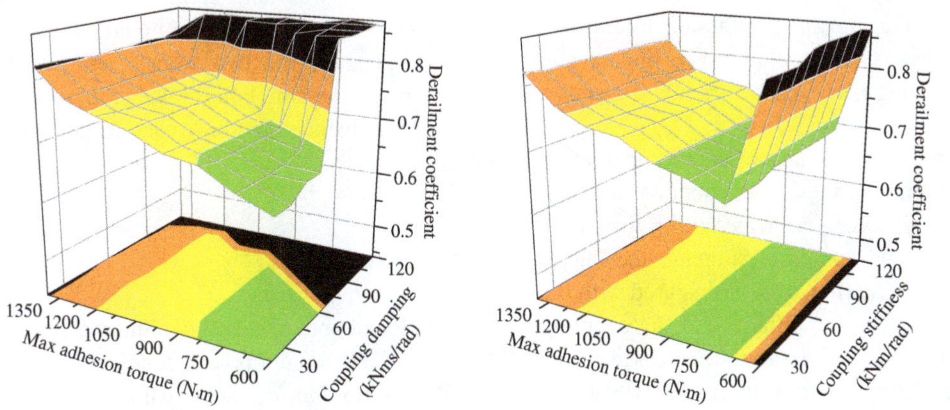

(a) (b)

Fig. 20 Influence of coupling parameter on the derailment coefficient of DCW: **a** Maximum adhesion torque $M_{\text{stick(max)}}$ versus coupling damping with coupling stiffness $K = 60$ kNm/rad; **b** Maximum adhesion torque $M_{\text{stick(max)}}$ versus coupling stiffness with coupling damping $d = 60$ kNms/rad

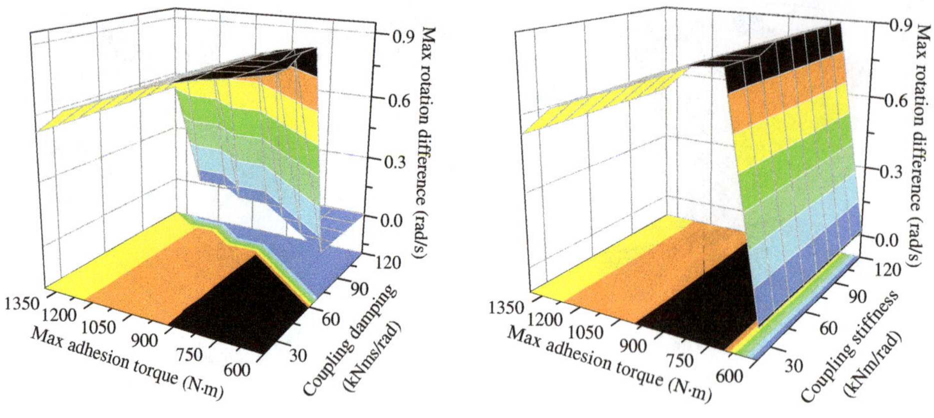

Fig. 21 Influence of coupling parameter on the maximum rotation difference of DCW: **a** Maximum adhesion torque $M_{stick(max)}$ versus Coupling damping with coupling stiffness $K = 60$ kNm/rad; **b** Maximum adhesion torque $M_{stick(max)}$ versus Coupling stiffness with coupling damping $d = 60$ kNms/rad

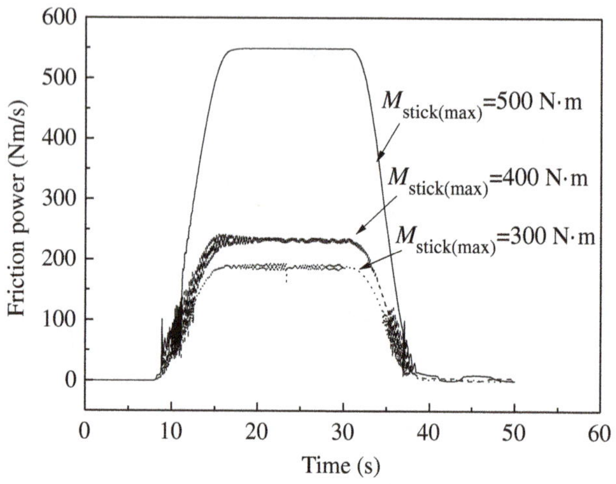

Fig. 22 Friction power

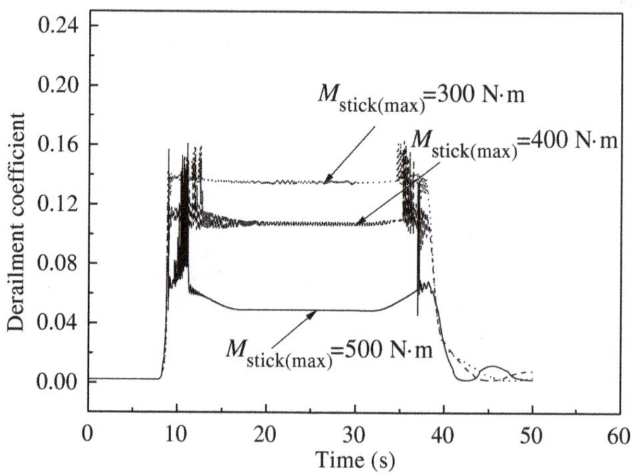

Fig. 23 Derailment coefficient

determines the work conditions of the differential. Therefore, the influence of the maximum adhesion torque and other coupling parameters of the clutch-type limited slip differential on the curving performance are investigated in this section.

Figure 18 illustrates the influence of the maximum adhesion torque, coupling stiffness, and coupling damping on the lateral wheel/rail force. With increasing the maximum adhesion torque and coupling damping, the lateral forces of wheel/rail increase (Fig. 18a). Compared with the coupling damping, however, the influence of the coupling stiffness is smaller (Fig. 18b). Due to the increased maximum adhesion torque, the torque difference between two wheels is more difficult to exceed the maximum adhesion torque, which causes that both wheels cannot rotate independently, and thus express more features of the conventional wheelset. As shown in Figs. 19 and 20, with increasing the maximum adhesion and coupling damping, the friction power and derailment coefficient increase. Meanwhile, the coupling damping dramatically reduces the relative speed of the two wheels, as shown in Fig. 21. According to the simulation results, the coupling stiffness has little influence on the dynamic performance of the DCW.

4 Discussions

As mentioned above, small maximum adhesion torque and small coupling damping are beneficial to improving the DCW's curving performance, which endue the DCW with properties of IRWs. However, too small maximum adhesion torque and coupling damping could deteriorate the DCW's self-steering performance in large radius curves and tangent lines. Generally, the maximum adhesion torque determines the work conditions of the DCW, and it depends on the wheel/rail adhesion conditions affected by many factors [19–22], such as normal load, sliding speed, temperature of the two bodies, contact geometry, weather conditions, and the presence of rain, snow, and dead leaves. On the other hand, with the reduction of the maximum adhesion torque, the friction power decreases (Fig. 22) and the derailment coefficient increases (Fig. 23). Therefore, a compromise should be achieved between running safety and wheel/rail wear.

5 Conclusions and future work

According to the simulation results, the DCW integrates both the features of the IRW and the conventional wheelset. In tight curves, the DCW can express the features of IRWs to achieve an improvement in the curving performance over the conventional wheelset. In tangent lines and large radius curves, the DCW has a self-steering capability as the conventional wheelset.

The study of coupling parameters shows that the maximum adhesion torque and coupling damping have a large influence on the dynamic behavior of DCW. With the increasing of the maximum adhesion torque and the coupling damping, the DCW tends to be a conventional wheelset. The maximum adhesion torque of the clutch-type limited slip device depends on the wheel/rail adhesion conditions.

However, in this paper we have only discussed the dynamic performance of single bogies, through which the maximum adhesion torque could not be determined and hence we cannot investigate how to control the maximum adhesion torque to adapt to different track conditions. Therefore, in the future research, the creep control will be studied to determine the maximum adhesion torque of clutch-type limited slip device with the full railway vehicle.

Acknowledgments This work was supported by the National Key Technology R&D Program of China (No. 2009BAG12A02), the National Basic Research Program of China (No. 2011CB711106), the Program for Innovative Research Team in University (No. IRT1178), the Program for New Century Excellent Talents in University (No. NCET-10-0664), and the National Key Technology R&D Program (No. 2009BAG12A01).

References

1. Kuba T, Lugner P (2012) Dynamic behaviour of tramways with different kinds of bogies. Veh Syst Dyn 50(S1):277–289
2. Shen G, Zhou J, Ren L (2006) Enhancing the resistance to derailment and side-wear for a tramway vehicle with independently rotating wheels. Veh Syst Dyn 44(S1):641–651
3. Garg VK, Dukkipati RV (1984) Dynamics of railway vehicle system. Academic Press, Toronto, pp 145–165
4. Dukkipati RV (1992) Independently rotating wheel system for railway vehicles: a state of the art review. Veh Syst Dyn 21(1):297–330
5. Ahmed AKW, Sankar S (1987) Lateral stability behavior of railway freight car system with elasto-damper coupled wheelset (part 2): truck model. Transm Autom Des 109(12):500–507
6. Ahmed AKW, Sankar S (1988) Steady-state curving performance of railway freight truck with damper-coupled wheelsets. Veh Syst Dyn 17(6):295–315
7. Chi M, Zhang W, Wang K, Zhang J (2003) Research on dynamic stability of the vehicle with coupled wheelsets. J Tongji Univ 31(4):464–468
8. Chi M, Wang K, Fu M, Ni W, Zhang W (2002) Analysis on wheel-rail lateral force of the bogie with independently rotating wheels for rear wheelsets. J Traffic Transp Eng 2(2):184–192
9. Satou E, Miyamoto M (1992) Dynamic of a bogie with independently rotating wheels. Veh Syst Dyn 20(1):519–534
10. Allotta B, Pugi L, Bartolini F, Cangioli F, Colla V (2010) Comparison of different control approaches aiming at enhancing the comfort of a railway vehicle. In: 2010 IEEE/ASME international conference on advanced intelligent mechatronics (AIM), Montreal, pp 676–681
11. Allotta B, Pugi L, Colla V, Bartolini F, Cangioli F (2011) Design and optimization of a semi-active suspension system for railway applications. J Mod Transp 19(4):223–232
12. Kaplan A, Hasselman TK, Short SA (1970) Independently rotating wheels for high speed trains. SAE Paper 700841
13. Dukkipati RV (1978) Dynamics of independently rotating wheelsets: a survey of the state of the art. Tech. Rep. LTR-IN-398, NRC Railway Laboratory
14. Jaschinski A, Chollet H, Iwnicki S, Wickens A, Von Würzen J (1999) The application of roller rigs to railway vehicle dynamics. Veh Syst Dyn 31(5–6):345–392
15. Gretzschel M, Bose L (2002) A new concept for integrated guidance and drive of railway running gears. In: Proceedings of the 1st IFAC conference on mechatronic systems, vol 1, Darmstalt, pp. 265–270
16. Goodall R, Mei TX (2005) Mechatronic strategies for controlling railway wheelsets with independently rotating wheels. In: Proceedings of the IEEE/ASME international conference on advanced intelligent mechatronicsn (AIM'05), vol 1, Como, Italy, pp. 225–230
17. Mei TX, Goodall RM (2001) Robust control for independently rotating wheelsets on a railway vehicle using practical sensors. IEEE Trans Control Syst Technol 9(4):599–607
18. Mei TX, Goodall RM (2003) Practical strategies for controlling railway wheelsets independently rotating wheels. J Dyn Syst Meas Control Trans ASME 125(3):354–360
19. Powell AJ, Wickens AH (1996) Active guidance of railway vehicles using traction motor torque control. Veh Syst Dyn 25(S1):573–584
20. Wickens AH (2009) Comparative stability of bogie vehicles with passive and active guidance as influenced by friction and traction. Veh Syst Dyn 47(9):1137–1146
21. Conti R, Meli E, Pugi L, Malvezzi M, Bartolini F, Allotta B, Rindi A, Toni P (2012) A numerical model of a HIL scaled roller rig for simulation of wheel–rail degraded adhesion condition. Veh Syst Dyn 50(5):775–804
22. Malvezzi M, Pugi L, Papini S, Rindi A, Toni P (2013) Identification of a wheel–rail adhesion coefficient from experimental data during braking tests. Proc Inst Mech Eng F 227(2):128–139

18

A comparative study on crash-influencing factors by facility types on urban expressway

Yong Wu · Hideki Nakamura · Miho Asano

Abstract This study aims at identifying crash-influencing factors by facility type of Nagoya Urban Expressway, considering the interaction of geometry, traffic flow, and ambient conditions. Crash rate (CR) model is firstly developed separately at four facility types: basic, merge, and diverge segments and sharp curve. Traffic flows are thereby categorized, and based on the traffic categories, the significances of factors affecting crashes are analyzed by principal component analysis. The results reveal that, the CR at merge segment is significantly higher than those at basic and diverge segments in uncongested flow, while the value is not significantly different at the three facility types in congested flow. In both un- and congested flows, sharp curve has the worst safety performance in view of its highest CR. Regarding influencing factors, geometric design and traffic flow are most significant in un- and congested flows, respectively. As mainline flow increases, the effect of merging ratio affecting crash is on the rise at basic and merge segments as opposed to the decreasing significance of diverging ratio at diverge segment. Meanwhile, longer acceleration and deceleration lanes are adverse to safety in uncongested flow, while shorter acceleration and deceleration lanes are adverse in congested flow. Due to its special geometric design, crashes at sharp curve are highly associated with the large centrifugal force and heavy restricted visibility.

Keywords Crash-influencing factors · Crash rates · Principal component analysis · Facility types · Urban expressway

Y. Wu (✉) · H. Nakamura · M. Asano
Department of Civil Engineering, Nagoya University,
C1-2(651) Furo-cho, Chikusa-ku, Nagoya 464-8603, Japan
e-mail: gerry_woo@126.com

1 Introduction

Improving traffic safety is a worldwide issue to be relieved urgently. Crash characteristics and their influencing factors, as the theoretical basis for safety improvement, may provide direction for policies and countermeasures aimed at smoothing hazardous conditions. For a better understanding of crash-influencing factors, researchers have continually sought ways through an extensive array of approaches, and the most prominent one is crash data analysis [1]. The conventional approaches have established statistical links between crash rate (CR) and its explanatory factors [2, 3]. In the analyses, traffic flows are generally represented by low-resolution data that is collected at a highly aggregated level, e.g., hourly or daily flows. Geometric features are primarily considered the hierarchy of radius or slope [4, 5]. Meanwhile, several studies have suggested that crashes are associated with the interaction of geometry, traffic flow, and ambient conditions [6]. However, most existing studies investigated the factors individually and the related CR models were developed based on single factor only. As a result, it is inadequate to identify the nature of individuals through aggregated analysis only, since the conditions preceding individual crashes are virtually different from each other [1].

Considering the insufficiency of CR analysis above, some studies have tried to identify crash characteristics at individual level, in an effort to predict crash risk on a real time basis [7–9]. Through these studies, the effect of traffic flow on crash risk has been well analyzed. In theory, the concept of real-time crash prediction exhibits huge promise for the application of proactive traffic management strategies for safety.

However, the combined effects of geometry, traffic flow, and ambient conditions on crashes still have not been well anayzed through the above studies. Furthermore, these papers primarily developed crash model for the whole traffic conditions, which may conflict with the fact that the influence of traffic flow on crashes may vary when traffic conditions change. In addition, even if crash characteristics are found out to be dependent on facility type that is composed of uniform segment individually, e.g., basic, merge, and diverge segments [2], the existing studies are focused on the entire route of intercity expressway without segmentation.

Another cause for the limited predictive performance of existing models is the inadequacy of analytic process [10]. As for statistical methods, the significance and independence of explanatory variables should be identified in advance for the reliability of statistics. Whereas, many previous studies paid little attention to this point and incorporated the potential influencing factors into crash modeling directly.

Urban expressway is one common type of separated highway with full control of access in large cities in Japan. Generally, it is composed of various facility types where geometric features and traffic characteristics are often different from each other. Correspondingly, crash characteristics and their influencing factors may also be different by facility type. In the meantime, compared to intercity expressway, crash characteristics and their related influencing factors of urban expressway are different [11]. Necessarily, urban expressway deserves to be analyzed independently and its crash characteristics should be identified based on specific facility type.

Given the problems of existing studies, the objective of this paper is to investigate crash characteristics based on CR models and their influencing factors by facility type on urban expressway. Meanwhile, the causes are identified by considering the interaction of geometry, traffic flow, and ambient conditions. Besides, geometric features are identified considering the driver-vehicle-roadway interaction. The significances of these factors affecting crashes are compared at different facility types using principal component analysis (PCA). Their influencing mechanisms are further discussed. In essence, this study can be regarded as a proactive analysis for crash risk prediction model in the future.

2 Study sites and datasets

2.1 Study sites

The test bed of this study is Nagoya Urban Expressway network (NEX) as shown in Fig. 1. Up to December 31, 2009, this network was about 69.2 km × 2 (two directions) in total length with over 250 ultrasonic detectors

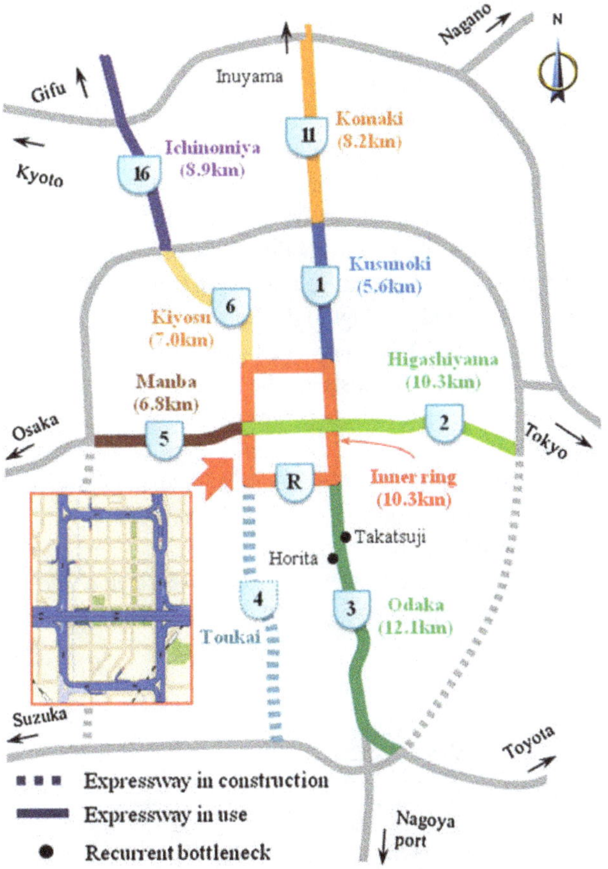

Fig. 1 Schematic map of NEX network (2009) (*Source* Nagoya Expressway Public Corp., modified by authors)

installed with an average spacing of 500 m (varied in 250–750 m) on mainline. Most routes are 4-lane roadways (2-lane/dir), except the inner ring (Route no. R) that is one-way roadway and where the number of lanes differs (2–5) with the change of ramp junctions. In the limited areas, such as the links of other routes to the inner ring, small curves are designed. In this network, two recurrent bottlenecks are located along Odaka line (Route no. 3).

Five databases are used in this study; (1) crash records with the occurrence time in minutes, the location in 0.1 km and the weather and pavement conditions; (2) detector data including traffic volume q, average speed v, and occupancy occ per 5 min; (3) geometric design and the location of detector in 0.01 km; (4) traffic regulation records for incidents (e.g., crash, working, and inclement weather) including the locations and periods of temporal lane and cross-section closures; and, (5) daily sunrise and sunset time records in Nagoya. Here, it is worth noting that detector data are processed for the whole cross section of each direction. The period of the data above is for 3 years (2007–2009) except for those on Kiyosu line (Route no. 6) that was opened from December 1, 2007.

2.2 Segmentation of facility types

Basic segment is extracted outside the 500 m up- and down-stream of ramp junctions considering the experience in Japan [12]. Correspondingly, merge or diverge segments are regarded as the sections inside the 500 m up- and down-stream of on- and off-ramps, respectively. The segmentation methods are shown in Fig. 2. Other than these segments, there is a special geometric design in NEX, curve with small radius. Figure 3 explains CR statistics dependent on radius. Obviously, compared to other segments, much higher CR exists in the curves with radius smaller than 100 m. Thus, these curves are defined as sharp curves and regarded as another distinct facility type of NEX. Given the limitation of segment samples available, basic, merge, and diverge segments, and sharp curve will be analyzed in this study.

The cross sections of inner ring are diverse and the length of individual layouts, i.e., 2-, 3-, 4-, or 5-lane, is not enough to be separately analyzed. Meanwhile, all of the sharp curves along Inner ring are 2-lane roadway. In this regard, only 2-lane segments are analyzed in this study and the geometric statistics by facility types are summarized in Table 1.

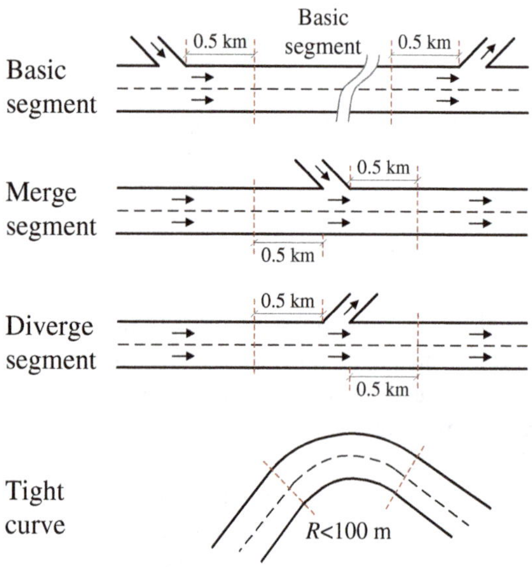

Fig. 2 Segmentation of facility types

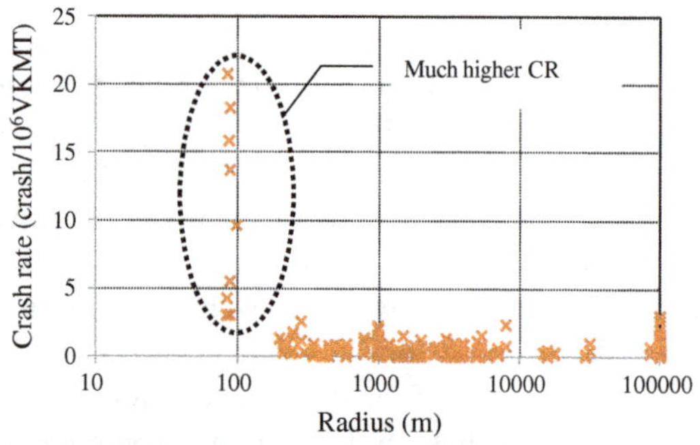

Fig. 3 Distribution of crash rate to radius

Table 1 Geometric statistics of facility types

Facility type	No. of segments	Total length (km)
Basic segment	38	56.6
Merge segment	28	20.9
Diverge segment	35	25.2
Sharp curve	9	2.5

3 Methodology

3.1 Data extraction

3.1.1 Detector data

In principle, detectors can count the number of vehicles at their locations only. In such case, the "coverage area" of detector is defined for estimating traffic conditions at crash locations through detector data. At basic segment, the boundary of two consecutive coverage areas is defined at the midpoint between two neighboring detectors. At merge and diverge segments, it is bounded at the ramp-junction point, and one segment can be divided into up- and down-stream areas. Each sharp curve can be matched with a single detector. Note that the time of crash is recorded by road administrators after the crash occurrence. In reality, it does not correspond to occurrence time exactly. For this reason, data within small time before crashes should be rejected to avoid mixing up crash-influencing and crash-influenced data. Therefore, the latest data at least 5 min before the recorded time are accepted after the exclusion of invalid data and the data within lane and section-closure intervals in advance.

3.1.2 Geometric features

Design consistency is the conformance of geometry of a highway with driver expectancy, and its importance and significant contribution to road safety is justified by understanding the driver–vehicle–roadway interaction [13] that may vary at individual locations in nature. In this regard, geometric variation in the upstream of crash location is proposed to reflect the effect of geometry on crashes. Considering the length of detector coverage area, the following variables in 500 m distance are extracted [12].

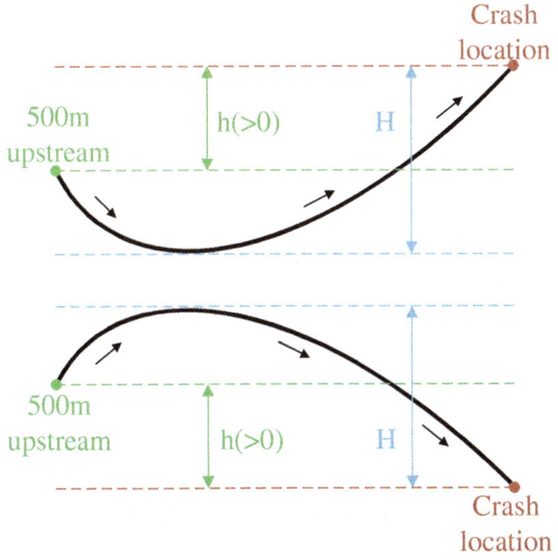

Fig. 4 Variation in road elevation

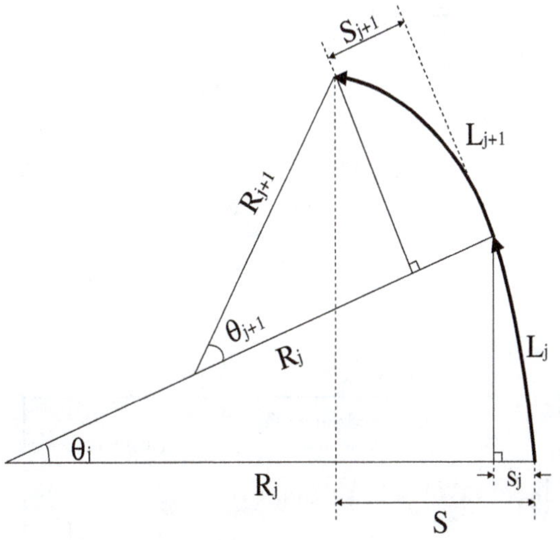

Fig. 5 Horizontal displacement

(1) Variation in road elevation h between the crash location and its 500 m upstream, and the maximum elevation difference H in this 500 m distance (Fig. 4).

(2) Horizontal displacement S. Radius is impossible to describe a section composed of various curves. On the other hand, centrifugal force is also associated with the horizontal displacement s in the direction of tangent to the curve j (Fig. 5). In such case, S in the 500 m distance (Σs_j) is adopted and calculated by the following equations.

$$\theta_j = \frac{L_j}{R_j} \quad (0 > ; \theta_j \leq \pi/2), \tag{1}$$

$$s_j = R_j(1 - \cos\theta_j), \tag{2}$$

where j is the ID of curve. R_j, θ_j, L_j, and s_j correspond to the radius, central angle, arc length, and horizontal displacement of curve j, respectively.

(3) Index of centrifugal force I_{CF}. Speed v always has a square relation with centrifugal force. This study designs I_{CF} ($I_{CF} = Sv^2$) to reflect the combined effect of speed v and horizontal displacement S, while it is not centrifugal force.

(4) Index of space displacement I_{SD}. I_{SD} ($I_{SD} = SH$) is used to reveal the comprehensive geometric features induced by horizontal and vertical variation in this study.

The geometric data above are collected every 0.1 km as crash is recorded in a unit of 0.1 km. Besides, these data are also extracted at the location of detector that is the common link between crash and detector data. Table 2 summarizes the process of data collection.

3.1.3 Ambient conditions

Common, prevailing, and uncontrolled environment and weather conditions are defined as ambient conditions. They are (1) ambient light classified into daytime and nighttime, which are the period from sunrise to sunset and from sunset to sunrise, respectively; (2) sunny, cloudy, and rainy weather conditions at the time of crash; (3) dry and wet pavement conditions at the location of crash; and, (4) day type on crash days including holiday and weekday. Here, holiday includes all weekends, and all national and traditional holidays like the Golden Week in May and the Obon Week in August in Japan.

3.1.4 Data matching

The related detector data, geometric features, and ambient conditions for individual crashes are matched as exemplified in Table 3. The crashes matching with invalid detector data and within lane and cross-section closure intervals are excluded in advance. As a result, a total of 1,591 crashes remain for the following analysis.

Table 2 Example of geometric variations collection

Route no.	Direction*	Kilopost**	h (m)	H (m)	S (m)	I_{SD} (m^2)
1	SB	0.0	−4.63	5.49	0.78	4.30
1	SB	0.1	−7.90	8.49	3.91	33.2
1	SB	0.2	−10.6	11.5	6.08	69.9
1	SB	0.21	−11.5	11.8	8.88	104.7
1	SB	0.3	−15.3	15.3	9.60	146.9
1	SB	–	–	–	–	–
1	SB	6.4	10.2	10.9	5.15	56.1

*SB South-bound

** 0.21: the Kilopost of detector #0101

Table 3 Examples of data matching for individual crashes

Crash ID	Traffic characteristics				Geometric features							Ambient conditions			
	q (veh/5 min)	v (km/h)	MR	DR	Facility type[a]	h (m)	H (m)	I_{CF} (km^3/h^2)	I_{SD} (m^2)	L_A^b (m)	L_D^b (m)	Light	Weather	Pavement	Day type
1	139	88.7	–	–	B	1.5	1.5	0	0	–	–	Day	Sunny	Dry	Holiday
2	96	95.0	–	0.02	D	0.5	1.3	302	45	–	220	Day	Sunny	Dry	Weekday
3	29	80.3	–	–	S	10.3	12.7	126	249	–	–	Night	Cloudy	Wet	Holiday
4	154	82.9	0.07	–	M	1.5	1.5	0	0	200	–	Day	Cloudy	Dry	Weekday

B, M, D, and S basic, merge, and diverge segments, and sharp curve, respectively, L_A length of acceleration lane at merge segment, L_D length of deceleration lane at diverge segment

3.2 Classification of traffic conditions

Congested flow, characterized by traffic oscillation, has different features from uncongested flow. It is necessary to make a distinction between two traffic regimes. Figure 6 shows the traffic volume–speed diagram at Horita on-ramp junction, one typical bottleneck in NEX. The speed of 60 km/h, corresponding to maximal flow is defined as the critical speed v_c that is used for classifying un- and congested flows [2, 14]. Besides, the corresponding value at another bottleneck (Takatsuji on-ramp junction) is also found out around 60 km/h.

The value of 60 km/h would be regarded as the related index at basic and diverge segments, since no bottleneck can be virtually found at both segments in NEX. At sharp curve, a threshold speed of 45 km/h is selected in general for classifying two traffic regimes based on traffic flow-speed diagram at Tsurumai curve (Fig. 7). The value is further checked at other sharp curves, and it is found out to be reliable for classifying un- and congested flows basically.

To reflect the variation in traffic characteristics, each traffic regime is further sub-classified. It is evident that speed has a high variance at low flow rates (see Figs. 6 and 7). Besides, occupancy is not a commonly used index. Thus, traffic density k calculated by Eq. (3) is proposed to be the measure of effectiveness to further classify the traffic conditions. In view of the number of crash samples available, the aggregation intervals of k are set as 10 and 30 veh/km for un- and congested flows, respectively.

$$k_{ei} = \frac{12 \times q_i}{v_i}, \tag{3}$$

where q_i and v_i denote traffic flow and average speed in 5 min # i, respectively. k_{ei} corresponds to the calculated traffic density in this 5 min.

3.3 Calculation of crash rate (CR)

CR for traffic condition n can be calculated by the following equation:

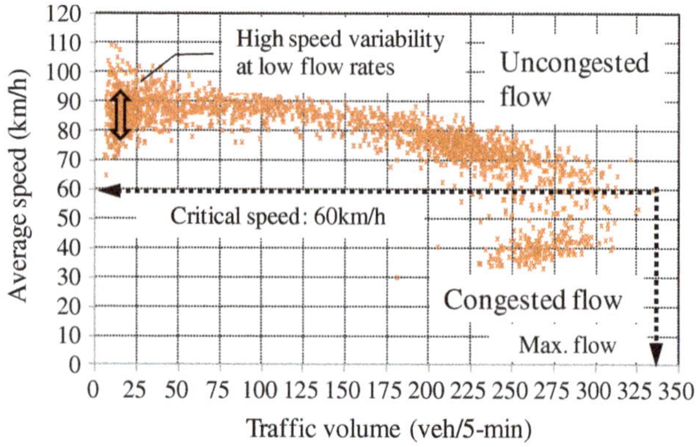

Fig. 6 q–v diagram at Horita on-ramp junction

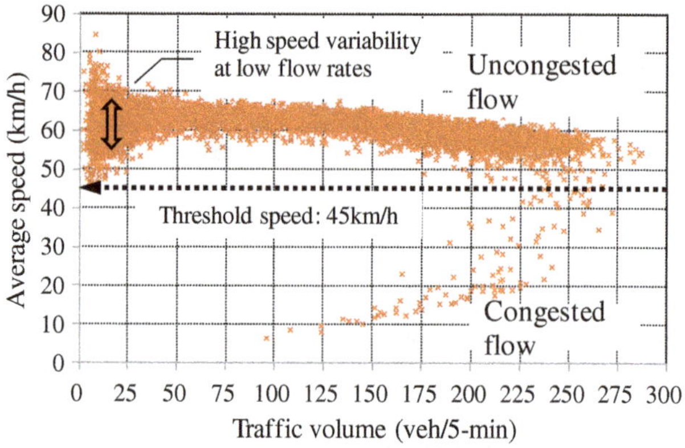

Fig. 7 q–v diagram in Tsurumai curve

$$CR_n = \frac{NOC_n \times 10^6}{\sum Q_{nl} L_l}, \tag{4}$$

where n and l are the ID of traffic condition and coverage area, respectively; NOC_n is the number of crashes for traffic condition n. $Q_{nl}L_l$ is the value of vehicle kilometers traveled (VKMT) in detector coverage area l for traffic condition n.

3.4 Principal component analysis (PCA)

PCA is a powerful tool for reducing a large number of observed variables into a small number of artificial variables that account for most of the variance in the dataset [15]. In general, through orthogonal transformation, a set of observations of possibly correlated variables can be converted into a set of values of linearly uncorrelated variables. Those converted values are defined as principal components. Technically, a principal component can be regarded to be a linear combination of optimally weighted observed variables [15]. As a result, the components are ranked in the order of accounting amount of total variance in the observed variables. Then, two criteria are generally available to select the number of component extracted: (1) 80 % rule, the extracted components should be capable to explain at least 80 % of the variance in the original dataset. (2) Eigen value rule, only components whose eigen values are over 1.0 can be retained.

4 Crash rate estimation models

In the following, the differences of crash characteristics by facility type are investigated by comparing CR models based on traffic conditions.

4.1 Uncongested flow

Figure 8 gives the CR tendency following traffic density k by facility type in uncongested flow. It is evident that sharp curve has a special characteristic compared to other segments. Its CR is the highest among four facility types at low-density stages. Then, the value follows a decreasing tendency to k. In contrast, the CR at other segments increases as k increases. Such phenomenon may be related to the design of small radius for sharp curve. Such geometric design can result in high centrifugal force that can act on the vehicle and tries to push it to the outside of the curve. Furthermore, higher speed may result in higher centrifugal force.

Regarding the differences at other segments, CR at merge segment increases rapidly at high-density stages and gets much higher compared to basic and diverge segments. The results of paired t-test at the three facility types in Table 4 also reveal that CR at basic/diverge segments is significantly lower than that at merge segment, while they are not significantly different from each other between basic and diverge segments. At merge segment, merging maneuvers can result in slow-down and lane-changing behaviors for mainline traffic. These interruptions may increase the possibility of vehicle conflicts. Such possibility can further increase with an increase in k.

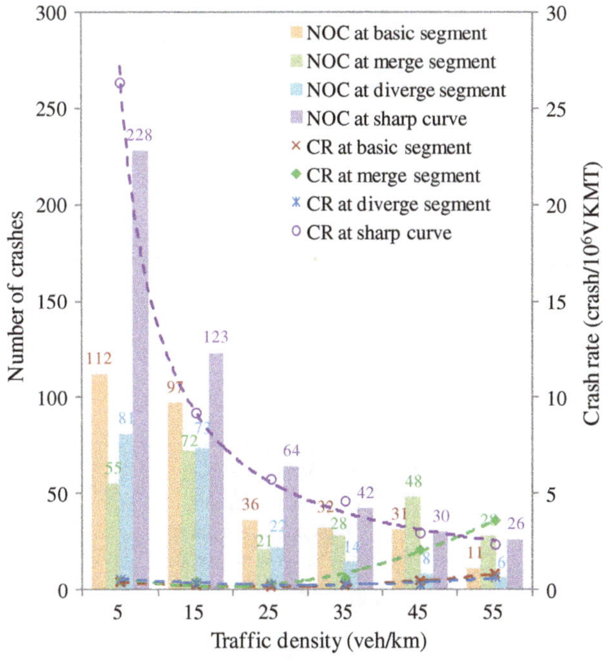

Fig. 8 CR comparison in uncongested flow

Table 4 *T*-test of CR in uncongested flow

Paired	t-Value	df	Sig.
Pair 1: basic and merge segments	−2.781	5	0.019
Pair 2: basic and diverge segments	−1.070	5	0.310
Pair 3: merge and diverge segments	2.320	5	0.043

Table 5 summarizes the CR regression models as function of k as well as the goodness-of-fit of models at four facility types. At sharp curve, the model is power function while they are quadratic functions at other facility types. All of the models and variables are significant at 95 % confidence level (not shown in Table 5). Regarding quadratic models, CR at merge segment is most sensitive to the increase in k, more than three times of CR increases as that at basic and diverge segments by the increase in one unit of k.

4.2 Congested flow

Figure 9 describes the differences of CR distribution to k by facility type in congested flow. It appears that CR follows increasing tendencies to k at four facility types. In contrast to other segments, sharp curve still has the highest CR in congested flow while no statistical regression model is developed at this facility type due to the limited crash samples. Since the differences of CR tendency at other segments are not clear in Fig. 9, a paired t-test is conducted as shown in Table 6. The results indicate that there is no

Table 5 CR regression models in uncongested flow

Facility type	Sample size	Model[a]
B	319 crashes	$CR = 6.81 \times 10^{-4}k^2 - 3.23 \times 10^{-2}k + 0.541$, $R^2 = 0.998$, $k(CR_{min}) = 24$
M	25 1 crashes	$CR = 2.55 \times 10^{-3}k^2 - 9.29 \times 10^{-2}k + 1.01$, $R^2 = 0.983$, $k(CR_{min}) = 20$
D	204 crashes	$CR = 5.68 \times 10^{-4}k^2 - 3.34 \times 10^{-2}k + 0.747$ $R^2 = 0.849$ $k(CR_{min}) = 28$
S	513 crashes	$CR = 132.2k^{-0.983}$, $R^2 = 0.992$

[a] $k\ (CR_{minimal})$: traffic density corresponding to the minimal CR

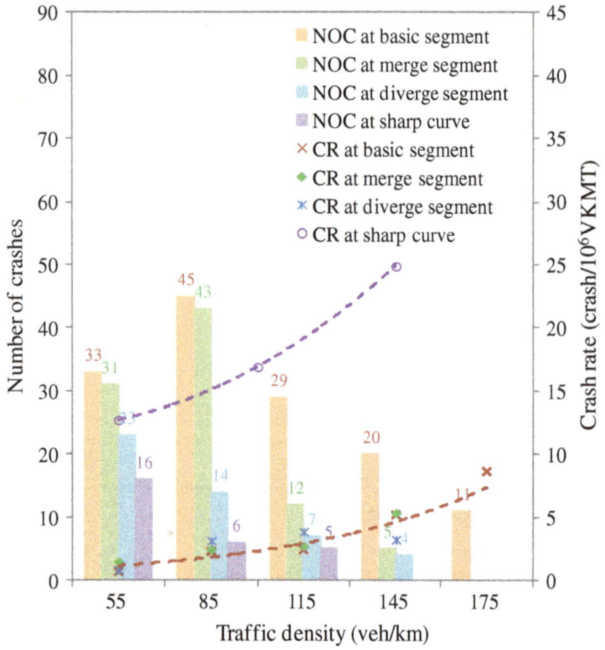

Fig. 9 CR comparison in congested flow

Table 6 *T*-test of CR in congested flow

Paired	*t*-Value	df	Sig.
Pair 1: basic and merge segments	−2.448	4	0.092
Pair 2: basic and diverge segments	−0.153	4	0.888
Pair 3: merge and diverge segments	0.325	4	0.767

significant difference of CR at basic, merge, and diverge segments. Such finding may imply that the effect of facility type on crashes is reduced in congested flow. For this reason, CR model is developed by combining the three facility types in order to increase the number of crash samples for reliability. As demonstrated in Table 7, an exponential function is adopted and it fits well to the combined CR tendency. The model and its variables are also significant at 95 % confidence level, while the results are not shown in Table 7.

Table 7 CR regression model in congested flow

Facility type	Sample size	Model
B + M + D	513 crashes	$CR = 4.87 \times 10^{-1} e^{0.0155k}$ $R^2 = 0.924$

5 Effects of influencing factors

The analyses above reveal that CR characteristics are different by facility type, which may be related to the different geometric designs and traffic characteristics. However, CR analysis is insufficient to examine a variety of factors by a single model. Instead, PCA is applied and the affecting mechanisms of individual factors are further investigated.

5.1 Introduction of variables

Table 8 explains individual variables combining with its type and some summary statistics. In nature, traffic flow diagram is two-dimensional, and k and v are used together to describe traffic conditions. As for geometric features, h, I_{CF}, and I_{SD} are picked out to reflect the vertical, horizontal, and comprehensive geometric variations, respectively. Dummy variables are referred to incorporate ambient conditions into PCA. A dummy variable usually takes 0 and 1. In this case, weather conditions (over 2 categories) are replaced by pavement conditions (only 2 categories), since two conditions are usually highly related to each other.

At merge and diverge segments, ramp traffic is a significant influencing factor on crashes [16]. This study employs ramp flow ratio to illustrate the interaction between ramp and mainline traffic. Merging ratio (MR) or diverging ratio (DR) is defined as the proportion of on- or off-ramp traffic out of the sum of ramp and mainline traffic, respectively. Meanwhile, the length of acceleration lane L_A or the length of deceleration lane L_D is adopted to reveal the space available provided for merging or diverging maneuvers, respectively.

Table 8 Introduction of individual variables

Variables	Statistics[a]		Description
	Max.	Min.	
k	238	0	Traffic density (veh/km)
v	141.0	4.7	Average speed (km/h)
MR	0.78	0.00	Merging ratio
DR	0.60	0.00	Diverging ratio
I_{CF}	1997	0.1	Index of centrifugal force (km^3/h^2)
h	14.8	0.1	Vertical variation (m)
I_{SD}	1112.7	0.0	Index of space displacement (m^2)
L_A	250	100	Length of acceleration lane (m)
L_D	432	100	Length of deceleration lane (m)
Pave	$f(1) = 27.7$ %		Equal to 1 if wet pavement, 0 otherwise
Light	$f(1) = 27.1$ %		Equal to 1 if nighttime, 0 otherwise
Day	$f(1) = 29.4$ %		Equal to 1 if holiday, 0 otherwise

f frequency

[a] Max./Min.: maximal and minimal values, respectively

5.2 PCA among various facility types

In essence, PCA rotates data by using a linear transformation. Consequently, only the monotonic loadings of factors can be reflected by this approach. For this reason, uncongested flow is further classified into low-and high-density conditions at approximately 25 veh/km in view of the value of k (CR_{min}) as shown in Fig. 10, since there are different monotonicities of CR model in two conditions. As a result, three traffic conditions are analyzed, i.e., low- and high-density uncongested flow as well as congested flow.

5.2.1 Low-density uncongested flow

Table 9 demonstrates PCA results at basic segment in low-density uncongested flow. In terms of the rules introduced in Sect. 3.4, four components are selected and all of the factors can explain at least 80 % of variance in the original dataset in terms of the value of cumulative percent.

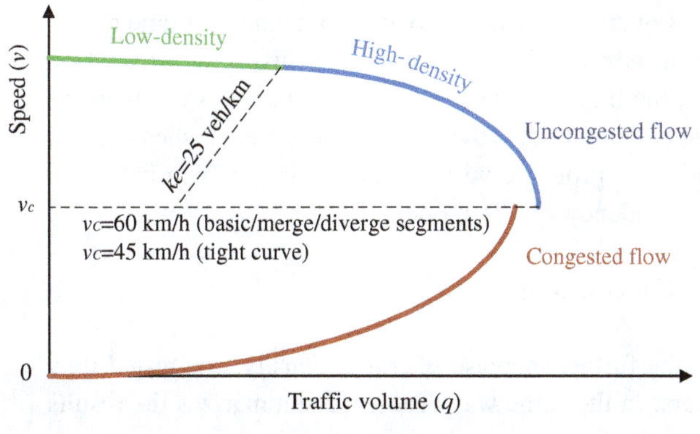

Fig. 10 Classification of traffic conditions

Table 9 PCA results at basic segments

Variables	Component			
	1st	2nd	3rd	4th
k_e	−0.194	**−0.852**	−0.119	0.103
v	0.285	0.182	**0.798**	−0.086
I_{CF}	**0.953**	0.005	0.053	−0.122
I_{SD}	**0.959**	0.011	−0.044	0.090
h	0.119	0.149	0.228	**0.973**
Pave	0.294	0.214	**0.783**	−0.095
Day	0.139	0.093	0.190	−0.468
Light	−0.164	**0.838**	−0.130	0.143
Initial Eigenvalue	2.12	1.54	1.37	1.12
Percent of variance	30.3	20.2	18.2	15.1
Cumulative Percent	30.3	50.5	68.7	83.8

The variables highly related to each component are in bold

In low-density uncongested flow, crashes at basic segment are found to be significantly associated with geometric variation (I_{CF} and I_{SD}), traffic density along with ambient light, speed coupled with pavement, and vertical variation h. Geometric variation is the 1st component, as great variation may result in frequent speed reduction. Accordingly, the difficulty for drivers to control vehicle behaviors increases. At low traffic density k, driver's attention is not high, and some discretionary behaviors may be operated. Such condition combining with the poor ambient light is possible to increase crash risk. Meanwhile, due to the reduced value of tire-pavement friction, high speed v combining with wet pavement can reduce the roadability. In such cases, k and v are two separate components, which can further demonstrate that both variables are not highly interrelated at low flow rate. In addition, vertical variation h has a positive loading because of the increased visibility restriction and the difficulty in maintaining vehicle behaviors for drivers.

Principal components at other segments are analyzed as shown in Table 10. The variables that are significantly related to each component are selected based on their loadings. For judging the relative significance of the same component by facility type, the percent of variance explained by each component is provided as well.

One difference at merge segment from basic segment is that MR combining with the length of acceleration lane L_A becomes a principal component. Meanwhile, day type is found to be significant. In terms of the percent of variance accounted by components, the significance of geometric variation gets lower in contrast to basic segment. Merging traffic is an important influencing factor, since it can induce interruption to mainline traffic. Such interruption may get stronger as MR increases. Besides, higher L_A can provide more space for ramp and mainline traffic to adjust for

Table 10 PCA results in low-density uncongested flow

Facility type (number of crash)	Item	Principal component									
		1st		2nd		3rd		4th		5th	
		F	L	F	L	F	L	F	L	F	L
Basic segment (225)	Component	I_{CF}	0.953	k	−0.852	v	0.798	h	0.973		
		I_{SD}	0.959	Light	0.838	Pave	0.783				
	Percent of variance	30.3		20.2		18.2		15.1			
Merge segment (140)	Component	I_{CF}	0.867	k	−0.841	MR	0.808	v	0.721	Day	0.874
		I_{SD}	0.901	Light	0.869	L_A	0.747	Pave	0.742		
	Percent of variance	20.1		18.8		15.4		13.9		11.9	
Diverge segment (167)	Component	DR	0.807	I_{CF}	0.948	k	−0.807	v	0.892	L_D	0.797
		h	0.845	I_{SD}	0.746	Light	0.860	Pave	0.848		
	Percent of variance	18.9		17.6		16.2		15.3		13.3	
Sharp curve (319)	Component	I_{SD}	0.854	I_{CF}	0.948	k	−0.820	Pave	0.929	Day	0.981
		h	0.978	v	0.753	Light	0.794				
	Percent of variance	23.7		17.5		16.4		13.6		12.7	

F factor, *L* loading

merging behaviors. Regarding the influence of day type on crashes, it may be related to the different vehicle compositions and driver populations between holiday and weekday, while such influence needs a further study to investigate vehicle behaviors at merge segment. As for geometric variation, on ramps in NEX are virtually allocated far from poor alignment like small curve. Thus, it is considered reliable that the significance of geometric variation affecting crashes is lower at merger segment compared to basic segment.

At diverge segment, the most significant difference from basic and merge segments is that the DR and the vertical variation h are related to the 1st component. Higher DR can significantly interrupt mainline traffic since it is necessary to pass through several lanes to move onto the deceleration lane for driving vehicles. Furthermore, higher h can make lane-changing maneuvers more difficult.

Generally, sharp curve has much worse design consistency compared to other segments. Crashes at sharp curve are found to be associated with poor vertical consistency (I_{SD} and h), high horizontal variation I_{CF} along with speed v, low traffic density k in nighttime, wet pavement, and holiday. In NEX, sharp curve is often designed to connect routes with different elevations. Thus, the vertical consistency is fairly poor. Smaller radius along with high v may cause notable centrifugal force. The affecting mechanisms of other component are similar to these at basic, merge, and diverge segments.

5.2.2 High-density uncongested flow

As traffic density increases, the inter-vehicle interaction gets more intensive. The corresponding results of PCA in high-density uncongested flow are summarized in Table 11. All of the components are of statistical significance.

In the case of high-density uncongested flow, it is distinct that traffic-related variables including k and v become an independent component, as a reflection of the increased interaction of vehicles. Furthermore, in terms of the value of loading, high density not low density is adverse to safety. The finding can further support the results of CR models: CR is decreasing to k in low-density uncongested flow, while it is increasing in high-density uncongested flow.

With respect to the differences by facility type, at merge segment, MR gets to be a factor related to the 1st component due to the increased interruption of ramp traffic with the increase of traffic density. During the variation in traffic conditions, the significance of DR becomes lower than geometric variation at diverge segment. However, in high-density uncongested flow, L_D is more important in contrast to low-density uncongested flow. Once a driver feels the difficulty for lane-changing maneuvers in diverging area, they may move onto the nearest lane to off-ramp in advance in the upstream of diverging area. As a result, the impact of lane-changing maneuvers on mainline traffic gets relatively low. In a sharp curve, crashes are still found to be probable with a decrease in k, which is similar to the tendency of CR model.

5.2.3 Congested flow

With the further increase of traffic density, congested flow appears. In the same way, Table 12 summarizes the results of PCA by facility type in congested flow.

Table 11 PCA results in high-density uncongested flow

Facility type (number of crash)	Item	Principal component									
		1st		2nd		3rd		4th		5th	
		F	L	F	L	F	L	F	L	F	L
Basic segment (94)	Component	I_{CF}	0.983	k	0.858	Day	0.776	h	0.925		
		I_{SD}	0.974	v	−0.885	Light	0.781				
	Percent of variance	28.9		21.8		17.1		14.4			
Merge segment (112)	Component	MR	0.818	k	0.936	L_A	0.842	Day	0.810	Pave	0.963
		I_{CF}	0.923	v	−0.899			Light	0.713		
		I_{SD}	0.960								
	Percent of variance	28.3		18.0		13.5		12.3		10.4	
Diverge segment (37)	Component	I_{CF}	0.970	DR	0.904	k	0.793	h	0.733	Pave	0.913
		I_{SD}	0.977	L_D	0.772	v	−0.704				
	Percent of variance	21.8		19.8		16.8		13.7		13.5	
Sharp curve (122)	Component	I_{CF}	0.723	k	−0.901	Pave	0.901	Day	0.869		
		I_{SD}	0.962	v	0.872						
		h	0.908								
	Percent of variance	33.2		21.9		15.3		13.9			

Table 12 demonstrates that the effect of traffic flow on crashes get more important in congested flow, compared to that in uncongested flow. Except merge segment, the significance of traffic flow affecting crashes is the highest. Based on the percent of variance, the influence of geometric design is further decreasing.

Regarding the differences by facility type, crashes at merge segment are found to be positively associated with smaller L_A, not higher L_A. For congested flow, smaller L_A may increase the difficulty of adequate speed adjustment for merging and lane-changing maneuvers. Besides, based on the loading of day type, weekday not holiday is a significant factor. It is likely related to higher percentage of heavy vehicles on weekday that may induce more frequent shockwave in congested flow. At diverge segment, as similar to merge segment, weekday is also a significant factor. Meanwhile, smaller L_D not higher L_D is adverse to safety. At sharp curve, poor ambient light can significantly restrict visibility, while visibility is critical for driving in small inter-vehicle spacing. Thus, ambient light becomes another important factor in congested flow compared to high-density uncongested flow.

From the analyses above, geometric features are found out to be the most significant influencing factor in uncongested flow. In this sense, the different CR characteristics by facility type in uncongested flow may be significantly associated with the variation in geometry. Poor design consistency induced by small radius is the potential cause for the highest CR in sharp curve. Ramp traffic can interrupt mainline traffic, and longer acceleration lane may provide longer interruption area. Both features can increase crash risk at merge segment. A lot of diverging traffic may move onto the lane nearest to deceleration lane in advance in the upstream of diverging area, since urban expressway carries a lot of commuters and many drivers are familiar with road structure. Hence, even if DR and L_D are found out as significant influencing factors, CR at diverge segment is not significantly higher than that at basic segment.

As traffic density increases, the effects of traffic-related variables increase and get more significant than geometry in congested flow. In this condition, once a breakdown initiates at bottlenecks, it can propagate to upstream section that may consists of several facility types, where traffic conditions are not significantly different. As a result, the difference of CR characteristics at basic, merge, and diverge segments is not significant. However, due to the heavily restricted visibility induced by the special geometric design, sharp curve still has higher CR than other facility types.

6 Conclusions and future work

This paper identified the different CR characteristics by facility type of Nagoya Urban Expressway. In uncongested flow, CR at basic, merge, and diverge segments appears convex downward to traffic density. In contrast, the value at sharp curve follows a decreasing tendency. In congested flow, CR at four facility types increases as traffic density increases. In both un- and congested flows, sharp curve has the worst safety performance in view of its highest CR among the four facility types. As for other segments, merge

Table 12 PCA results in congested flow

Facility type (number of crash)	Item	Principal component									
		1st		2nd		3rd		4th		5th	
		F	L	F	L	F	L	F	L	F	L
Basic segment (138)	Component	k	0.950	I_{CF}	0.842	h	0.699	Day	0.942		
		v	−0.947	I_{SD}	0.743	Pave	0.814				
	Percent of variance	26.0		20.3		18.9		15.4			
Merge segment (91)	Component	MR	0.871	k	0.883	I_{CF}	0.672	L_A	−0.871	Day	−0.933
		h	0.790	v	−0.884	Light	0.745	Pave	0.716		
		I_{SD}	0.825								
	Percent of variance	22.1		18.3		14.6		13.9		10.1	
Diverge segment (48)	Component	k	0.868	I_{CF}	0.849	DR	0.854	Pave	0.721	Light	0.951
		v	−0.801	I_{SD}	0.879	L_D	−0.783	Day	−0.716		
				h	0.859						
	Percent of variance	21.9		19.6		15.4		13.8		11.6	
Sharp curve (27)	Component	k	−0.772	I_{CF}	0.909	Pave	0.799	Day	0.912		
		v	0.950	I_{SD}	0.875		0.949				
				h	0.971	Light					
	Percent of variance	30.5		26.6		22.5		11.1			

segment has higher CR compared to the basic and diverge segments in uncongested flow. Comparatively, CR at three facility types is not significantly different in congested flow.

The causes of the differences were further investigated by focusing on traffic conditions and considering the interaction of geometry, traffic flow, and ambient conditions. Generally, geometric features are the most significant factors in uncongested flow. With the increase of traffic density, the effects of traffic-related variables increase and become most significant in congested flow. For ramp traffic, the significance of MR affecting crashes is on the rise as mainline flow increases. In contrast, the significance of DR gets decreasing. In addition, higher L_A and L_D are adverse to safety for uncongested flow, while smaller L_A and L_D are adverse for congested flow. Crashes at sharp curve are highly associated with the after effects of its special geometric design, such as large centrifugal force and heavy restricted visibility.

The potential benefits of integrating these findings in safer geometric design and traffic control are numerous. The analysis can provide a basis for geometric audit for safety regarding design consistency. Meanwhile, based on the estimated CR models, road administrators can easily image the safety performance with the variation of traffic conditions at a given facility type. Furthermore, PCA results may help prioritize countermeasures and further estimate the safety performance of an adopted countermeasure.

For more accurate analysis of crash characteristics, data in smaller time window, e.g., 1 min even 30 s, are highly recommended to improve the reliability of statistics, since crash occurrence is significantly associated with the short-term turbulence of traffic flow [1]. Furthermore, it is better to examine the effect of inter-lane interaction on crashes if the lane-based data is available. In this study, ramp traffic is found out to play a significant role for safety at merge and diverge segments. Thus, a microscopic analysis on driver behavior is needed. In essence, PCA is a qualitative analysis and the results are insufficient for applying specific countermeasures for a given case. Future studies are expected to acquire the quantitative effects of various influencing factors on crashes.

Acknowledgments The authors gratefully acknowledge the support of Nagoya Expressway Public Corporation for the data provision and other helps for this study.

References

1. Abdel-Aty M, Pande A (2007) Crash data analysis: collective versus individual crash level approach. J Saf Res 38(5):581–587
2. Wu Y, Nakamura H, Asano M (2012) A comparative study on crash rate characteristics at different intercity expressway facility

types. In: Proceedings of the 46th IP meeting of Japan Society of Civil Engineering (JSCE), Saitama University, 2–4 Nov 2012
3. Golob TF, Recker WW (2004) A method for relating type of crash to traffic flow characteristics on urban freeways. Transp Res Part A 38(1):53–80
4. Rengarasu TM, Hagiwara T, Hirasawa M (2009) Effects of road geometry and cross section variables on traffic accidents using homogeneous road segments. Transp Res Rec 2102:34–42
5. Shively TS, Kockelman K, Daminen K (2010) A bayesian semi-parametric model to estimate relationships between crash counts and roadway characteristics. Transp Res Part B 44(5):699–715
6. Bajwa S, Warita H, Kuwahara M (2010) Effects of road geometry, weather and traffic flow on safety. In: Proceedings of the 15th international conference of Hongkong Society for transportation study, The Hong Kong University of Science and Technology, Hong Kong, 11–14 Dec 2010
7. Abdel-Aty M, Pemmanaboina R (2006) Calibrating a real time traffic crash prediction model using archived weather and ITS traffic data. IEEE Trans Intell Transp Syst 7(2):167–174
8. Caliendo C, Guida M, Pasis A (2007) A crash-prediction model for multilane road. Accid Anal Prev 39(4):657–670
9. Christoforou Z, Cohen S, Karlaftis MG (2011) Identifying crash types propensity using real-time traffic data on freeways. J Saf Res 42:43–50
10. Hossain M, Muromachi YA (2012) A bayesian network based framework for real-time crash prediction on basic freeway segments of urban expressways. Accid Anal Prev 45:373–381
11. Wu Y, Nakamura H, Asano M (2013) A comparative study on crash characteristics between urban and intercity expressway basic segments. In: Selected proceedings of the 13th word conference on transport research, 15 pp
12. Hikosaka T, Nakamura H (2001) Statistical analysis on relationship between crash rate and traffic flow condition in basic expressway sections. In: Proceedings of the 21st Japan Society of Traffic Engineering Meeting, Tokyo, 29–30 Oc 2001. (in Japanese)
13. Ng JC, Sayed T (2004) Effects of geometric design consistency on road safety. Can J Civ Eng 31:218–227
14. Shawky M, Nakamura H (2007) Characteristics of breakdown phenomena in Japan urban expressway merging sections. In: Proceedings of TRB 86th annual meeting, Washington DC, 21–25 Jan 2007
15. Orme JG, Combs-Orme T (2009) Multiple regression with discrete dependent variables. Oxford University Press, New York
16. Zhou HG, Chen HY, Zhao JG et al (2010) Operational and safety performance of left-sides off-ramps at freeway diverge areas. In: Proceedings of TRB 89th annual meeting, Washington DC, 10–14 Jan 2010

19

A modification of local path marginal cost on the dynamic traffic network

Zheng-feng Huang · Gang Ren · Li-li Lu · Yang Cheng

Abstract Path marginal cost (PMC) is the change in total travel cost for flow on the network that arises when time-dependent path flow changes by 1 unit. Because it is hard to obtain the marginal cost on all the links, the local PMC, considering marginal cost of partial links, is normally calculated to approximate the global PMC. When analyzing the marginal cost at a congested diverge intersection, a jump-point phenomenon may occur. It manifests as a likelihood that a vehicle may unsteadily lift up (down) in the cumulative flow curve of the downstream links. Previously, the jump-point caused delay was ignored when calculating the local PMC. This article proposes an analytical method to solve this delay which can contribute to obtaining a more accurate local PMC. Next to that, we use a simple case to calculate the previously local PMC and the modified one. The test shows a large gap between them, which means that this delay should not be omitted in the local PMC calculation.

Z. Huang (✉)
Faculty of Maritime and Transportation, Ningbo University, Ningbo 315211, China
e-mail: huang321000@gmail.com

G. Ren · L. Lu
Jiangsu Key Laboratory of Urban ITS, Southeast University, Nanjing 210096, China

G. Ren · L. Lu
Jiangsu Province Collaborative Innovation Center of Modern Urban Traffic Technologies, Nanjing 210096, China

Y. Cheng
Department of Civil and Environmental Engineering, University of Wisconsin, Madison, WI 53705, USA

Keywords Transportation network · Path marginal cost · Cumulative flow curve · Dynamic traffic · System optimization

List of symbols

- Q_a^t The capacity of cell a at time interval t
- N_a^t The maximum number of vehicles that can be presented in cell a at time interval t
- n_a^t The vehicle occupancy of cell a at time interval t
- S_a^t The sending flow from cell a at time interval t
- R_a^t The receiving flow to cell a at time interval t
- $y_{a,b}^t$ The transmission flow from cell a to b at time interval t
- y_a^t The outflow from cell a at time interval t, expressed as $y_a^t = \sum_{b \in I_a^+} y_{a,b}^t$
- β_a^t The vehicle occupancy heading for branch cell a from upstream adjacent diverge cell divided by all the vehicle occupancy in this upstream diverge cell at time interval t
- I_a^- The upstream cell set of cell a
- I_a^+ The downstream cell set of cell a

1 Introduction

Path marginal cost (PMC) is the change in total travel cost for flow on the network that arises when time-dependent path flow changes by 1 unit. In the fields of transport economy and intelligent transportation, PMC has remained the normally computed value for finding the congestion toll [1, 2] or the system-optimal dynamic traffic [3, 4]. However, until now, no method can calculate the marginal cost on all the links after the perturbation of unit vehicle. So generally, different types of local PMC are used to approximate the global PMC. Based on cumulative flow curve, Ghali et al. [5] provided a sound analytical formulation for marginal cost on each link along the path where new vehicle was added. These link marginal costs were summed up as a local PMC. However, the link interactions were not considered. Shen et al. [6] proposed a perturbation propagation time method to modify the marginal cost, where the interaction of sequential links was considered. More recently, Qian et al. [7] stated that link interactions on congested diverge links may present a jump-point feature. However, they have not provided an approach to take the jump-point caused delay into the local PMC. Aforementioned definition of PMC can be clearly presented in Fig. 1.

We take Fig. 1 to explain jump-point phenomenon. It is the fluctuation of cumulative vehicles caused by vehicle sequence at link 1 and the rounding calculation procedure for diverging flows. Jump-point phenomenon is related to the dynamic traffic loading method. Generally, simulation methods are selected to load dynamic traffic, because the actual travel time, which is needed to calculate local PMC, can be obtained by traffic simulation. Among them, the cell transmission model (CTM) proposed by Daganzo [8] is relatively an accurate dynamic traffic simulation method, because not only physical queue but also the feature of traffic shockwave is considered. When employing this method to calculate the transmission flow at diverge intersection, a rounding operation is implemented to guarantee the integer formality for flow. If the upstream diverge link is congested, the rounding operation combined with vehicle sequence information could make the delayed vehicle at each time interval not always head for the same downstream branch link as the additional vehicle move toward. For instance, suppose a new vehicle is inserted at the congested upstream cell (a presentation of link segment) at time interval t, and the vehicle heading for another downstream branch cell (not the same as the direction of the additional vehicle) is delayed by unit time interval after the rounding calculation for transmission flow. Then, this delay makes the cumulative flow curves in the branch cells lift up (or down) by unit vehicle at time interval t. This phenomenon of jumping up and down may continue until the vanishment of upstream bottleneck. Qian et al. [7] names it as a jump-point phenomenon. In their sense, it seems difficult to identify each jump point at the branch cells. Thus, they ignore the jump-point caused delay in the calculation of local PMC. However, it may lead to problems of stability with the application like iterative system-optimal dynamic traffic assignment methods that incorporate calculation of local PMC, such as the method of successive average.

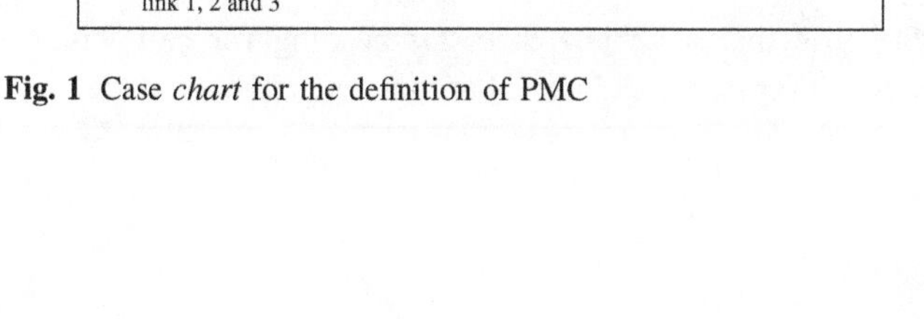

Fig. 1 Case *chart* for the definition of PMC

This article provides a modified local PMC for diverge cells. Specifically, calculate its key component, which is the delay generated by jump point. In the first section, CTM at diverge cell is reviewed. In the second section, the calculation of local PMC considering jump-point phenomenon is given. In the third section, using a simple diverge network, we compare the results of the previously local PMC and our modified one.

2 Review of CTM for diverge cells

Let the length of each time interval be identical and equal to the free-flow time on each link at diverge intersection. Then, the links in Fig. 2a can be converted to a cell network including three diverge cells in Fig. 2b. Although only two branch links are shown here for convenience, the analysis of more than two branch links is similar.

Because sending and receiving flow, transmission flow, and vehicle occupancy are key variables in the simulation process, we review their formulae at time interval t in advance.

When the backward wave propagation speed is assumed to be the free-flow speed, the formulae for receiving and sending flows of cell a can be described by

$$\begin{aligned} S_a^t &= \min\{Q_a^t, n_a^t\}, \\ R_a^t &= \min\{Q_a^t, N_a^t - n_a^t\}, \end{aligned} \tag{1}$$

where n_a^t can be obtained from the traffic simulation iteration of previous time interval $t-1$.

The transmission flow formula for flow from cell a to cell b is denoted by

$$\begin{aligned} y_a^t &= \min\{S_a^t, \min\{R_b^t/\beta_b^t | b \in I_a^+\}\}, \\ y_{a,b}^t &= \beta_b^t \cdot y_a^t, \forall b \in I_a^+, \end{aligned} \tag{2}$$

If β_b^t is equal to zero, the corresponding term inside the brace should be deleted.

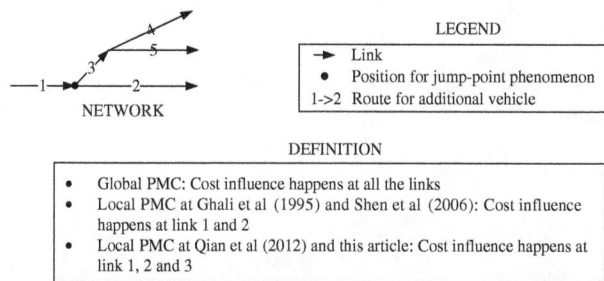

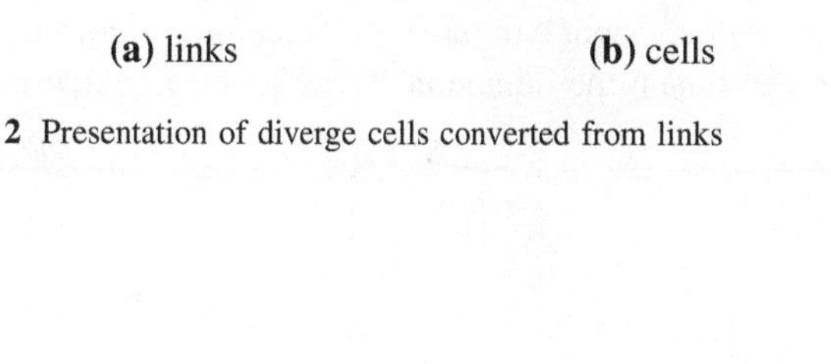

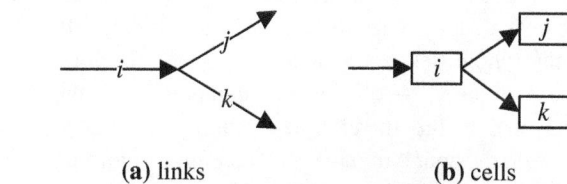

Fig. 2 Presentation of diverge cells converted from links

Equation (3) is used to update the vehicle occupancy of cell i:

$$n_a^{t+1} = n_a^t + \sum\nolimits_{b\in I_a^-} y_{b,a}^t - y_a^t. \quad (3)$$

Note that transmission flow in Eq. (2) should be rounded during each time interval to ensure that the flow and vehicle occupancies are integers.

3 Calculation of local PMC

We use the example network shown in Fig. 2 to analyze the local PMC caused by additional vehicle through cell i and j. The local PMC is defined to consist of two types of additional cost. One is unmodified marginal cost, which is equal to the previously local PMC that puts the jump-point phenomenon aside; the other is J-P cost, which is specifically used to describe the jump-point caused delay. Qian et al. [7] introduced these two parts. We borrow their theory in the follows. A modification is that the J-P cost trend when uncongested downstream cell is different from their analysis. Finally, we provide the method to attain the J-P cost.

3.1 Unmodified marginal cost

We depict unmodified cost according to three types of traffic condition: (1) cell i is uncongested; (2) cell i is congested, and cell j is uncongested; (3) cell i and cell j are congested.

If the first type occurs, the insertion of additional vehicle at cell i will not cause extra delay to its following vehicles. Therefore, the unmodified cost generated for the vehicles at the diverge cells is identical to the travel time of the additional vehicle, which is equal to the free-flow time passing through cell i and j.

If the second type occurs, it means that flow perturbation only occurs at cell i not j. We depict the unmodified cost generated at cell i here (shown in Fig. 3a). At the beginning of flow-perturbation time interval (the arrival time of additional vehicle), the cumulative arrival flow of cell i lifts up by 1 unit. Until the queue-vanishing time interval t_i^C, can the following vehicles not be influenced by the flow perturbation anymore. Therefore, only those vehicles that arrive among the time range $[\tau, t_i^C]$ are delayed by the flow perturbation. Each vehicle is delayed by l, which is the inverse of traffic capacity (or discharging rate) at cell i. In terms of the whole delayed vehicles $M_2 - M_1$, the total delay time will be $t_i^D - t_i^B$. Another component of the unmodified cost is the travel time spent by the added vehicle, which is equal to $t_i^B - \tau$. Therefore, summing them up can obtain the unmodified cost in cell i, which is equal to $t_i^D - \tau$.

If the third type occurs (shown in Fig. 3a, b), the flow propagation process should be considered. It is required that all the diverge cells should be taken as a whole to calculate the unmodified cost which possesses two components. The first component is the travel time of additional vehicle through cell i and j, which is equal to $t_j^B - \tau$. The second component is the delay for the vehicles $M_4 - M_3$ (Note that $M_4 - M_3$ is part of $M_2 - M_1$ in Fig. 3 although the vertical coordinate intervals may be different for convenience) and $M_5 - M_4$, which is equal to $t_j^D - t_j^B$. In other word, we can take cells i and j as a single virtual cell to obtain the unmodified cost which is equal to $t_j^D - \tau$. Therefore, there is a hidden assumption that no vehicle toward cell k is influenced by the additional vehicle. However, the vehicle sequence information combined with the rounding calculation in the dynamic traffic simulation may make some vehicles toward cell k be delayed, which is a jump-point phenomenon explained in the subsequent subsection.

3.2 J-P cost

The jump-point phenomenon may occur when cell i is congested. However, the analysis processes for the second and third types are the same; thus, we would only show the analysis for simplicity.

In the first place, we should explain the jump-point phenomenon graphically. When an additional vehicle is added to cell i, we should compute Eq. (2) after each time interval to obtain the number of vehicles in the queue of cell i heading for each branch cell. After rounding the number to the nearest integer, we may discover a changed outflow. For instance, it is possible that for a specific time horizon $\left[t^A, t^{A'}\right]$ of the period following time interval τ, the arrival rate of cell j may lift up (down) by 1 unit, whereas a corresponding decrease (increase) may occur in cell k (shown in Fig. 3). These time horizons are named jump points.

Subsequently, show jump-point phenomenon from the analytical explanation. Suppose that the number of vehicles in queue on cell i heading for cell j during time interval t is denoted as $x_{i,j}^t$. So the traffic ratio β_j^t is expressed by

$$\beta_j^t = x_{i,j}^t \Big/ \left(x_{i,j}^t + x_{i,k}^t\right). \quad (4)$$

As a special case of Eq. (2), transmission flow is expressed as

$$y_{i,j}^t = y_i^t \cdot x_{i,j}^t \Big/ \left(x_{i,j}^t + x_{i,k}^t\right),$$
$$y_{i,k}^t = y_i^t \cdot x_{i,k}^t \Big/ \left(x_{i,j}^t + x_{i,k}^t\right). \quad (5)$$

Suppose that a vehicle at cell i heading for cell j is postponed for unit time interval at time interval $t - 1$ when

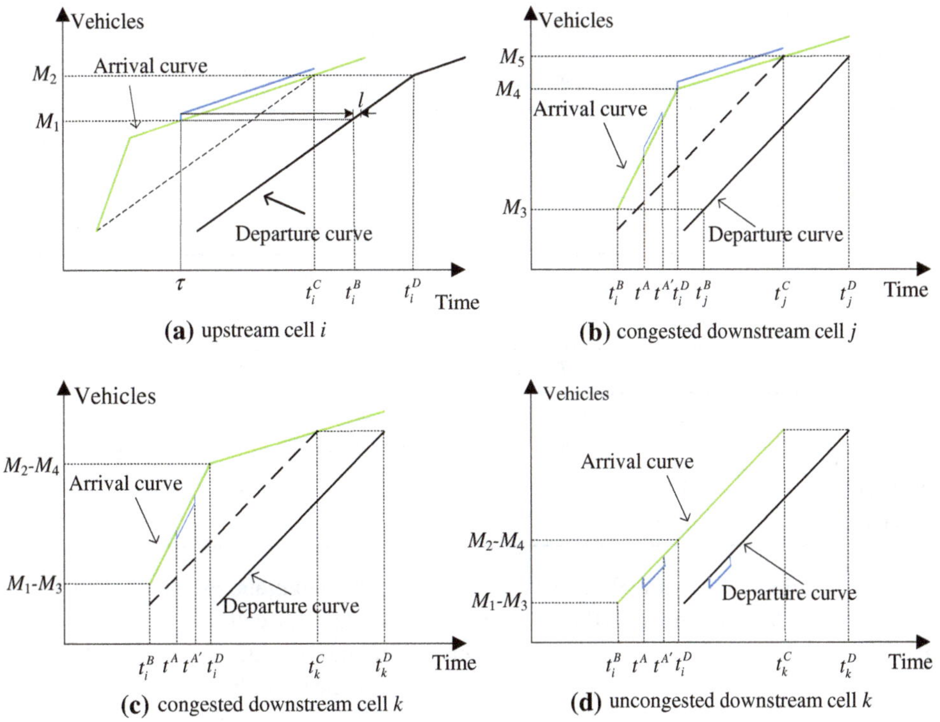

Fig. 3 Cumulative flow curves

compared to its previously simulated position. Then, the delayed vehicle will be added to the next time interval. The transmission flow at time interval t can be expressed:

$$y_{i,j}^{t}{}' = y_i^t \cdot \left(x_{i,j}^t + 1\right) \Big/ \left(x_{i,j}^t + x_{i,k}^t + 1\right),$$
$$y_{i,k}^{t}{}' = y_i^t \cdot x_{i,k}^t \Big/ \left(x_{i,j}^t + x_{i,k}^t + 1\right). \tag{6}$$

Numerically, we round the transmission flow to the nearest integer. In the rounding operation, although some abnormal transmission flow may not make the equation $y_{i,j'}^t + y_{i,k'}^t = y_i^t$ be satisfied, we can avoid this shortcoming by adjusting rounding rule. To describe the jump-point phenomenon clearly, we do not discuss this rounding rule here. If the cumulative flow of cell j or k at time interval $t-1$ is the same with the previously simulated results, but the cumulative flow at time interval t after calculating Eq. (6) is different to the corresponding simulated results, it is indicated that the jump point is generated at time interval t. The lifted-down vehicle is postponed to the next time interval $t+1$ and be added to the vehicle occupancy of that time. The above jump point can keep the "jumping" shape if the following transmission flow is unchanged compared to the previously simulated transmission flow at time interval $t+1$. Totally, there could be multiple jump points, which depend on the demands from both paths queued on cell i.

We use Fig. 3b, c, d to explain different types of J-P cost. Regarding cell j, the unit flow is assumed to be lifted up at time horizon $\left[t^A, t^{A'}\right]$, so its J-P cost should be added by $t^{A'} - t^A$. This added part can be explained in other word: a vehicle joins in the queue line of cell j with unit time interval ahead of original time interval, whereas its outflow time is unchanged. Regarding cell k, a reduction of one inflow vehicle in the corresponding time horizon is assumed. If cell k is congested, the J-P cost would be decreased by $t^{A'} - t^A$, whose explanation is similar to the former one. If cell k is not a bottleneck, the J-P cost would be increased by $t^{A'} - t^A$. The reason is that there is an assumption that the free-flow travel time at cell k cannot be shortened. Therefore, when the vehicles travel through the uncongested cell k with a free-flow travel time, the time interval of outflow will be changed to the same step size with the one of inflow. This could lead to the increase of J-P cost. Undoubtedly, unit time interval would be saved if one inflow vehicle is increased to uncongested cell k during certain time interval. It is noted that the delay modification to uncongested cell k here differs from Qian et al. [7]'s argument which deemed it unchanged.

To obtain the J-P cost during congested time horizon [t_i^B, t_i^D], we should prepare the merging queue rule in the upstream intersections in priority. It can obtain the vehicle sequence information or rank the position of each vehicle in the upstream diverge cell, which can help to determine the ratio of the vehicle occupancy from different routes. Assuming that the merging queue rule is given, we can use an analytical method to estimate the J-P cost at the diverge

Table 1 List of conditions and parameters and J-P cost for downstream cell *a time interval*

Condition of cell $a \in \{I_i^+\}$	Uncongested and vehicle added	Uncongested and vehicle erased	Uncongested and vehicle unchanged	Congested and vehicle added	Congested and vehicle erased	Congested and vehicle unchanged
θ_t^a	−1	−1	−1	1	1	1
δ_t^a	1	−1	0	1	−1	0
Trend of J-P cost	−1	1	0	1	−1	0

Table 2 Comparison of inflow at downstream cells

Time interval	T	$T+1$	$T+2$	$T+3$	$T+4$	$T+5$
Inflow at cell j not consider jump point	3	4	4	4	3	0
Inflow at cell k not consider jump point	3	2	2	2	1	0
Inflow at cell j consider jump point	4	4	4	4	2	0
Inflow at cell k consider jump point	2	2	2	2	2	0

intersection. In the first place, we define two variables. The assignment of these variables and the subsequent J-P cost calculation are described below.

3.2.1 Congestion level variable

Assume that θ_t^a represents the congestion level of cell $a(\forall a \in I_i^+)$ at time interval t; 1 signifies bottleneck and −1 signifies no bottleneck. The "no bottleneck" also accords to the free-flow speed state in the CTM. Flow profile with respect to density is a trapezoidal pattern, which guarantees that no bottleneck occurs at most of the low occupancy states.

3.2.2 Perturbation variable

Assume that perturbation variable δ_t^a represents a change in the aspect of cumulative vehicles for cell $a(\forall a \in I_i^+)$ at time interval t. The assigned number 1 (0 or −1) represents 1 vehicle exceeding (no change compared to or 1 vehicle less than) the previously simulated cumulative vehicles. Next, we acquire the perturbation variable time interval by time interval using the information including the perturbation variable and the postponed vehicle at previous time interval and the postponed vehicle at current time interval.

First define γ_t^a to describe the relation between path and cell a, and initialize this variable and δ_t^a with 0 for different time intervals and cells; then perform the following steps to assign δ_t^a beginning from the time interval t_i^B:

Step 1 Add the previous perturbation variable δ_{t-1}^a to the current perturbation variable δ_t^a, if t is larger than t_i^B;

Step 2 Identify the route of the new entering vehicle at cell i of the current time interval, which is postponed from the previous time interval or just the additional vehicle we added for the PMC calculation; and assign γ_t^a with 1 if this vehicle is heading for cell a;

Step 3 Round the outflow of cell i; based on the vehicle sequence in the queue, identify the upcoming postponed vehicle compared to the previously simulated vehicle occupancy; update γ_t^a with $\gamma_t^a = \gamma_t^a - 1$, if the delayed vehicle is heading for cell a;

Step 4 $\delta_t^a = \delta_t^a + \gamma_t^a$;

Step 5 let $t = t + 1$ and perform sequential steps 1–4; it will not be completed until the time is out of range $[t_i^B, t_i^D]$.

3.2.3 Formula for J-P cost

We use $\sum_{t\in[t_i^B,t_i^D]} \sum_{a\in I_i^+} \theta_t^a \cdot \delta_t^a$ to calculate the J-P cost.

To test the validation of our formula for J-P cost, we will do a comparison here. We list the most probable six cases of cell condition in the first row of Table 1, where different traffic conditions and vehicle variations compared to previously simulated cumulative vehicles are given. The parameters of our method in accord with them can be shown in the middle rows of Table 1. The trend of J-P cost can be known directly from the cell condition and listed at the last line. Fortunately, their corresponding trend of J-P cost is the same to the one by our analytical formula. Thus, it indicates that our calculation method can attain the accurate J-P cost.

4 Case study I

We give a case (shown in Fig. 2) to describe the distinct difference between the previous and modified PMC. Because the jump-point phenomenon may occur only when cell i is congested, whatever the state of downstream cell

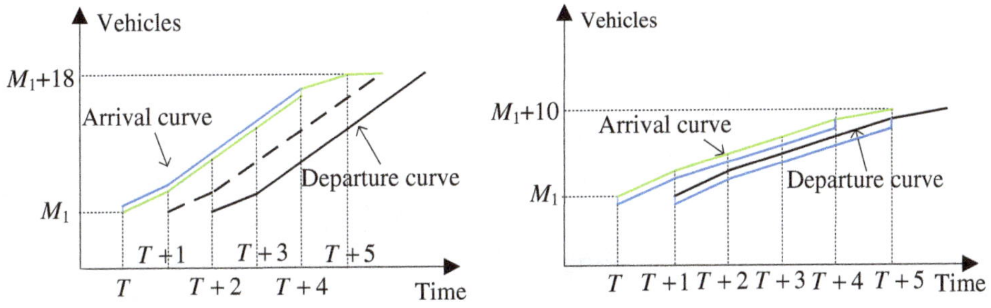

(a) congested cell j **(b)** uncongested cell k

Fig. 4 Cumulative flow curve when cell i is congested

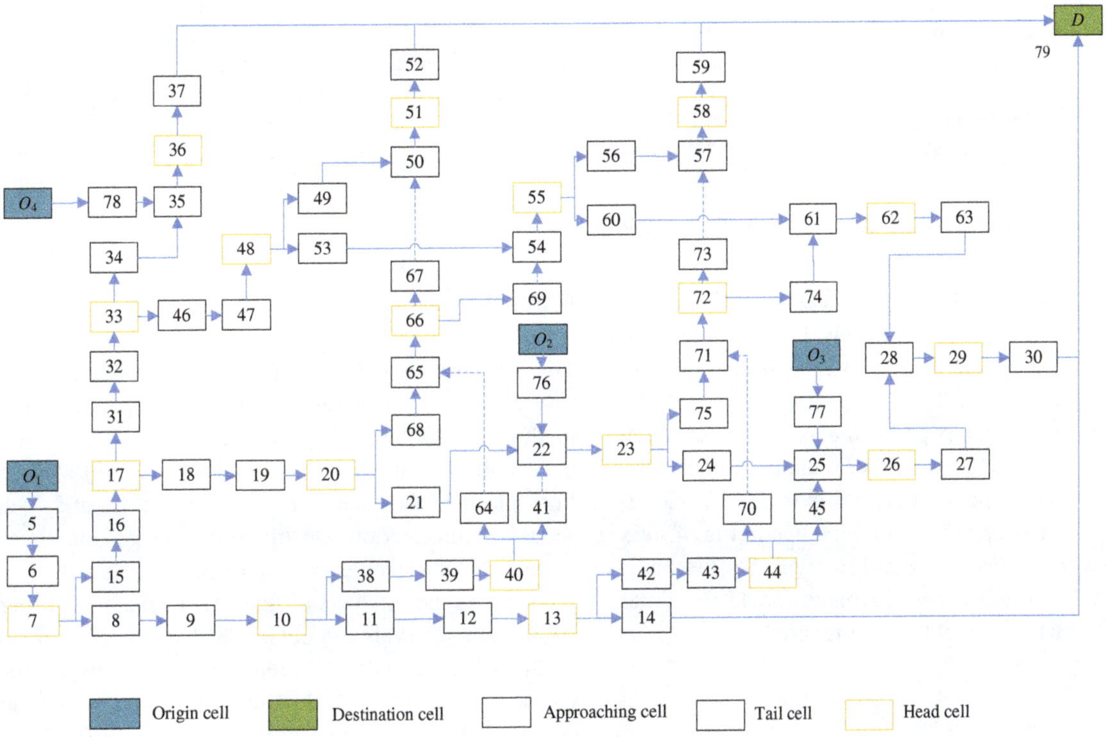

Fig. 5 Cell network

j and k. We select the condition that cell j is congested and k is uncongested as an example for clearly explanation. Other conditions are listed as follows.

- The unit time interval is set equal to 3 s;
- The capacity at cell i is set to be six vehicles per time interval;
- The receiving flow of cell j and the maximal vehicle occupancy of cell i are assumed to be large enough to let $y_i^t = S_i^t = 6$ be possible;
- The vehicle occupancies toward different diverge cells at cell i at T are $x_{i,j}^T = 5$ and $x_{i,k}^T = 4$;
- The input flow at cell i toward cell j and k before $T + 3$ is always equal to 4 and 2 per time interval separately; then the input flow at cell i is stopped at $T + 3$ and the following time intervals;
- Assume that the arrival time interval of additional vehicle at cell i is $T - 2$, and use T to replace its departure time interval t_i^B.

If the jump-point phenomenon is ignored, the previously simulated inflows at cell j and k still work. The only difference is that a vehicle is added to the inflow of cell j when the bottleneck is vanished. We can use Eq. (7) to obtain the inflow from T to $T + 5$ at cell j and k, which are shown in the second and third rows of Table 2.

$$y_{i,j}^t = S_i^t \cdot x_{i,j}^t \Big/ \left(x_{i,j}^t + x_{i,k}^t\right). \tag{7}$$

Because cell j is congested, we let the travel time of arrival vehicle among the time interval from T to $T + 5$ be equal to 2 units. The shape of their flow arrival curves can

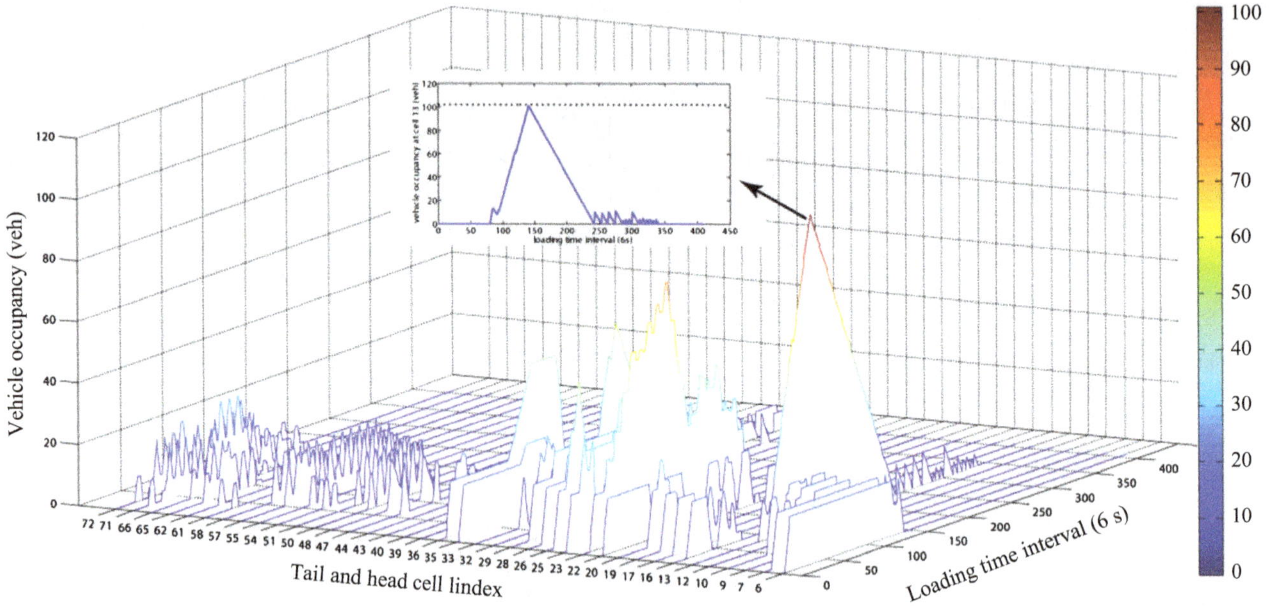

Fig. 6 Vehicle occupancy of tail/head cell changing with time interval

be shown in Fig. 4 (green line). In this case, $T + 7$ accords to the time interval t_j^D of Fig. 4b. So, the previous PMC (or unmodified cost) is equal to nine, which is the subtract result of $T + 7$ and $T - 2$.

If we consider the jump point, the vehicle occupancy would be $x_{i,j}^T = 6$ and $x_{i,k}^T = 4$. After rounding calculation, the inflow is shown in the fourth and fifth rows of Table. 2. The changed curve is depicted with blue line in Fig. 4. It is obviously that the J-P cost is equal to eight. So, the modified PMC is 17. The J-P cost accounts for 47 % of the modified PMC, which should not be ignored.

5 Case study II

To illustrate the feasibility of applying the method in larger network Fig. 5, we use the following cell network to calculate the system-optimum dynamic traffic assignment (SO-DTA) by using method of successive averages that embed least PMC searching. Qian et al. [7] demonstrated the feasibility of using this method to solve SO-DTA problem.

Varied cells make up for the network. The origin and destination and approaching cells have the same length with basic cell; the length of other cells equals six basic cells. We assume all the vehicles depart origins cells toward the same destination. The assigned occupancies changing with the time are shown in Fig. 6. No spillback occurs in the cells, indicating that the SO-DTA is reasonable.

6 Conclusions

The J-P cost caused by jump-point phenomenon is considered in the calculation of local PMC. An analytical method for solving J-P cost is proposed. This treatment contributes to obtaining a more accurate local PMC. In the first place, based on the historical data and vehicle sequence information, we figure out congestion state and perturbations valuables at each time interval; then, summing up all the products of these two valuables during the whole time intervals can obtain the J-P cost. A comparison of the J-P cost under different cell conditions shows the correctness of our method. A case study shows that the J-P cost may make up a high proportion of local PMC, which should not be ignored.

The application of the modified local PMC in the congestion charging is beyond our study scope. But, it can be realized in theory. For two parallel routes between an OD pair, the actual route costs rather than marginal ones are equal under UE assumption. Theoretically, we can charge toll for the link in the larger PMC route to transfer congested flow to achieve SO network flows as managers desire to see. Congestion toll case with this method in a small network has been tested by Qian et al. [9]. In practice, other newly developed congestion toll related techniques such as tradable credits, flat toll, and tactical waiting [10, 11] are more effective, because the ideally time-varying fine toll is hard to solve. However, the comparison of PMC in different routes can still serve as a measurement method for these practical charging types.

References

1. De Palma A, Lindsey R (2011) Traffic congestion pricing methodologies and technologies [J]. Transp Res Part C 19(6):1377–1399
2. Zhong R, Sumalee A, Maruyama T (2012) Dynamic marginal cost, access control, and pollution charge: a comparison of bottleneck and whole link models [J]. J Adv Transp 46(3):191–221
3. Rakha H, Tawfik A (2009) Traffic networks: dynamic traffic routing, assignment, and assessment. In: Asakura Y (ed) Encyclopedia of complexity and systems science. Springer, New York, pp 9429–9470
4. Chow AH (2009) Dynamic system optimal traffic assignment—a state-dependent control theoretic approach [J]. Transportmetrica 5(2):85–106
5. Ghali MO, Smith MJ (1995) A model for the dynamic system optimum traffic assignment problem [J]. Transp Res Part B 29(3):155–170
6. Shen W, Nie Y, Zhang HM (2006) Path-based system optimal dynamic traffic assignment models: formulations and solution methods. In: Chootinan P (ed) In intelligent transportation systems conference. ITSC'06 IEEE, Toronto, pp 1298–1303
7. Qian ZS, Shen W, Zhang HM (2012) System-optimal dynamic traffic assignment with and without queue spillback: its path-based formulation and solution via approximate path marginal cost [J]. Transp Res Part B 46(7):874–893
8. Daganzo CF (1995) The cell transmission model, part II: network traffic [J]. Transp Res Part B 29(2):79–93
9. Qian ZS, Michael H (2011) Computing individual path marginal cost in networks with queue spillbacks [J]. Transp Res Rec: Journal of the Transportation Research Board 2263(1):9–18
10. Xiao F, Shen W, Michael Zhang H (2012) The morning commute under flat toll and tactical waiting [J]. Transp Res Part B 46(10):1346–1359
11. Xiao F, Qian ZS, Zhang HM (2013) Managing bottleneck congestion with tradable credits [J]. Transp Res Part B 56:1–14

Permissions

All chapters in this book were first published in JMT, by Springer; hereby published with permission under the Creative Commons Attribution License or equivalent. Every chapter published in this book has been scrutinized by our experts. Their significance has been extensively debated. The topics covered herein carry significant findings which will fuel the growth of the discipline. They may even be implemented as practical applications or may be referred to as a beginning point for another development.

The contributors of this book come from diverse backgrounds, making this book a truly international effort. This book will bring forth new frontiers with its revolutionizing research information and detailed analysis of the nascent developments around the world.

We would like to thank all the contributing authors for lending their expertise to make the book truly unique. They have played a crucial role in the development of this book. Without their invaluable contributions this book wouldn't have been possible. They have made vital efforts to compile up to date information on the varied aspects of this subject to make this book a valuable addition to the collection of many professionals and students.

This book was conceptualized with the vision of imparting up-to-date information and advanced data in this field. To ensure the same, a matchless editorial board was set up. Every individual on the board went through rigorous rounds of assessment to prove their worth. After which they invested a large part of their time researching and compiling the most relevant data for our readers.

The editorial board has been involved in producing this book since its inception. They have spent rigorous hours researching and exploring the diverse topics which have resulted in the successful publishing of this book. They have passed on their knowledge of decades through this book. To expedite this challenging task, the publisher supported the team at every step. A small team of assistant editors was also appointed to further simplify the editing procedure and attain best results for the readers.

Apart from the editorial board, the designing team has also invested a significant amount of their time in understanding the subject and creating the most relevant covers. They scrutinized every image to scout for the most suitable representation of the subject and create an appropriate cover for the book.

The publishing team has been an ardent support to the editorial, designing and production team. Their endless efforts to recruit the best for this project, has resulted in the accomplishment of this book. They are a veteran in the field of academics and their pool of knowledge is as vast as their experience in printing. Their expertise and guidance has proved useful at every step. Their uncompromising quality standards have made this book an exceptional effort. Their encouragement from time to time has been an inspiration for everyone.

The publisher and the editorial board hope that this book will prove to be a valuable piece of knowledge for researchers, students, practitioners and scholars across the globe.

List of Contributors

Liang Ye
Transport Planning and Research Institute, Ministry of Transport of China, Room 1109, building 2, Jia 6 Shuguangxili, Chaoyang, Beijing 100028, China

Ying Hui
School of Transportation Engineering, Tongji University, Shanghai 201804, China

Dongyuan Yang
School of Transportation Engineering, Tongji University, Shanghai 201804, China

G. Y. Zhou
State Key Laboratory of Traction Power, Tribology Research Institute, Southwest Jiaotong University, Chengdu 610031, China

J. H. Liu
State Key Laboratory of Traction Power, Tribology Research Institute, Southwest Jiaotong University, Chengdu 610031, China

W. J. Wang
State Key Laboratory of Traction Power, Tribology Research Institute, Southwest Jiaotong University, Chengdu 610031, China

G. Wen
State Key Laboratory of Traction Power, Tribology Research Institute, Southwest Jiaotong University, Chengdu 610031, China

Q. Y. Liu
State Key Laboratory of Traction Power, Tribology Research Institute, Southwest Jiaotong University, Chengdu 610031, China

Ahmed Mohamed Semeida
Civil Engineering Department, Specialization of Transportation and Traffic Engineering, Faculty of Engineering, Port Said University, Port Said, Port Fouad 42523, Egypt

Dajing Zhou
Key Laboratory of Magnetic Levitation Technologies and Maglev Trains (Ministry of Education of China), Southwest Jiaotong University, Chengdu 610031, Sichuan, China
Superconductivity and New Energy R&D Center, Southwest Jiaotong University, Chengdu 610031, Sichuan, China

Jiaqing Ma
Key Laboratory of Magnetic Levitation Technologies and Maglev Trains (Ministry of Education of China), Southwest Jiaotong University, Chengdu 610031, Sichuan, China
Superconductivity and New Energy R&D Center, Southwest Jiaotong University, Chengdu 610031, Sichuan, China

Lifeng Zhao
Key Laboratory of Magnetic Levitation Technologies and Maglev Trains (Ministry of Education of China), Southwest Jiaotong University, Chengdu 610031, Sichuan, China
Superconductivity and New Energy R&D Center, Southwest Jiaotong University, Chengdu 610031, Sichuan, China

Xiao Wan
Key Laboratory of Magnetic Levitation Technologies and Maglev Trains (Ministry of Education of China), Southwest Jiaotong University, Chengdu 610031, Sichuan, China
Superconductivity and New Energy R&D Center, Southwest Jiaotong University, Chengdu 610031, Sichuan, China

Yong Zhang
Key Laboratory of Magnetic Levitation Technologies and Maglev Trains (Ministry of Education of China), Southwest Jiaotong University, Chengdu 610031, Sichuan, China
Superconductivity and New Energy R&D Center, Southwest Jiaotong University, Chengdu 610031, Sichuan, China

Yong Zhao
Key Laboratory of Magnetic Levitation Technologies and Maglev Trains (Ministry of Education of China), Southwest Jiaotong University, Chengdu 610031, Sichuan, China
Superconductivity and New Energy R&D Center, Southwest Jiaotong University, Chengdu 610031, Sichuan, China
School of Materials Science and Engineering, University of New South Wales, Sydney, NSW 2052, Australia

Jie Jiang
School of Automobile and Traffic Engineering, Jiangsu University, Zhenjiang 212013, China

Yu-lin Chang
School of Automobile and Traffic Engineering, Jiangsu University, Zhenjiang 212013, China

Na Wu
State Key Laboratory of Traction Power, Southwest Jiaotong University, Chengdu 610031, China

Jing Zeng
State Key Laboratory of Traction Power, Southwest Jiaotong University, Chengdu 610031, China

Tao Li
Dept. of Remote Sensing and Geospatial Information Engineering, Southwest Jiaotong University, Chengdu 610031, Sichuan, China
Institute of Space and Earth Information Science, Chinese University of Hong Kong, Hong Kong, China

Guoxiang Liu
Dept. of Remote Sensing and Geospatial Information Engineering, Southwest Jiaotong University, Chengdu 610031, Sichuan, China

Hui Lin
Institute of Space and Earth Information Science, Chinese University of Hong Kong, Hong Kong, China

Rui Zhang
Dept. of Remote Sensing and Geospatial Information Engineering, Southwest Jiaotong University, Chengdu 610031, Sichuan, China
Institute of Space and Earth Information Science, Chinese University of Hong Kong, Hong Kong, China

Hongguo Jia
Dept. of Remote Sensing and Geospatial Information Engineering, Southwest Jiaotong University, Chengdu 610031, Sichuan, China

Bing Yu
Dept. of Remote Sensing and Geospatial Information Engineering, Southwest Jiaotong University, Chengdu 610031, Sichuan, China

Hongcheng Gan
Transportation Center, Northwestern University, Evanston, IL, USA
Department of Transportation Engineering, University of Shanghai for Science and Technology, Shanghai, China

Xin Ye
Civil Engineering Department, California State Polytechnic University, Pomona, CA, USA

Hamed Pouryousef
Civil and Environmental Engineering Department, Michigan Technological University, 1400 Townsend Drive, Houghton, MI 49931, USA

Pasi Lautala
Civil and Environmental Engineering Department, Michigan Technological University, 1400 Townsend Drive, Houghton, MI 49931, USA

Thomas White
Transit Safety Management, 3604 220th Pl SW, Mountlake Terrace, WA 98043, USA

M. Alqhatani
School of Civil, Environmental and Chemical Engineering, RMIT University, GPO Box 2476, Melbourne, VIC 3001, Australia

S. Setunge
Ministry of Higher Education, Riyadh, Saudi Arabia

S. Mirodpour
School of Civil, Environmental and Chemical Engineering, RMIT University, GPO Box 2476, Melbourne, VIC 3001, Australia

Bing Yu
Department of Remote Sensing and Geospatial Information Engineering, Southwest Jiaotong University, 111 North 1st Section, 2nd Ring Road, Chengdu 610031, Sichuan, China

Guoxiang Liu
Department of Remote Sensing and Geospatial Information Engineering, Southwest Jiaotong University, 111 North 1st Section, 2nd Ring Road, Chengdu 610031, Sichuan, China

Rui Zhang
Department of Remote Sensing and Geospatial Information Engineering, Southwest Jiaotong University, 111 North 1st Section, 2nd Ring Road, Chengdu 610031, Sichuan, China

Hongguo Jia
Department of Remote Sensing and Geospatial Information Engineering, Southwest Jiaotong University, 111 North 1st Section, 2nd Ring Road, Chengdu 610031, Sichuan, China]

Tao Li
Department of Remote Sensing and Geospatial Information Engineering, Southwest Jiaotong University, 111 North 1st Section, 2nd Ring Road, Chengdu 610031, Sichuan, China

Xiaowen Wang
Department of Remote Sensing and Geospatial Information Engineering, Southwest Jiaotong University, 111 North 1st Section, 2nd Ring Road, Chengdu 610031, Sichuan, China

Keren Dai
Department of Remote Sensing and Geospatial Information Engineering, Southwest Jiaotong University, 111 North 1st Section, 2nd Ring Road, Chengdu 610031, Sichuan, China

Deying Ma
Department of Remote Sensing and Geospatial Information Engineering, Southwest Jiaotong University, 111 North 1st Section, 2nd Ring Road, Chengdu 610031, Sichuan, China

Ramakrishna VADDE
Texas A&M University-Kingsville, Department of Civil Engineering MSC 194, Kingsville, Texas 78363, USA

Dazhi SUN
Texas A&M University-Kingsville, Department of Civil Engineering MSC 194, Kingsville, Texas 78363, USA

Joseph O. SAI
Texas A&M University-Kingsville, Department of Civil Engineering MSC 194, Kingsville, Texas 78363, USA

Mohammed A. FARUQI
Texas A&M University-Kingsville, Department of Civil Engineering MSC 194, Kingsville, Texas 78363, USA

Pat T. LEELANI
Texas A&M University-Kingsville, Department of Civil Engineering MSC 194, Kingsville, Texas 78363, USA

Meiwu AN
Saint Louis County Department of Highways and Traffic and Public Works, 121 South Meramec Avenue, 8th Floor Saint Louis, Missouri 63105, USA

Mei CHEN
Department of Civil Engineering, University of Kentucky, Lexington, KY 40506-0281, USA

Lindong WANG
China Academy of Railway Sciences, Locomotive and Car Research Institute, Beijing 100081, China

Qiang HUANG
China Academy of Railway Sciences, Locomotive and Car Research Institute, Beijing 100081, China

Yuqing ZENG
China Academy of Railway Sciences, Locomotive and Car Research Institute, Beijing 100081, China

Fengtao LIN
China Academy of Railway Sciences, Locomotive and Car Research Institute, Beijing 100081, China

Fazhi LI
Key Laboratory of Road and Traffic Engineering of the Ministry of Education, School of Transportation Engineering, Tongji University, Shanghai 201804, China

Zhengyu DUAN
Key Laboratory of Road and Traffic Engineering of the Ministry of Education, School of Transportation Engineering, Tongji University, Shanghai 201804, China

Dongyuan YANG
Key Laboratory of Road and Traffic Engineering of the Ministry of Education, School of Transportation Engineering, Tongji University, Shanghai 201804, China

Lu Sun
Department of Civil Engineering, The Catholic University of America, Washington, DC 20064, USA
International Institute of Safe, Intelligent and Sustainable Transportation & Infrastructure, Southeast University, Nanjing 210096, China

Xingwen Wu
State Key Laboratory of Traction Power, Southwest Jiaotong University, Chengdu, China

Maoru Chi
State Key Laboratory of Traction Power, Southwest Jiaotong University, Chengdu, China

Jing Zeng
State Key Laboratory of Traction Power, Southwest Jiaotong University, Chengdu, China

Weihua Zhang
State Key Laboratory of Traction Power, Southwest Jiaotong University, Chengdu, China

Minhao Zhu
State Key Laboratory of Traction Power, Southwest Jiaotong University, Chengdu, China

Yong Wu
Department of Civil Engineering, Nagoya University, C1-2(651) Furo-cho, Chikusa-ku, Nagoya 464-8603, Japan

Hideki Nakamura
Department of Civil Engineering, Nagoya University, C1-2(651) Furo-cho, Chikusa-ku, Nagoya 464-8603, Japan

Miho Asano
Department of Civil Engineering, Nagoya University, C1-2(651) Furo-cho, Chikusa-ku, Nagoya 464-8603, Japan

Zheng-feng Huang
Faculty of Maritime and Transportation, Ningbo University, Ningbo 315211, China

Gang Ren
Jiangsu Key Laboratory of Urban ITS, Southeast University, Nanjing 210096, China
Jiangsu Province Collaborative Innovation Center of Modern Urban Traffic Technologies, Nanjing 210096, China

Li-li Lu
Jiangsu Key Laboratory of Urban ITS, Southeast University, Nanjing 210096, China
Jiangsu Province Collaborative Innovation Center of Modern Urban Traffic Technologies, Nanjing 210096, China

Yang Cheng
Department of Civil and Environmental Engineering, University of Wisconsin, Madison, WI 53705, USA

www.ingramcontent.com/pod-product-compliance
Lightning Source LLC
Chambersburg PA
CBHW080650200326
41458CB00013B/4805
* 9 7 8 1 6 8 2 8 5 0 9 5 4 *